Studies in Computational Intelligence 457

For further volumes:
http://www.springer.com/series/7092

Ngoc Thanh Nguyen, Bogdan Trawiński,
Radosław Katarzyniak, and Geun-Sik Jo (Eds.)

Advanced Methods for Computational Collective Intelligence

Editors
Ngoc Thanh Nguyen
Institute of Informatics
Wrocław University of Technology
Wrocław
Poland

Radosław Katarzyniak
Institute of Informatics
Wrocław University of Technology
Wrocław
Poland

Bogdan Trawiński
Institute of Informatics
Wrocław University of Technology
Wrocław
Poland

Geun-Sik Jo
Department of Computer Science
& Information Engineering
INHA University
Incheon
Korea

ISSN 1860-949X e-ISSN 1860-9503
ISBN 978-3-642-43379-5 ISBN 978-3-642-34300-1 (eBook)
DOI 10.1007/978-3-642-34300-1
Springer Heidelberg New York Dordrecht London

Softcover reprint of the hardcover 1st edition 2013

Printed on acid-free paper

Springer is part of Springer Science+Business Media (www.springer.com)

Preface

Collective intelligence has become one of major research issues studied by today's and future computer science. Computational collective intelligence is understood as this form of group intellectual activity that emerges from collaboration and competition of many artificial individuals. Robotics, artificial intelligence, artificial cognition and group working try to create efficient models for collective intelligence in which it emerges from sets of actions carried out by more or less intelligent individuals. The major methodological, theoretical and practical aspects underlying computational collective intelligence are group decision making, collective action coordination, collective competition and knowledge description, transfer and integration. Obviously, the application of multiple computational technologies such as fuzzy systems, evolutionary computation, neural systems, consensus theory, knowledge representation etc. is necessary to create new forms of computational collective intelligence and support existing ones.

Three subfields of application of computational technologies to support forms of collective intelligence are of special attention to us. The first one is semantic web treated as an advanced tool that increases the collective intelligence in networking environments. The second one covers social networks modeling and analysis, where social networks are this area of in which various forms of computational collective intelligence emerges in a natural way. The third subfield relates us to agent and multiagent systems understood as this computational and modeling paradigm which is especially tailored to capture the nature of computational collective intelligence in populations of autonomous individuals.

The book consists of 35 extended chapters which have been selected and invited from the submissions to the 4th International Conference on Computational Collective Intelligence Technologies and Applications (ICCCI 2012) held on November 28–30, 2012 in Ho Chi Minh City, Vietnam. All chapters in the book discuss theoretical and practical issues connected with computational collective intelligence and related technologies and are organized into six parts. Part I consists of three chapters in which authors discuss applications of computational intelligence technologies to modeling and usage of semantic web and ontology applications to solving real problems as well as natural language processing. Part II is composed of five chapters devoted to social networks and e-learning. Five chapters in Part III cover both methodological and practical problems related to agent and multiagent systems. Part IV of the book contains seven chapters in which theoretical and practical aspects of data mining methods including clustering, association rules, and time series as well as their application to face recognition are presented. Part V of the book is focused on soft computing. It encompasses eight chapters in which theoretical and practical aspects of evolutionary algorithms, probabilistic reasoning, artificial neural networks, and fuzzy systems are studied. The book completes with Part VI in which seven chapters pertain optimization and control methods including routing algorithms, management models and systems, and fuzzy control.

The editors hope that the book can be useful for graduate and Ph.D. students in Computer Science, in particular participants in courses on Soft Computing, Multiagent Systems, and Data Mining. This book can be also useful for researchers working on the concept of computational collective intelligence in artificial populations. It is the hope of the editors that readers of this volume can find many inspiring ideas and use them to create new cases of intelligent collectives. Many such challenges are suggested by particular approaches and models presented in individual chapters of this book. The editors hope that readers of this volume can find many inspiring ideas and influential practical examples and use them in their future work.

We wish to express our great attitude to Prof. JanuszKacprzyk, the editor of this series, and Dr. Thomas Ditzinger from Springer for their interest and support for our project.

The last but not least we wish to express our great attitude to all authors who contributed to the content of this volume.

November 2012

Ngoc Thanh Nguyen
Bogdan Trawiński
Radosław Katarzyniak
Geun-Sik Jo

Contents

Part I

Semantic Web and Ontologies

Recognizing and Tagging Vietnamese Words Based on Statistics and Word Order Patterns

Hieu Le Trung[1], Vu Le Anh[2], Viet-Hung Dang[1], and Hai Vo Hoang[2]

[1] Duy Tan University, Da Nang, Vietnam
[2] Hoa Sen University, Da Nang, Vietnam

Abstract. In Vietnamese sentences, function words and word order patterns (*WOPs*) identify the semantic meaning and the grammatical word classes. We study the most popular WOPs and find out the candidates for new Vietnamese words (NVWs) based on the phrase and word segmentation algorithm [7]. The best WOPs, which are used for recognizing and tagging NVWs, are chosen based on the support and confidence concepts. These concepts are also used in examining if a word belongs to a word class.

Our experiments were examined over a huge corpus, which contains more than 50 million sentences. Four sets of WOPs are studied for recognizing and tagging nouns, verbs, adjectives and pronouns. There are 6,385 NVWs in our new dictionary including 2,791 new noun-taggings, 1,436 new verb-tagging, 682 new adj-taggings, and 1,476 new pronoun taggings.

1 Introduction

Vietnamese, like many languages in Southeast Asia, is an analytic (or isolating) language. As such its grammar highly relies on word orders and sentence structure rather than morphology (word changes through inflection). A Vietnamese word may consist of one or more syllables. The lexicon is an open and dynamic system. There are many words popularly used but never listed in any recent dictionaries. This paper focuses on recognizing these new words and their corresponding word classes.

The first characteristic of our approach is utilizing the *systemic functional grammar* introduced by Halliday and colleagues [6]. Halliday's theory set out to explain how spoken and written texts construct meanings and how the resources of language are organised in open systems and functionally bound to meanings. C.X.Hao and other Vietnamese linguistics made a revolution in Vietnamese grammar when they pointed out that Vietnamese grammar must be considered as an systematic functional grammar [5,2,10]. They proved that function words and the word order patterns (*WOPs*) identify the grammar function and the classes of the other words in a phrase.

The second characteristic of this paper is the *word segmentation algorithm* introduced in our prior work based on statistics [7]. Under analysis and statistical data, Vietnamese is proved to be not really chaotic. Vietnamese sentences are

N.T. Nguyen et al. (Eds.): *Adv. Methods for Comput. Collective Intelligence*, SCI 457, pp. 3–12.
DOI: 10.1007/978-3-642-34300-1_1

the combination of phrases that usually appear together. By applying the word segmentation algorithm, we found that, the number of these phrases is only less than 500,000. These phrases can be grouped into classes, which based on WOPs.

The complete list of WOPs can be built based on the functional grammar and the word segmentation algorithm. The most popular word orders are listed by the algorithm before being manually tagged. With the linguists's support, all WOPs are determined by applying functional grammar rules. WOPs then can be used for recognizing and tagging words.

In this paper, we will study four sets of WOPs for recognizing nouns, verbs, adjectives and pronouns. The relationship between WOPs and word classes are examined with the support and confidence concepts. Only the best WOPs are chosen due to their confidence with a given word class. We also introduce the confidence and support concepts between a given word and a set of WOPs. They are used for determining if the word belongs to a word class.

Our first contribution is the process of constructing the list of WOPs based on the word segmentation algorithm and functional grammar. The second is the confidence and support concepts for examining the relationships among the tags, WOPs and words. They are used for choosing the best WOPs which recognize words tagged by the given taggers. The last contribution includes the list of WOPs and the list of new words for four taggers noun, verb, adjective and pronoun. All experiments have been done on the huge database with high accuaracy.

The organization of the rest is: Section 2 mentions the related works. The basic definitions and mathematical concepts are described in section 3. The algorithms for NVW recognition, WOPs construction are studied in section 4. Section 5 is about the experiments and the main results. Section 6 concludes the paper.

2 Related Works

Vietnamese word class classification, the fundamental problem of language research, has been argued and reached to many similar points in terms of creteria, the number and names of word classes. Nevertheless, these creteria have not unanimous due to different detailed creteria of Vietnamese researchers [3,5].

Agreeing with Nguyen H.C., we classify word types based on two creteria, "grammar function" [1] to see if a word can be the center of a phrase, and "composing ability" to see if it can be the auxiliary word building up the phrase. However, it should be noted that, the rules of these creteria in language text books are listed as several examples. Meanwhile, our rules system is contructed with programs on huge database based on word order patterns.

Most of research teams working on tagging Vietnamese word types start with the set of word types, which is admitted by many language experts (nouns, verbs, adjective, adverb, prep., ...), and then categorize words into new subtypes with more detailed creteria due to the purposes and methods of their own.

Nguyen Q.C. et. al. [8,9] constructed a note system with 48 tags of word types on 10 limited domains on entries. The VnPOSTag [14] has 14 tags and 1 undefined tag; the VnQtag [16] has 48 tags and 1 underfined tag while VCL system has 13 main tags with 29 sub-tags. Dien D. et.al. [11] built their tagging system by mapping English tag set. To our knowledge, previous works' authors mainly focused on the methodologies of solving tagging problem, but not on building up a tag set that can be utilized by research communities. Moreover, word type exchanging phenominon in Vietnamese was not considered thoroughly in tagging problem [8] in computer science.

Algorithms for POS tagging focus on 3 main approaches: based on grammar principles, on statistical probability and on both of them.

Based on grammar priciples, Nguyen Q.C., Phan T.T. et al. [8,9] built a system that combines tri-gram pos-tagger and writing style pos-tagger. The system consists of 270 rules and 48 tags, extracted from more than 70.000 words. The result accuracy is 80% with statistics based method and is 90% with statistics and grammar based method.

Conducting research based on statistics based method, Nguyen T.M.H. et al. proposed tri-gram QTAG [16]. QTAG was trained and tested with texts of 63.732 words. The best accuracy resutled is 94% with the set of 9 tags and 85% with the set of 48 tags.

Based on Machine Learning tools, models of CRF and MEM, Phan X.H. and Nguyen C.T. et al.[15] created VLSP pos-tagger, which was trained from 10.000 sentences and the label set of Viet Treebank. CRFs give the best accuracy of 90.4% while MEM gives 91.03%. There have also been other related research groups, such as Oanh T.T. et al. conducting vnPOS in 2008, Dien D. et al. using hybrid TBL method to build up the language database (LDB) of bilingual English-Vietnamese dictionary, consisting of more than 500.000 sentences [11].

Generally, the above research activities are developed independently and there are no comparisons under common criteria. Most of them are examined and built based on small LDBs with the goal of characterizing several features, testing methods, and therefore are not enough to represent all the characteristics of language usage, tags as well as grammar structures of Vietnamese. The achievements so far are still at lab's work with low applicability and low inheritability. They do not fullfil the need in reality for a strong fundamental base where Vietnamese language processing can thrive.

3 Basic Definitions and Mathematics Concepts

Basic concepts are introduced in subsection 3.1. WOPs are studied in subsection 3.2. The confidence and support concepts are studied in subsection 3.3.

3.1 Basic Concepts

Syllables, Words and Sentences. The set of Vietnamese syllables is represented by Σ_s. Vietnamese words consist of one or more syllables. The set of Vietnamese words is represented by $\mathcal{W} \subseteq \Sigma_s^*$. Each sentence is a sequence of words. The set of Vietnamese sentences is represented by $\mathcal{L} \subseteq \mathcal{W}^*$.

Tags represent classes of word types. The finite set of tags is denoted by Σ_T. $\Theta : \mathcal{L} \mapsto \Sigma_T^*$ is the POS tagging function in which each sequence of words $q = w_1 \dots w_l$ is tagged by a sequence of tags $\Theta(q) = t_1 \dots t_l$ $(t_i \in \Sigma_T)$. That means, w_i is tagged by t_i in the sequence of words q. For each $t \in \Sigma_T$, $w \in \mathcal{W}$, we define two sets $W(t) := \{w \in \mathcal{W} \mid \exists q \in \mathcal{L}_s : w \text{ is tagged by } t \text{ in } q\}$ and $T(w) := \{t \in T \mid \exists q \in \mathcal{L} : w \text{ is tagged by } t \text{ in } q\}$.

Corpus is the finite set of sentences represented by $\mathcal{C} \subseteq \mathcal{L}$.

Tagged Corpus. $\mathcal{T} = \{p = w_1[t_1] \dots w_l[t_l] \mid w_i \in \mathcal{W},\ t_i \in \Sigma_T\}$ is a finite set of tagged phrases. Tagged phrase $p = w_1[t_1] \dots w_l[t_l]$ is a sequence of word/tag pairs, in which word w_i is tagged t_i in p, for al $i = \overline{1, l}$.

Dictionary $\mathcal{D} = \{(w, t) | w \in \mathcal{W}\ , t \in \mathcal{T}\ (w)\}$ is a finite set of word/tag pairs, where $w \in \mathcal{W}$ is a *new Vietnamese word* (NVW) to dictionary $\mathcal{D}$, if $\nexists t \in \Sigma_T$: $(w, t) \in \mathcal{D}$.

3.2 Word Order Pattern

Frequently word order (FWO). In the previous work [7], we have shown that: (i) Vietnamese is not chaotic, each sentence is a combination of *frequently word orders* (FWO); (ii) The list of frequently word orders can be determined by applying unsupervised learning processing with some linking function. In this paper, we assume that the list of FWOs is computed already. Our prior work should be referenced for more details and better reading of this work. FWO is tagged manually and is denoted by $\mathcal{C}$.

Word order pattern (WOP). WOPs occur frequently in language. Hence, WOPs can be determined by using FWO. We define that a WOP is a sequence of tags which are used for tagged some FWO. The finite set of WOPs is represented by $\mathcal{P}$.

A word $w \in \mathcal{W}$ *matches* a WOP $p \in \mathcal{P}$ on the corpus $\mathcal{C}$ if $\exists q \in \mathcal{C}$ which contains w and tagged by p. Suppose $\mathrm{W}_\mathcal{C}(p) = \{w \in \mathcal{W} | w \text{ matches by } p \text{ on } \mathcal{C}\}$. The *popular (or support)* of p for tagger $t \in \mathcal{T}$ over $\mathcal{C}$ is defined as follows:

$$Sup_\mathcal{C}(p, t) := \frac{|\mathrm{W}_\mathcal{C}(p) \cap W(t)|}{|W(t)|}$$

We have $Sup_\mathcal{C}(p, t) \leq 1$, and $Sup_\mathcal{C}(p, t)$ can be considered as the probability that a word tagged by t will match p.

Suppose $Q \subseteq \mathcal{P}$ is a set of WOPs and $\mathrm{W}_\mathcal{C}(Q) = \{w \in \mathcal{W} | w \text{ matches by some } p \in Q \text{ on } \mathcal{C}\}$. The *popular (or support)* of p for tag $t \in \mathcal{T}$ over $\mathcal{C}$ is defined as follows:

$$Sup_\mathcal{C}(Q, t) := \frac{|\mathrm{W}_\mathcal{C}(Q) \cap W(t)|}{|W(t)|}$$

$q = w_1 \dots w_l \in \mathcal{C}$ *is able to be tagged* by a WOP $p = t_1 \dots t_l \in \mathcal{P}$ if $t_i \in T(w_i), i = 1, \dots, l$ Suppose $\mathrm{L}_\mathcal{C}^{all}(p) = \{q \in \mathcal{C} \mid q \text{ is able to be tagged by } p\}$, $\mathrm{L}_\mathcal{C}(p) = \{q \in \mathcal{C} \mid q \text{ is tagged by } p\}$. The confidence of p is defined as follows:

$$Conc(p) := \frac{|\mathrm{L}_C(p)|}{|\mathrm{L}_C^{all}(p)|}$$

We have $Conc(p) \leq 1$. In the case $Conc(p) \sim 1$ we can use p to recognize the tagger of any word matching p.

4 WOPs and NVW Recognition

The process of WOP construction is as follows:

Algorithm: WOPs construction
input: Huge number of sentences
output: Set of WOPs
1. *Computing set of FWOs from huge number of sentences .*
2. *Tagging FWO manually.*
3. *Listing WOPs which is the tager of some WOP.*
4. *Computing the confidence of WOPs.*
5. *Ignoring WOPs whose confidence are not higher than* min_{con}.

FWOs construction is an unsupervised process which is described in our previous work [7]. Each FWO is tagged manually using many taggers represented for different types of function words. From the tagged FWOs, the list of WPOs are computed. It should be noted that the confidence of a WOP is strongly dependent of the number of its function word tags. Hence, after step 5, only the WOPs which contains many function word tags are chosen. It is predictable that these WOPs are mentioned in grammar works [5,1]. The min_{con} is chosen by the experiments described in section 5.

The different sets of WOPs are determined for recognizing different word classes: noun, verb, adjective and pronoun. The details of the process are:

Algorithm: Constructing WOPs for recognizing NVWs
input: Tagger t and the set of WOPs, Q, which contains t.
output: The set of WOPs, $P \subseteq Q$, is used for recognizing NVW with tag t.
1. *Sorting* Q *as the list* $p_1, p_2, \ldots, p_n$ *by the confidence decrease .*
2. $P = \emptyset$, $i = 1$.
3. repeat $P = P \cup \{p_i\}$, $i = i + 1$
4. until $Sup_C(P, t) = 1 \vee i = n + 1$
5. return P

The first algorithm guarantees that all WOP are confident, while the second algorithm guarantees that a set of WOP are fully supported for recognizing NVWs.

The algorithm for checking if a word w is a word class t is straightforward. Suppose P is the set of WOPs using for the recognition. The confidence and support values of w and P are computed using formula described in section 3. If these values are greater than predefined thresholds $(\theta_{sup}, \theta_{con})$ then they are recognized.

Finally, in the case we want to whether a sequence of syllables is a NVW or not, for each tag t we will examine if it belongs to word class t. The examination is based on the set of WOPs using for t recognition. If w passes all examinations, it is a NVW.

5 Experiments and Main Results

The environment of our experiments is mentioned in subsection 5.1 where the corpus or the POS tagset will be described. Subsection 5.2 represents the main results of the experiments on funtional words and WOPs. The results of NVW recognition is shown in subsection 5.3.

5.1 Experiment Environment

Corpus, FWOs and Tagged Corpus. Our database of sentences is collected from 250.034 articles in the TuoiTre online newspaper and more than 9.000 novels in VNThuquan website. After applying data normalization (fixing the code fonts and repairing spelling mistakes of syllables) and sentence segmentation [7], the initial corpus has 790.565 FWOs with 3.488.698 syllables. The tagged corpus is generated from 122.967 FWOs (about 15.5%) including 418.112 Vietnamese word/tag pairs. Each chosen FWO contains at least 2 words and must have unique POS tagging solution. The tagged corpus will be used for WOP construction.

Dictionary, POS system. Our initial dictionary is built from two ditigital sources: (i) http://vi.wiktionary.org/wiki/ and (ii) http://tratu.soha.vn/. There are 31, 127 words in the dictionary. Our POS tagset is proposed from the linguists. It has 40 tags, with 5 main tags for content words ([NN], [VB], [Adj], [NPpp] and [NPpl]) and 35 tags for function words. Descriptions, the number of words and occurrences for all tags of our system are shown in Table 1.

5.2 Function Words and WOPs System

Experiments have been done on function words and important results are presented in this Subsection.

- *The number of function words is not big but they occur very frequently.* There are 2,061 function words in our tagged corpus database (about 6.6%), but the percent of their occurence is up to 48.5%. In addition, the statistics indicate that for every content word, there exists at least one function word that frequently appears nearby it.
- *Most of function words have a unique tag.* There are 1,950 words (94.6%) that have only one tag.

These statistics confirm again that function words play a very important role in Vietnamese grammar.

Table 1. The POS tagset

No.	Tag	Description	NoW	Num	No.	Tag	Description	NoW	Num
1	Adj	Adjective	7.033	29.373	21	NNv	Verbal Nouns	52	4.031
2	ADn	Adverb behind Verb	83	7.387	22	PNde	Demonstrative Pronouns	40	6.796
3	ADp	Adverb before Verb	107	21.757	23	PNpl	Pronouns of Place	64	1.243
4	CC	Coordinating conjunction	6	5.326	24	PNtm	Pronouns of Time	13	5.486
5	CD	Subordinating conjunction	54	8.645	25	PNir	Interrogative Pronouns	35	6.741
6	DPc	Indefinite Pronouns	33	2.689	26	PNps	Personal Pronouns	108	17.264
7	DPe	Predeterminer	12	11.517	27	PNob	Object Pronouns	62	12.477
8	INpo	Posessive Prepositions	4	8.694	28	PNpo	Posessive Pronouns	225	7.749
9	INpl	Prepositions of Place	14	12.581	29	PNrp	Replexive Pronouns	56	5.811
10	INtm	Prepositions of Time	8	1.346	30	PNrl	Relative Pronouns	56	2.782
11	INrs	Prepositions of Result	3	810	31	PNtm	Pronouns of Time	194	8.453
12	INpt	Prepositions	6	15.542	32	PNlv	Pronouns of Level	32	6.214
13	JDn	Adverb behind Adjective	19	1.777	33	NPpp	Proper Noun of People	3.788	6.235
14	JDp	Adverb before Adjective	20	5.275	34	NPpl	Proper Noun of Place	2.367	3.741
15	NBc	Cardinal Number	106	3.399	35	RB	Interjection	220	4.248
16	NBe	Ordinal Number	134	12.341	36	ID	Idiom	465	523
17	NBn	Nomial Number	125	9.084	37	VBe	Specific Verb	7	6.054
18	NN	Noun	13.842	96.063	38	VBm	Modal Verb	66	10.974
19	NNc	Collective Noun	64	9.050	39	VBa	Auxilliary Verb	28	25.654
20	NNe	Specific Nouns	5	578	40	VB	Verb	9.354	80.219

WOPs system. A total WOPs from the tagged corpus includes 3,042 patterns. Table 2 shows some WOPs and their numbers of occurrences (Num), values of support (Sup) and confidence (Conf). We found that:

- *The confidence of a WOP strongly depends on the number of function words and content words in that WOP.* There are 82 and 315 WOPs whose confidence is higher than 0.9 and 0.7 respectively. These WOPs have only one content word. The confidence of a WOP is increased if the number of function words is increased. The reason of these results is that function words determine the functional grammar of the neighbour words.
- *More than 80% words of a given tag match a WOP recognizing that tag.* The average of support value of WOPs is 0.75. It implies that *"Using WOPs is a good solution for recognizing NVW"*.

With $min_{cof} = 0.9$, we have 82 WOPs (see Table 3). These WOPs are refined again for recognizing nouns, verbs, adjectives and proper nouns. Finally, we have: (i) 31 WOPs (support = 0.68) for recognizing nouns (ii) 25 WOPs (support = 0.74) for recognizing verbs (iii) 14 WOPs (support = 0.86) for recognizing adjectives (i) 12 WOPs (support = 0.71) for recognizing proper nouns.

5.3 NVW Recognition

With $min_{cof} = 0.9$, we run the recognition algorithm over our untagged corpus again. There are 6,385 NVWs recognized including 2,791 new nouns, 1,436 new

Table 2. List WOPs and Number Occurrences, Support, Confidence

WOPs	Num	Sup	Conf
[PNps][VB][NBe]	648	0.03	1.00
[VB][INpt][DPe]	566	0.03	1.00
[INpo][NP]	420	0.02	0.96
[VBa][Adj][PNps]	782	0.02	0.91
[INpl][NBc][NN]	613	0.02	0.9
[NP][ADp][VB]	477	0.02	0.89
[NN][PNde][ADp]	2.077	0.1	0.84
[NBe][NN]	478	0.09	0.84
[PNde][JDp][Adj]	625	0.05	0.75
[PNde][VBe][Adj]	479	0.06	0.74

Table 3. Statistics of WOPs

No.	min_{con}	N.WOPs	N.NN	S.NN	N.VB	S.VB	N.Adj	S.Adj	N.NP	S.NP
1	0.95	38	14	0.63	11	0.67	7	0.75	6	0.65
2	0.9	82	31	0.68	25	0.74	14	0.86	12	0.74
3	0.85	133	59	0.71	36	0.79	21	0.89	17	0.79
4	0.8	198	71	0.79	59	0.85	38	0.92	32	0.83
5	0.75	237	84	0.85	71	0.91	45	0.97	37	0.90
6	0.7	315	113	0.95	87	0.97	61	0.99	54	0.94

Table 4. List of WOPs and FWOs of NVW *trình_duyệt*

No.	WOPs	N.M	FWOs
1	[DPe][NN][INpo]	18	những[DPe] trình_duyệt[NN] của[INpo]
2	[NN][INpo][PNps]	12	trình_duyệt[NN] của[INpo] bạn[PNps]
3	[VBe][NBe][NN]	8	là[VBe] một[NBe] trình_duyệt[NN]
4	[NN][PNde][ADp]	5	trình_duyệt[NN] này[PNde] đã[ADp]
5	[DPe][NN][PNde]	5	những[DPe] trình_duyệt[NN] khác[PNde]
6	[DPc][DPe][NN][ADp]	4	hầu_hết[DPc] các[DPe] trình_duyệt[NN] đều[ADp]

verbs, 682 new adjectives and 1,476 new proper nouns. The accuaracy is 100%. The explanation is functional grammar of WOP.

Let us consider an example. *"trình_duyệt"* is recognized as a new noun. Here are top 6 of WOPs in which *"trình_duyệt"* matches. (i) WOP1 with 18 times (ii) WOP2 with 12 times (iii) WOP3 with 8 times,... The matching FWOs is shown in Table 4. Clearly, with these evidences *"trình_duyệt"* is definitely a new-noun.

6 Conclusion

The function words and the WOPs identify the grammar functions and the word classes of the other words in Vietnamsese sentences. The WOPs can be constructed by applying the word segmentation algorithm and functional grammar. We determined the list of most popular word orders and then tagged them manually. Based on that corpus and the linguists' support, we fulfilled the list of WOPs.

The new words of a given word class can be found by using the built set of WOPs. The best of WOPs for the word class recognition are determined by the confidence and the support concepts. The confidence is defined by the conditional probability formula to measure the quality of a WOP to the word class. The support is determined by the populartity of the WOP over the word class. The best set of WOPs is the set of the most confident WOPs and the support of the set must be greater than the support threshold. Similarly, we also define the confidence and support concepts between a word and a given set of WOPs. They measure the quality and the quantity of the fact that the word belongs to the word class. Hence, they are used for examining if a word belongs to a given word class.

We have examined our theory over a huge corpus, which contains more than 50 million sentences to extract the most popular Vietnamese word orders and compute the statistical data. We introduce 4 sets of WOPs for recognizing the noun, verb, adjective and pronoun. Finally, we have recognized 6,382 NVWs in our new untagged corpus, including 2,791 new nouns, 1,436 new verbs, 682 new adjectives, 1,476 new proper nouns. Our tagged corpus can be updated online by our system.

We plan to apply WOPs in POS tagging and grammar tree analyzing problems. These ideas come from the fact that Vietnamese grammar depend strongly on the WOPs. Moreover, that the word string is considered as the basic unit for examining the grammar will make the process more accurate and less time consuming than the approach examining word by word. Obviously, if we have the mapping which maps each popular word string to a WOP then our new algorithms can analyze the grammar tree more efficiently than the classical ones.

References

1. Hao, C.X.: Vietnamese: Draft of Functional Grammar. Education Publisher, Hanoi (2004) (in Vietnamese)
2. Hao, C.X.: Vietnamese - Some Questions on Phonetics, Syntax and Semantics. Education Publisher, Hanoi (2000) (in Vietnamese)
3. Ban, D.Q.: Vietnamese Grammar. Education Publisher, Hanoi (2004) (in Vietnamese)
4. Chu, M.N., Nghieu, V.D., Phien, H.T.: Linguistics Foundation Vietnamese. Education Publisher, Hanoi (1997) (in Vietnamese)
5. Thuyet, N.M., Hiep, N.V.: Components of Vietnamese Sentence. Hanoi National University Publisher (1998) (in Vietnamese)

6. Halliday, M.A.K.: Introduction to functional grammar, 2nd edn. Edward Arnold, London (1994)
7. Le Trung, H., Le Anh, V., Le Trung, K.: An Unsupervised Learning and Statistical Approach for Vietnamese Word Recognition and Segmentation. In: Nguyen, N.T., Le, M.T., Świątek, J. (eds.) ACIIDS 2010, Part II. LNCS (LNAI), vol. 5991, pp. 195–204. Springer, Heidelberg (2010)
8. Chau, Q.N., Tuoi, T.P.: A Pattern-based Approach to Vietnamese Key Phrase Extraction. In: Addendum Contributions of the 5th International IEEE Conference on Computer Sciences - RIVF 2007, pp. 41–46 (2007)
9. Chau, Q.N., Tuoi, T.P.: A Hybrid Approach to Vietnamese Part-Of-Speech Tagging. In: Proceedings of the 9th International Oriental COCOSDA Conference (O-COCOSDA 2006), Malaysia, pp. 157–160 (2006)
10. Hai, L.M., Tuoi, P.T.: Vietnamese lexical functional grammar. In: The 1st International Conference on Knowledge and Systems Enginnering, pp. 168–171 (2009)
11. Dien, D., Kiem, H., Toan, N.V.: Vietnamese Word Segmentation. In: The Sixth Natural Language Processing Pacific Rim Symposium, Tokyo, Japan, pp. 749–756 (2001)
12. Ha, L.A.: A method for word segmentation in Vietnamese. In: Proceedings of Corpus Linguistics 2003, Lancaster, UK (2003)
13. Phuong, L.H., Huyên, N.T.M., Roussanaly, A., Vinh, H.T.: A Hybrid Approach to Word Segmentation of Vietnamese Texts. In: Martín-Vide, C., Otto, F., Fernau, H. (eds.) LATA 2008. LNCS, vol. 5196, pp. 240–249. Springer, Heidelberg (2008)
14. Oanh, T.T., Cuong, L.A., Thuy, H.Q., Quynh, L.H.: An Experimental Study on Vietnamese POS Tagging. In: International Conference on Asian Language Processing, IALP 2009, pp. 23–27 (2009)
15. Nguyen, C.T., Nguyen, T.K., Phan, X.H., Nguyen, L.M., Ha, Q.T.: Vietnamese word segmentationwith CRFs and SVMs: An investigation. In: Proceedings of the 20th Pacific Asia Conference on Language, Information and Computation (PACLIC 2006), Wuhan, CH (2006)
16. Huyen, N.T.M., Luong, V.X., Phuong, L.H.: A case study of the probabilistic tagger QTAG for Tagging Vietnamese Texts. In: Proceedings of the First National Symposium on Research, Development and Application of Information and Communication Technology, Vietnam (2003)

A Combination System for Identifying Base Noun Phrase Correspondences

Hieu Chi Nguyen

Ho Chi Minh University of Industry,
12 Nguyen Van Bao, Go Vap District, Ho Chi Minh City, Vietnam
nchieu@hui.edu.vn

Abstract. Bilingual Base Noun Phrase extraction is one of the key tasks of Natural Language Processing (NLP). This task is more challenges for the pair of English-Vietnamese because of the lack of available Vietnamese language resources such as robust NLP tools and annotated training data. This paper presents a bilingual dictionary-, a bilingual corpus- and knowledge-based method to identify Base Noun Phrase correspondences from a pair of English-Vietnamese bilingual sentences. Our method identifies anchor points of the Base Noun Phrase in English sentence, and then it performs alignment based on these anchor points. Our method not only overcomes the lack of resources of Vietnamese but also improves the performance of miss-alignment, null-alignment, overlap and conflict projection of the existing methods. The proposed technique can be easily applied to other language pairs. Experiment on 35,000 pairs of sentences in the English-Vietnamese bilingual corpus showed that our proposed method can obtain the accuracy of 78.5%.

Keywords: Base Noun Phrase, anchor points, BasedNP pairs.

1 Introduction

The extraction of bilingual Base Noun Phrases (BaseNP) in the Bilingual corpus is an important task for many natural language processing systems, especially; it is the key task in Machine Translation systems. In the extraction of bilingual Base Noun Phrases, identifying BaseNP correspondences plays an important role. However, one of the primary problems for NLP researchers working in new languages or new domains is a lack of available annotated data (like Penn Treebank). A collection of data is neither easy nor cheap. While previous empirical and Machine Learning methods for identifying BaseNP correspondences of pairs of resource-rich languages such as English-French of Kupiec [2] (1993), English-German of Koehn [14] (2003), English-Japanese of Wantanabe et al. [7] (1999), Japanese-Korean of Hwang et al [18]. (2004), and English-Chinese of Wang et al. (2002)[13], Yonggang Deng [19] (2005) have been quite completed, it is still a challenge for a pair of English-Vietnamese. The challenge of the task becomes greater due to absence of robust NLP tools (POS tagging, morphology analyses, NP chunking, etc) and annotated training data for Vietnamese (*Vietnamese Treebank*).

In this paper, we present a new combination system called "Identifying Base Noun Phrase Correspondences" from a pair of English-Vietnamese sentences. While the

N.T. Nguyen et al. (Eds.): *Adv. Methods for Comput. Collective Intelligence*, SCI 457, pp. 13–23.
DOI: 10.1007/978-3-642-34300-1_2

existing approaches which align chunks such as Kupiec [2] (1993), Wang et al. (2002)[13] or project cross-language such as Yarowsky et al. [8][9] (2001), Riloff et al. [12] (2002), acquire BaseNP correspondences after getting results of word alignment, our method identifies anchor point (left and right bound of BaseNP) of transformed source sentence on the target sentence language and align word in them. In word alignment, we limit the search domain by a beam threshold. The use of this approach aims at improving the quality of misalignment or null alignment, overlap and conflict.

Furthermore, our model can perform the extraction on a pair of sentences and solve the lack of available English-Vietnamese bilingual corpus. This model does solve not only the problem of English-Vietnamese BaseNP extraction but can be also applied to pairs of other languages which are similar to English-Vietnamese. The experimental results with English- Vietnamese Base Noun Phrase Extraction show that our model can obtain 78.5% (precision), 73% (recall) on the 35,000 pairs of bilingual sentences.

The remain of this paper is organized as follows: section 2 shows the related works, section 3 presents our method to identify BaseNP Correspondences, our experimental results are discussed in section 4. Finally, section 5 is our conclusion.

2 Related Works

In previous works, some researchers proposed approaches and methods that make use of annotated data and tools for resource-rich languages to overcome the annotated resource shortage in other languages, e.g. Vietnamese [8][9]. Yarowsky proposed methods *"that use automatically word-aligned raw bilingual corpus to project annotations from resource-rich language to arbitrary second language"*. The methods apply to French, Chinese, Spanish, and Czech to induce POS-taggers, base noun phrase brackets, base named-entity taggers. However, Yarowsky has not yet solved the problem of aligned to null, overlap and conflict.

Besides Yarowsky's works, there are some previous works presented in [17] (2003), [14] (2003), [11] (2004), etc. D. Dien proposed a method that directly projects POS-annotations from the English side to Vietnamese side via available word alignments. However, D. Dien has just performed 1000 words POS-tagging and word-alignment. Koehn developed a corpus for a model translating and extracting bilingual noun phrase. The model was applied to the English and German languages, which are tool- and resource-rich. Resnik and Hwa [9] (2002) proposed a projection model with assumptions about the existing of two dependent parse trees and a Direct Correspondence Assumption (DCA).

Kupiec [2] (1993) proposed a method to extract bilingual noun phrases using statistical analysis of the co-occurrence of phrases. The Kupiec's method is an instance of a general approach to statistical estimation, represented by the Expectation Maximization (EM) algorithm. Wang et al. (2002)[13] proposed a bilingual chunking model for structure alignment, which comprises three integrated components: chunking models of both languages and cross-constraint; using chunk as the structure. The crossing constraint requests a chunk in one language that corresponds at most to one chunk in the other language.

Our surveying previous works concludes that aligned to null, cross-reference and conflict problems are still the open problems in current research of English-Vietnamese

NLP. This paper introduces our combination system that helps to invert word order in noun phrases of the source language (to suit those of target language and aligns word on the anchor points for cross-reference and conflict problems solving). Furthermore, our heuristics is proposed to solve the empty word-alignment.

3 Our Proposed Methodology

*** Task *description***

Table 1. Examples of BaseNP pairs in a pair of bilingual sentences

A pair of bilingual sentences	Aligned BaseNP pairs
Se: "This is the first year5 since transportation7 deregulation in 1980 that we have had a dramatic and broad-based upturn in perceived transportation22 rates", said Bernard Alone, a transportation28 logistics professor at Ohio State in Columbus.	[*This*: day]; [*the first year*: name đầu tiên]; [*transportation deregulation*: bãi bỏ quy định về vận chuyển]; [*We*: chúng tôi]; [*a dramatic and broad-based upturn*: một sự tăng trưởng ấn tượng và trên bình diện rộng]; [*perceived transportation rates*: tỷ suất vận chuyển nhận biết được]; [Bernard LaLonde: Bernard LaLonde]; [*a transportation logistics professor*: một giáo sư vận chuyển tiếp vận]; [*Ohio State*: bang Ohio]; [*Columbus*: Columbus]
Sv: "Đây là năm3 đầu tiên kể từ khi bãi bỏ quy định về vận chuyển10 năm11 1980 rằng chúng tôi đã có một sự tăng trưởng đầy ấn tượng và trên bình diện rộng trong tỷ suất vận chuyển27 nhận biết được", theo Bernard LaLonde, một giáo sư vận chuyển35 tiếp vận ở bang Ohio tại Columbus.	

Give a pair of English-Vietnamese bilingual sentences $S_e = w_e^1 w_e^2 ... w_e^u$ and $S_v = w_v^1 w_v^2 ... w_v^t$, where u and t is the sentence length of English and Vietnam sentence. A sequence of NP chunks can be represented as:

$$B_e = [w_e^1 w_e^2]^1 ... [w_e^{k-2} w_e^{k-1} w_e^k]^i ... [w_e^{u-1} w_e^u]^m = n_e^1 ... n_e^i ...$$

and $B_v = [w_v^1 w_v^2]^1 ... [w_v^{k-2} w_v^{k-1} w_v^k]^i ... [w_v^{u-1} w_v^u]^m = n_v^1 ... n_v^i ...$

Where, n_e^i denotes the ith NP chunk of S_e, and n_v^i denotes the ith NP chunk of S_v. Finding n_v^i correspondence of n_e^i.

**** Comparison of the structure of English and Vietnamese noun phrase***

According to Quirk and Green Baum [1] (1990), the positions for elements in a noun phrase in English are as follows: (1) determiners (predetermine and/or determiner), (2) numerals (ordinal, cardinal), (3) descriptive adjectives (in the order: opinion, size, shape, color), (4) adjectival of place of origin or nationality, (5) adjectival nouns (or noun adjuncts), (6) the head noun, (7) adjectival phrases, (8) adjectival predicates, and (9) adjectival clauses.

But according to the part-of-speech result of English, the basic noun phrase patterns are found to be as follows:

EbaseNP → Pre-determiner + Determiner + Numberals + Modifier + head noun
Modifier → Adjectives* + Adjectival
Adjectival → present participlel past participle I nounl ∍
EBaseNP → EBaseNP + conj + EbaseNP

This result agrees with Marcu's comments (1995) "*The goal of the BaseNP chunks was to identify essentially the initial portions of non-recursive noun phrases up to the head including determiners but not including post modifying prepositional phrases or clauses*". Like this, a BaseNP includes the elements to be shown in Fig 1.

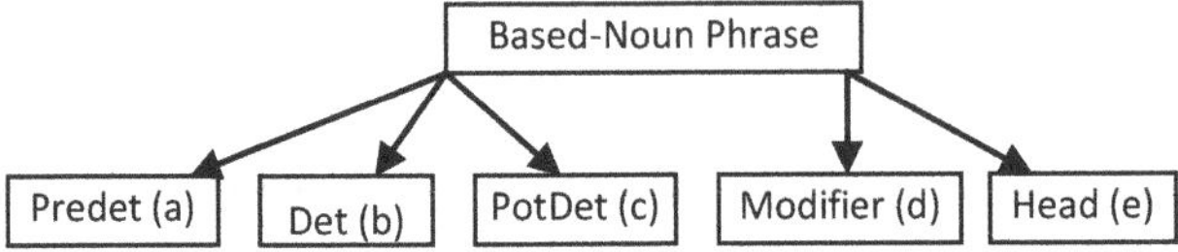

Fig. 1. Based-Noun Phrase elements in English

These structures correspond with first six positions in Quirk and Green Baum [1]. According to Nguyen Tai Can (1999)[6], the positions of elements in a noun phrase in Vietnamese are as follows: (1) pre-determiner (2) numerals (plural particles and/or cardinal numerals), (3) identifier, (4) classifiers, (5) the head noun, (6) Adjectival (noun, verb), (7) Adjectives (of shape and/or color, ordinal numeral), (8) phrases, (9) possessives, (10) included clauses, and (11) post-determiners.

However, according to a BaseNP is a simple noun phrase that does not contain another noun phrase recursively, Vietnamese BaseNP cuts down (8) phrases, (10) clauses. So they are found to be as follows:

VBaseNP→ Predetermine + Numerals + identifier + classifier + head noun + Modifier + Other
Modifier → Adjectival + Adjectives*
Other → possessive + post-determiner
VBaseNP → VBaseNP + conj + VbaseNP

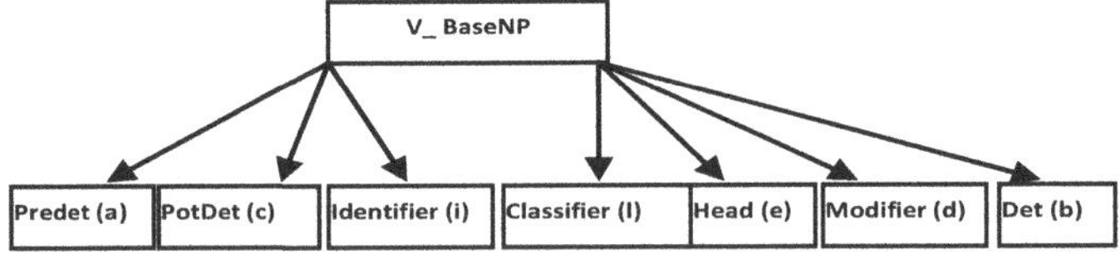

Fig. 2. Based-Noun Phrase elements in Vietnamese

Certain differences may be noted between English Base Noun Phrases and Vietnamese Base Noun Phrases:

While predetermine and determiner occurs only in the first positions in a Base Noun Phrases in English, Vietnamese determiners are of two kinds: those which occur in the fourth position in a BaseNP (herein called "*classifier*"), and those in the eleventh position in a BaseNP (herein called "*post-determiner*"). Vietnamese has, in addition, a separate position for the identifier *cái*, which functions very much like the determiner *the* in English.

While numerals (both ordinal and cardinal) occur in the second position in a BaseNP in English, cardinal numerals in Vietnamese fill the first position, while ordinal numeral BaseNPs occupy the seventh position. In addition, Vietnamese also has plural particles, which occur in the first position in a BaseNP when the plural reference of the BaseNP needs to be emphasized (The notable exception is Nguyen Tai Can (1999)[6], who claims that *một*, *những* and *các* are articles in Vietnamese).

While different kinds of adjectives occur in the 3rd position in a BaseNP in English, the adjectives in Vietnamese occur in the 7th position in a Vietnamese BaseNP.

Table 2. Our approach's POS & symbol

Symbol	English NPhrase	Vietnamese NPhrase
a	predetermine	predetermine
b	determiner	possessives, post-determiners
c	numerals	numerals
d	adjectives, adjectival nationality, adjectival nouns	adjectivaux (noun, verb), phrases, adjectives, clauses
e	head noun	head noun
f	adjectival phrases, adjectival predicates, adjectival clauses	*null*
i	*null*	identifier
l	*null*	classifiers
In our research: a = {PDT}; b = {DT, PRP$}; c = {CD}; i = {CA, PL}; l = {CL}; d = {VBG, VBN, JJ+, JJR, JJS, NN, NNS, NNP, NNPS}; e = {NN, NNS, NNP, NNPS}; Hybrid's POS = Penn Treebank's POS ∪ {CA, CL, PL}		

**** System Architecture***

The Combination System for Identifying Base Noun Phrase Correspondences on a pair of bilingual sentences in this paper is based on the projection methodology of Yarowsky et al. [8][9] (2001), with some important improvements. First, we determine anchor points of a replacing base Noun Phrase of the English sentence (left and right bound of base Noun Phrase). Translating the Vietnamese Noun Phrase into English word by word (using a Parse Tree) forms the replacing base Noun Phrase. After that, we find the corresponding anchor points by the DictAlign methodology on S.J. Ker and J.S. Chang [5] (1997) with a δ threshold. However, instead of S.J. Ker and J.S. Chang's dislocation, δ threshold is determined by an experiment to solve the conflict in word alignment. The system architecture is shown in Figure 3. The result of experiment on the δ threshold is shown in Figure 4. We explored the δ threshold on three types of corpus. Series 1 is built on the Patterns of English of H.V. Buu [4] (1996). Series 2 is built on the English for Compute science of T.V. Hung [3] (1995). Series 3 is built on the WSJ1.

There are 100 pairs of bilingual sentences on a type. The outcome of the exploration shows that the maximal precisions of BaseNP extraction of the three types of corpus are different. After experimenting, we choose 2.7 for δ threshold.

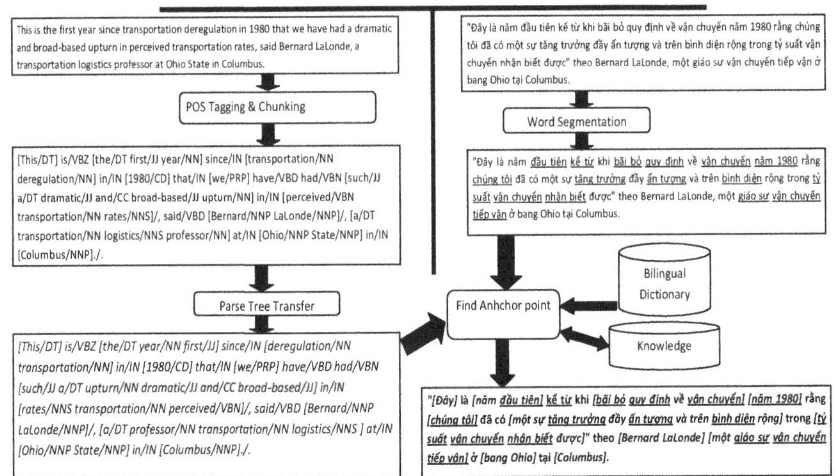

Fig. 3. System Architecture for identify BaseNP correspondence on a pair of English-Vietnamese bilingual sentences

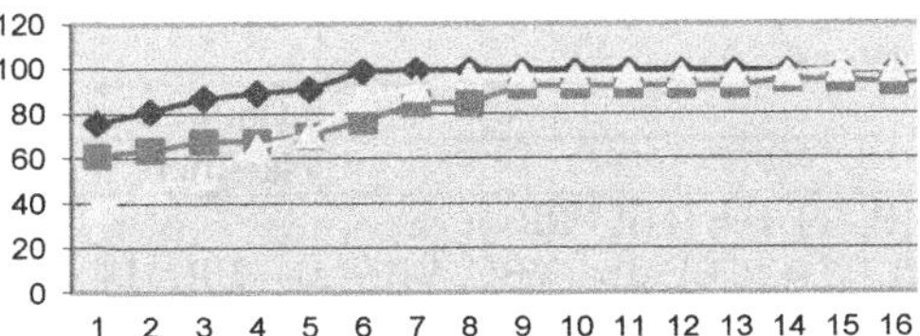

Fig. 4. Effect of the δ values to the precision

Table 3. Experimented values

No	δ	No	δ	No	δ
1	*0.5*	8	*1.7*	15	***2.7***
2	*0.7*	9	*1.8*	16	*2.9*
3	*0.9*	10	*1.9*	17	*3.5*
4	*1.2*	11	*2*	18	*4.5*
5	*1.4*	12	*2.1*	19	*5.5*
6	*1.5*	13	*2.2*	20	*6.5*
7	*1.6*	14	*2.5*	21	*7.5*

We apply the English resource and English-Vietnamese comparison knowledge to build the bilingual BaseNP identification and extraction model as follows:

1. To Align sentences in the parallel corpus
2. To tag words in sentences with POS information, analyze English sentences to recognize noun phrases (H. Cunningham et al., 2002[10]),

3. To rearrange the word order of noun phrases in English sentences according to those of Vietnamese sentences and determine the anchor points.
4. To segment Vietnamese words [20]
5. To Align Words on the anchor points
6. To extract English-Vietnamese BaseNP

**** Proposed Algorithms***

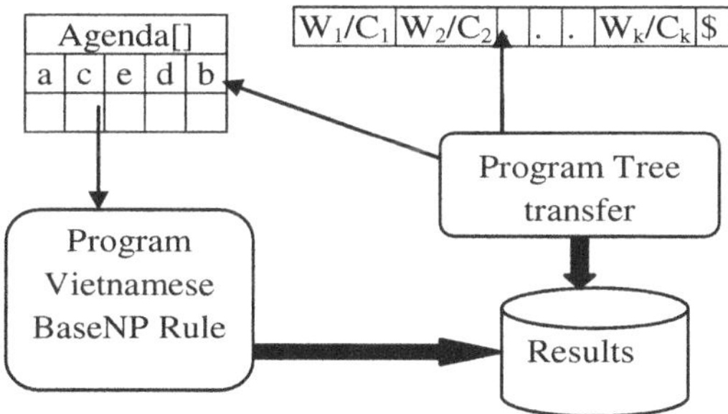

Fig. 5. Parse tree transfer and Vietnamese BaseNP rule generation Model

Figure 5 illustrates how the model works for transferring the order of words in English to Vietnamese and generating Vietnamese BaseNP rules.

By observation Figure 6, we see that there are two types of cross alignment: local and global. The former can be solved by changing the order of the English Noun Phrase to that of the Vietnamese. The latter is done by performing beam search (limit search space). We realize that the curtail thing for solving global ambiguity is the start searching point and the searching space. If these two things can be controlled, the problem of global cross alignment will be solved. Therefore we present the concept *anchorL*(i,j_L) and *anchorR*(i,j_R), where, i is an anchor point in the English sentence; j_L and j_R are correspondence anchor points in Vietnamese sentence; d_L and d_R are a search range; that can be represented as:

$$j_L = \text{BeamSearch}(\text{NPk}, p_L, d_L) \tag{1}$$

$$j_R = \text{BeamSearch}(\text{NPk}, p_R, d_R) \tag{2}$$

$$d_L = |\neg \text{NP}_k| * \delta + c \tag{3}$$

$$d_R = |\text{NP}_k| * \delta + c \tag{4}$$

here:

$|\text{NP}_k|$ is the number of words in the English BaseNP, $k=1...m$ (m is the number of BaseNP in the English sentence),$|\neg \text{NPk}|$ is the number of words which do not belong to English BaseNP (count from the right bound of NP_k-1 to left bound of NP_k), c, δ are thresholds for BaseNP length ratio in our system (we choose c=2 and δ= 1.6 ÷ 2.7 depend on type of bilingual corpus as Figure 4 and Table 3).

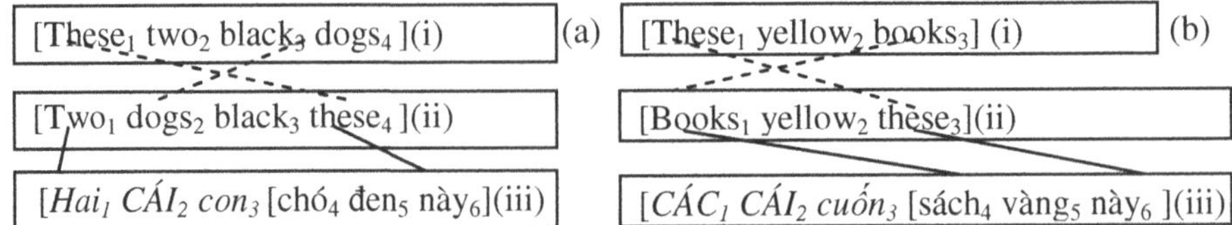

Fig. 6. Problematic BaseNP projection

As can be seen in Figure 6a, if we use anchor point technique without transferring the Parse Tree in aligning (i) with (iii) we only get "*chó đen này*" instead of getting "*hai CÁI con chó đen này*". However, if we align (ii) with (iii), we receive [*hai cái con chó đen này*] as in Figure 6b, the "*CÁC$_1$ CÁI$_2$ cuốn$_3$*" was extracted automatically by using characters of Vietnamese language.

4 Experiment

*** *Bilingual Corpus***

In our experiment, we use the CADASA corpus which is built from the English-Vietnamese bilingual books and translation from WSJ corpus (CoNLL-2000) as in Table 4 ([16], 2003). This corpus contains 35,475 pairs of bilingual sentence (about 476,000 English words and 636,000 Vietnamese words).

Table 4. Statistic of the average length of sentence

No	Bilingual Corpus	Pairs of sentences	Length of sentence	
			English	Vietnamese
1	Children's Encyclopedia	6,118	9	11
2	Network Encyclopédie	15,948	17	25
3	Life in Australia	2,144	8	12
4	Stories	10,034	10	11
5	WSJ corpus	1,235	27	36
	Total:	**35,479**	**14**	**18**

*** Evaluation Methodology**

The result is evaluated in terms of the corresponding BaseNP extraction precision (*Pre*), recall (*Rec*) ([20], 2005) and Alignment Error Rate (AER) ([15], 2003), as defined in the following:

$$\text{Precision}(B,B') = \frac{|B \cap B'|}{|B'|} \quad (5)$$

$$\text{Recall}(B,B'') = \frac{|B \cap B''|}{|B''|} \quad (6)$$

$$\text{AER}(B,B') = 1 - \frac{2*|B \cap B''|}{|B| + |B''|} \quad (7)$$

Where:

B' = English BaseNPs are produced by Gate
B = Correspondence BaseNPs assessed by human on B'
B'' = Correspondence BaseNPs are assessed by human

*** Results**

This section describes the experimental setup and performance evaluation of the proposed model to identify and extract corresponding English-Vietnamese BaseNP by two ways. In the first way, we evaluate the model automatically on gold standard and the result is shown in Table 5. In the second way, we randomly selected 2,700 aligned sentences from 35,000 pairs of bilingual sentences in our bilingual corpus for the testing corpus. The remaining 32,300 pairs of bilingual sentences are used for the training corpus (add semantics for bilingual dictionary in dictionary alignment). After that, ten people test the base NP alignment manually using the technique of Och and Ney (2003), as shown in Table 6. The result is evaluated in BaseNP precision, recall and alignment error rate using (5), (6) and (7) formulas, as shown in section 4.2.

Table 5. Result evaluated by computing with threshold $\delta = 2.7$ on Gold Standard

Pairs of sentences	Words of English	Words of Vietnamese	Pre	Rec	AER
1000	16,218	24,789	89%	83.7%	9.1%

Table 6. Result evaluated by human with threshold $\delta = 2.7$ on our bilingual corpus

No	Bilingual Corpus	Pairs of sentences	Pre	Rec	AER
1	Children's Encyclopedia	600	81.8%	76.1%	13.6%
2	Network Encyclopedia	1,000	79.2%	74.1%	14.8%
3	Life in Australian	200	78.4%	73.7%	15.1%
4	Stories	800	76.5%	70.4%	17.4%
5	WSJ	100	75.7%	69.9%	17.7%
	Total:	**2,700**	**78.5%**	**73%**	**15.6%**

5 Conclusion

Nowadays automatic English-Vietnamese Noun Phrase Extraction from a Bilingual corpus has been considered a crucial task for many applications in NPL. Language pairs such as English-French, English-Chinese have great sources of parallel corpus, annotated data and NPL tools and therefore a lot of researches on these have been done with prosperous results. However, this is not the case of English-Vietnamese BaseNP Extraction.

In this paper, we have shown a method of acquiring bilingual BaseNP using English NP chunking, English Vietnamese comparison knowledge and projection method via aligning anchor points of Yarowsky [8][9]. We applied such NLP technologies as POS tagging, NP chunking of the English language and word segmentation of the Vietnamese language. To acquire corresponding anchor points of the Vietnamese language, we change the word order of English BaseNP according to that of Vietnamese BaseNP and only do word alignment on anchor points with a threshold to determine range of search in word alignment based on the Anchor point Alignment Model. This model includes two stages DictAlign and ClassAlign based on Ker and Chang [5] ; others are proposed by us. We achieved *78.5%* in precision on 35,000 pairs of sentences in the English-Vietnamese bilingual corpus. The proposed technique can be easily applied to other language pairs.

References

1. Quirk, R., Greenbaum, S.: A University Grammar of English. Longman Group Limited, London (1990)
2. Kupiec, J.: An Algorithm for finding Noun phrase Correspondences in Bilingual Corpora. In: Proceedings of the 31st Annual Meeting on Association for Computational Linguistics, Columbus, Ohio, USA, pp. 17–22 (1993)
3. Hung, T.V.: Enlish for Computerscience, Printed in Ho Chi Minh City, Vietnam (1995)
4. Buu, H.V.: Patterns of English, Printed in Ho Chi Minh City, Vietnam (1996)
5. Ker, S.J., Chang, J.S.: A Class-based Approach to Word Alignment. Computational Linguistics 23(2), 313–343 (1997)
6. Can, N.T.: Vietnamese syntax grammar, Printed in Hanoi, Vietnam (1999)
7. Wantanabe, H., Kurohashi, S., Aramaki, E.: Finding Structural Correspondences from Bilingual Parsed Corpus, IBM Research, Tokyo Research Laboratory (1999)
8. Yarowsky, D., Ngai, G.: Inducing Multilingual POS Taggers and NP Bracketers via Robust Projection across Aligned Corpora. Johns Hopkins University Baltimore, MD (2001)
9. Yarowsky, D., Ngai, G., Wicentowski, R.: Inducing Multilingual Text Analysis Tools via Robust Projection across Aligned Corpora. In: Proc. of NAACL 2001 (2001)
10. Cunningham, H., Maynard, D., Bontcheva, K., Tablan, V.: GATE: A framework and graphical development environment for robust NLP tools and applications. In: Proc. of the 40th Anniversary Meeting of the Association for Computational Linguistics (2002)
11. Hwa, R., Resnik, P., Weinberg, A., Kolak, O.: Evaluating Translational Correspondence using Annotation Projection. In: The Proceedings of the 40th Anniversary Meeting of the Association for Computational Linguistics (2002)

12. Riloff, E., Schafer, C., Yarowsky, D.: Inducing Information Extraction Systems for New Languages via Cross-Language Projection. In: Proceedings of the 19th International Conference on Computational Linguistics, COLING 2002 (2002)
13. Wang, W., Zhou, M.: Structure Alignment Using Bilingual Chunking. In: The 19th International Conference on Computational Linguistics, COLING 2002 (2002)
14. Koehn, P.: Noun Phrase Translation. Ph.D. dissertation, University of Southern California (2003)
15. Och, F.J., Ney, H.: A Systematic Comparision of Various Statistical Alignment Models. Association for Computational Linguistics (2003)
16. Rebecca, Vickes, S.: The Fahasa/Heinemann Illustrated Encyclopedia, vol. 1,2,3 (2003)
17. Dien, D., Kiem, H.: POS-Tagger for English-Vietnamese Bilingual Corpus. In: HLT-NAACL 2003 Workshop (2003)
18. Hwang, Y.S., Paik, K., Sasaki, Y.: Bilingual Knowledge Extraction Using Chunk Alignment. In: PACLIC 18, December 8-10. Waseda University, Tokyo (2004)
19. Deng, Y.: Bitext Alignment for Statistical Machine Translation. Ph.D. dissertation, Johns Hopkins University, Baltimore, Maryland (2005)
20. Chau, Q.N., Tuoi, T.P., Tru, H.C.: Vietnamese Proper Noun Recognition. In: Proceedings of the 4th IEEE International Conference on Computer Sciences Research, Innovation and Vision for the Future, Ho Chi Minh City, Vietnam (2006)

Towards the Web in Your Pocket: Curated Data as a Service

Stuart Dillon[1], Florian Stahl[2], and Gottfried Vossen[2,1]

[1] University of Waikato Management School, Hamilton, New Zealand
stuart@waikato.ac.nz

[2] ERCIS, University of Münster, Münster, Germany
{florian.stahl,gottfried.vossen}@ercis.de

Abstract. The Web has grown tremendously over the past two decades, as have the information needs of its users. The traditional "interface" between the vast data resources of the Web and its users is the search engine. However, search engines are increasingly challenged in providing the information needed for a particular context or application in a comprehensive, concise, and timely manner. To overcome this, we present a framework that does not just answer queries based on a pre-assembled index, but based on a subject-specific database that is curated by domain experts and dynamically generated based on vast user input.

Keywords: Data curation, digital curation, curation process, Data as a Service, Web in the Pocket, information provisioning.

1 Introduction

Over the past 20 years, the volume of data stored electronically has become superabundant and the Web is said to be the biggest collection of data ever created [3]. Moreover, accessible data on the Web, whether created by computers, by users, or generated within professional organizations, continue to grow at a tremendous pace [19,22,28]. Social networks like Facebook, search engines like Google, or e-commerce sites like Amazon, generate and store new data in the TB range every day. Due to the emerging usage of cloud computing [2], this trend will not only continue, but accelerate over the coming years, as more data is permanently stored online, is linked to other data, and is aggregated in order to form new data [1]. As a consequence, the question arises as to how to best source the precise information required[1] at a given moment or in a particular context, be it for personal or professional use. The concept of a "Web in the Pocket" proposed in this paper is a possible solution.

For the past 15 years, "search" has been the tool of choice for garnering Web information; the underlying data is extracted using a search engine [9,8,24,7]. But while (dynamic, on-demand) search is preferable to (static) directories in

[1] We here use *data* and *information* interchangeably, although we consider information as the result of extracting "usable" data by a user.

N.T. Nguyen et al. (Eds.): *Adv. Methods for Comput. Collective Intelligence*, SCI 457, pp. 25–34.
DOI: 10.1007/978-3-642-34300-1_3

the face of an exploding Web, the point has been reached where search, even though nowadays available in many facets, may no longer be appropriate in many situations. Having data relevant to specific areas or applications selected, quality-checked and integrated ("curated") in advance as well as made available as a configurable, adaptable, and updatable service can provide a competitive advantage in many situations [25,15,33,26]. While recent data marketplaces [32] go a step in this direction at least for specific applications, the idea presented in this paper is to have access to a service that has been configured precisely to a user's needs and budget and that composes and delivers corresponding data in a topic-centric way.

We envision that this data service can even be made available offline on both stationary and mobile devices, the latter resulting in a "Web in your pocket" (WiPo). A simple way to view the WiPo idea is to think of complete context- or application-specific data being "pushed" to the user when required; this data has been collected according to user input, and has then been processed in various ways (explained below) to ensure completeness, accuracy, quality, and up-to-dateness. This is in contrast to the current search model, where information is dynamically "pulled" by the user, often only in rudimentary ways. A further key feature of WiPo is that it will generate data from a range of public and private sources. The way to achieve quality data is to exploit curation, where data curation refers to the long-term selection, cleansing, enrichment, preservation and retention of data and to provide access to them in a variety of forms. It is known, for example, from museums and has already been successfully applied to scientific data (amongst other by the British Digital Curation Centre, see www.dcc.ac.uk). An application built on curated data can help solve the identified information problems, as raw data will become less useful because of its availability in sheer quantities. We consider WiPo to be a data centric application that provides a comprehensive overview and detailed information on a given topic area through a configurable service.

As will be seen, WiPo can be useful to both business (i.e., professional) and non-business (i.e., private) users. Since the information needs of these two groups are likely to be different, we will treat them as separate cases. Further, information may be required at different frequencies and at differing levels of quality. For instance, stock prices are data that are usually needed with a relatively high frequency and quality, whereas other information may be required less frequently or where quality is less important. Quality (reliability) of information is normally increased as the number of data sources employed increases. We term this "broadness" (a term suggested to us by Jim Hendler). While both frequency and broadness values are likely to be continuous (i.e., on a continuum), for simplicity we take a bimodal approach. As shown in Figure 1, there are four possible classifications for each of the business and non-business scenarios, represented as a four-field matrix. In the following, we present sample use cases for all eight possible scenarios resulting from the four-field matrix and the two application areas of *businesses* and *non-business*.

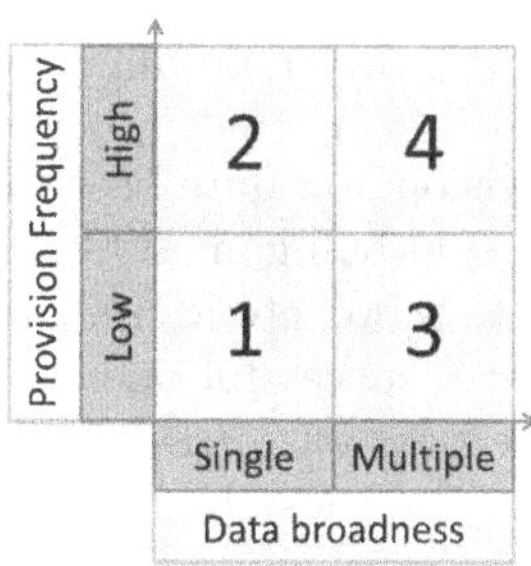

Fig. 1. Classification Matrix

Business Cases

1. Low frequency of data provision (one-off), low data broadness (single source): A company wishes to determine the cost of airline lounge membership for its frequently flying CEO.
2. High frequency of data provision, low data broadness: A manufacturing firm wishes to keep abreast with the cost of steel on the spot market. One (reliable) source is satisfactory; however, the data needs to be provided on a continuous basis.
3. Low frequency of data provision, high data broadness (multiple sources): A company is considering changing its sole provider of office supplies. It has already been through a tendering process, but wishes to source external information about providers.
4. High frequency of data provision, low data broadness: It is essential that firms have a good understanding of the marketplace(s) in which they operate. They need to know what others in the supply chain are doing and what their customers are doing. They need to know about external (e.g., political or legal) events that may change operating conditions. This information might come from a variety of sources including official indexes, market reports, business commentators, as well as internal data sources for comparison and benchmarking, and has to be up to date.

Non-business Cases

1. Low frequency of data provision (one-off), low data broadness (single source): An air traveler is searching for a one-night layover in an arbitrary city. They wish to know "What is the cheapest hotel within 5 km of the international terminal?"
2. High frequency of data provision, low data broadness: A small-time investor has a number of shares and stocks. He/she wants to be able to view the current share price of the shares that they own on demand.
3. Low frequency of data provision, high data broadness (multiple sources): A hospital patient has recently been diagnosed with a genetic condition that others in his or her extended family have also suffered from. The patient has

some information (from the hospital and their family) about the condition, but also wishes to gain a better understanding of alternative treatments and of local support groups.
4. High frequency of data provision, low data broadness: Many people regularly bet on sporting events, e.g., football games. Gamblers can bet on a range of things such as results, first to score, etc. Here, the more up-to-date information one can obtain, the more successful one can be. These users will need extensive statistics over the past few seasons, weather forecasts for games, team and player news (e.g., injuries) to help them improve their betting returns.

We will further explore the case of the hospital patient later as an example of how WiPo is intended to work. The rest of the paper is structured as follows: Section 2 outlines related work. Section 3 elaborates on how to provide curated data, starting with an overview, then highlighting the speciality of user input, and finally discussing the process of curation. Future work is outlined in Section 4.

2 Related Work

The Web in the Pocket is, in a sense comparable, to a materialized data warehouse [20] that is made portable, and to Fusion Cubes as suggested in [1] for situational data. More related to our work is research on search engines. Cafarella et al. [11] examine how search engines can index pages in the deep Web. Building data marts or services on given deep Web sources is described by Baumgartner et al. [4]. Generally speaking, however, retrieval and indexing of documents is no longer the problem it used to be, though far from being solved in its entirety [3].

There are several issues to address in order to fully satisfy current and future information needs. Dopicha [16] found that it was technically impossible to meaningfully answer queries such as "return all pages that contain product evaluations of fridges by European users". A similar conclusion was reached by Ceri [13]. To solve this, Dopichaj advocates the Semantic Web, whose idea is to enhance text on the Web by semantic information to make it machine understandable. Ceri [13], on the other hand, proposed the so called *Search Computing* (SeCo) framework. A detailed description of an architecture for SeCo is described in [5]. Briefly, a query to a SeCo search engine is processed by a query optimizer that determines suitable search services to which it then sends subqueries. Campi et al. [12] describe how service marts (e.g., as suggested in [4]) can be built and registered with the framework. Search service results are then joined and displayed to users. In a subsequent step, users have the opportunity to modify their queries. This is referred to as *liquid* query processing [6]. In order to realize this framework, two new user groups need to be created: providers of data offering data as a services and developers building search services based on data services.

One of the few approaches combining curation and Web Information Retrieval (IR) has been proposed by Sanderson et al. [30] who suggest (focusing on a single domain) a process similar to life science data consolidation as described in

[21]. In life science, for instance, different institutions contribute to a nucleotide sequence databases; to ensure consistency data are exchanged on a nightly bases [21]. Sanderson et al. [30] apply this procedure to content retrieval by sending predefined nightly queries to pre-registered services. Relevant information is harvested and temporarily stored. The information retrieved is then audited by data curators who decide upon its relevance; only relevant data is kept. Moreover, care is taken to prevent harvesting the same data more than once. A similar harvest-and-curate approach was suggested by Lee et al. [23] who outline ideas to enhance their ContextMiner (see `contextminer.org` for a tool offering contextual information to data) by making it scalable. However, beyond these sources there is apparently no in-depth research on combining data curation and IR, and even those do not appear to have the potential to serve information needs on a large scale in a sophisticated way. In any data curation process humans are needed at some point, which is why the education and qualification of data curators is also intensively discussed in the literature [18,31,27].

3 The WiPo Approach

3.1 Overview

The overall process view of the WiPo approach is generic and applicable to many use cases. The concrete application design, however, is use-case dependent, i.e., the individual steps of the process have to be implemented considering the purpose of the application and the specific needs of the user. Figure 2 presents the basic process underlying WiPo in the form of a Petri net [34], in which activities are depicted as rectangles and states are depicted as circles; the important rule is that states and activities must strictly alternate. A state (of the underlying data) can be thought of as objects from a database, a document containing certain information, or of collections of such items (e.g., a list of selected sources after the Source Selection activity). A double lined box indicates that this particular activity has a refinement, i.e., is composed of other activities at a lower level of abstraction. At the high level shown in Figure 2 the overall process consist of steps *Input Specification*, resulting in a list of potential sources and pre-filters; *Source Selection*, resulting in a list of sources; *Data Mining*, resulting in raw data; *Past Filters*, reducing the raw data to "relevant" raw data; and *Curation*. These sub-processes will be explained in more detail in the following natural language description.

The first step is receiving user input. In this step it is specified what the data service should be looking for. An input specification usually is a list of links, keywords, or documents, or a combination thereof. Due to the importance of this step, Section 3.2 elaborates further on input specification and presents a refinement of this step. In the case of the hospital patient, the input could be bookmarks of medical Web sites, documents describing cures and treatments, or clinical trial results. After this pre-processing step the most appropriate sources are determined based on standard algorithms and heuristics (e.g., [3,7]). This explicitly includes discovery of more recent documents of the same class as the

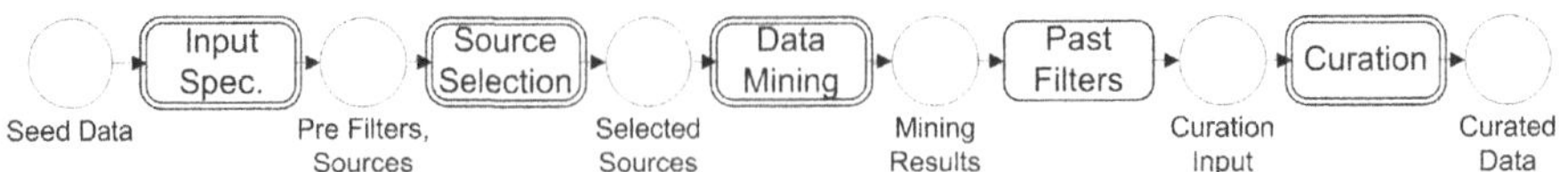

Fig. 2. WiPo process overview as a Petri net

ones provided, e.g., more recent medical publications in our sample case. Web sources can be categorized in two ways: A source may be either structured or unstructured. Concerning access to it, it may be either publicly available, semi-public (i.e., only available to subscribers), or private (i.e. only available to a given user). The task of selecting sources also entails the setting of pre-filters (e.g., filters of time if only recent documents are relevant). Furthermore, it is strongly connected to the action of meta-data management which focuses on storing information about the various sources.

The next step is data mining or information retrieval. Using techniques from these well-established fields (e.g., [36], relevant information is extracted from the predefined sources. Depending on the degree of structure the sources provide, the complexity of this process can range from very simple to extremely complex. Following data mining, a post-filter can be applied in order to reduce the potentially vast amount of mining results to a more manageable size, containing only most relevant information. Considering our patient example, mining could be restricted to a medical journal database as well as a collection of sites such as `uptodate.com`; a pre-filters would be time (since only recent documents are of interest); a post-filter can limit the results to the patient's specific condition.

The mining process is followed by curation, which we define as *making data fit for a purpose*. To ensure the "fitness" of the data, a domain expert will be involved in this step, which can depend on the particular knowledge domain, the professionalism of the data service provider, or even on the price the user is willing to pay for quality information. For the former option, a medical doctor as domain expert could ensure that the content of the received documents is factually true, and design a process to integrate various different medical sources in a timely manner. Curation itself consists of four subtasks: a) imposing data quality by finding trustworthy sources; b) adding value; c) adding data lineage information and d) integration of different data sources, including additional sources and files given by the user, as well as tasks such as data cleansing and data fusion. Fusion also includes joining mined results with the initial information provided by the user. When curation is completed, the data will be written to the curated database from where the data service collects, collates and presents them — appropriately visualized by the WiPo application — to the user. The entire setup just described is presented in Figure 3 starting from User Input. Thin arrows indicate a relationship between two items and block arrows represent the flow of data within the process.

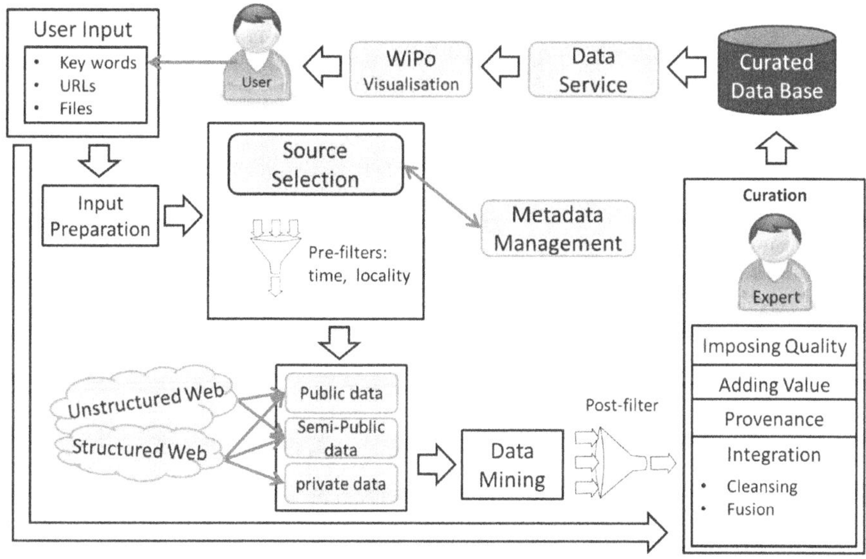

Fig. 3. WiPo Architecture

3.2 Incorporating User Input

In most of the established search engines such as Google or Bing, keywords and search strings, maybe even regular expressions or SQL-like queries, are used to interpret what a user is searching for. Those are then matched against an index of pre-crawled Websites to retrieve relevant documents [35]. Most of them allow the search to be restricted to a specific domain or top-level domain; others can be given a predefined list of URLs and restrict search to that list. The "history" of search engines has seen numerous approaches to capture a user's intuition. However, there is — to the best of our knowledge — no approach incorporating personal, offline data with externally, Web-sourced data.

In practical terms, using a suitable interface, the user can upload documents, supply a list of relevant links and potentially also keywords. WiPo will provide a dedicated language interface for this purpose, for which we envision usage of an XML standard along the lines of DITA (Darwin Information Typing Architecture). Input Analysis will then convert the given documents to a homogeneous file format in order to apply a generic classifier. The given URLs will be crawled and also fed into the classifier. In that way from both the given documents and URLs the essence is extracted in the form of topics or additional keywords. These mined keywords as well as the user-supplied keywords are then used to choose appropriate sources from a list of all available sources (which might also be extended based on user supplied URLs). Furthermore, relevant other dimensions such as time or location are determined to generate a list of sources including pre-filters. This process is shown in Figure 4.

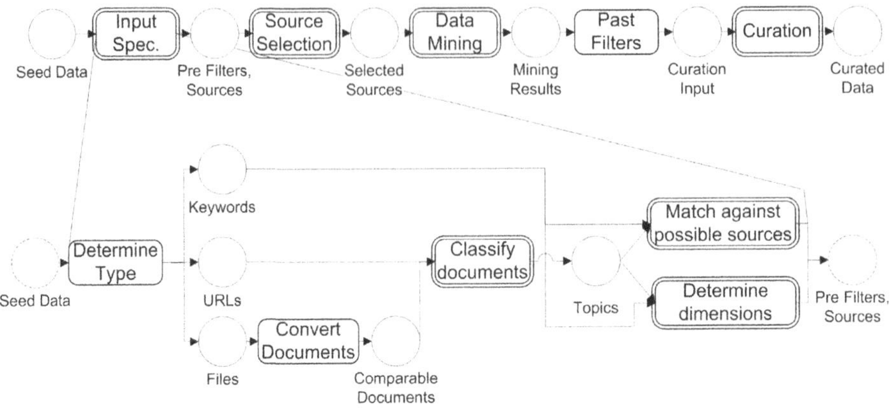

Fig. 4. Detailed view on Input Specification

3.3 Bringing Data Curation to IR

A research stream that thus far has only received little attention in the area of Web IR is data curation. The original idea of "digital" curation was first discussed in the library and information sciences, where it still vastly resides. It is focused on huge scientific data sets of physical, biological, or astronomical data with the aim of preserving data gained through scientific experiments for later usage [14,27]. To achieve this goal the idea is to clean and update data to new technical standards in order to make them accessible for future use [17,25,29]. In library science, electronic repositories storing data and information are referred to as institutional repositories [31,14]. Research conducted in that domain can also be beneficial when developing the central data repository as basis of a new Web search utility. Also relevant in this context is data *provenance* brought into the discussion of data curation by (amongst others) Buneman et al. [10], who are highly involved in the British Digital Curation Centre. Data provenance or lineage refers to the action of making the history of some piece of data available in a way that its origin can be traced and modifications reconstructed. This is in particular important w.r.t. crediting the right people when using their data (compare the action of referencing works by other authors). Also knowing the origin of a source can arguably increase trust and support reproducibility.

4 Conclusions and Future Work

We have introduced a framework that combines established methods from the fields of information retrieval, Web search and library sciences. These combined together offer a unique approach which we term *Web in the Pocket* (WiPo). We believe this offers an opportunity for a revolutionary new approach that will, in time, supersede the current, and somewhat limited, search-based approach for information retrieval. This paper has focused on the presentation of an overall conceptual model for WiPo as well as a basic architecture. Two aspects of

WiPo require further investigation and clarification, namely curation and input specification. In Section 3.2 we showed how we envision input specification; the consequential next step is to work on a concrete implementation. In 3.3 we briefly summarized the concept of curation, and this also needs a more detailed exploration within the context of the WiPo approach.

References

1. Abello, A., et al.: Fusion Cubes: Towards Self-Service Business Intelligence. To Appear in Journal on Data Semantics (2013)
2. Armbrust, M., et al.: A view of cloud computing. CACM 53(4), 50–58 (2010)
3. Baeza-Yates, R., Raghavan, P.: Chapter 2: Next Generation Web Search. In: Ceri, S., Brambilla, M. (eds.) Search Computing. LNCS, vol. 5950, pp. 11–23. Springer, Heidelberg (2010)
4. Baumgartner, R., Campi, A., Gottlob, G., Herzog, M.: Chapter 6: Web Data Extraction for Service Creation. In: Ceri, S., Brambilla, M. (eds.) Search Computing. LNCS, vol. 5950, pp. 94–113. Springer, Heidelberg (2010)
5. Bozzon, A., Brambilla, M., Ceri, S., Corcoglioniti, F., Gatti, N.: Chapter 14: Building Search Computing Applications. In: Ceri, S., Brambilla, M. (eds.) Search Computing. LNCS, vol. 5950, pp. 268–290. Springer, Heidelberg (2010)
6. Bozzon, A., Brambilla, M., Ceri, S., Fraternali, P., Manolescu, I.: Chapter 13: Liquid Queries and Liquid Results in Search Computing. In: Ceri, S., Brambilla, M. (eds.) Search Computing. LNCS, vol. 5950, pp. 244–267. Springer, Heidelberg (2010)
7. Bozzon, A., Brambilla, M., Ceri, S., Fraternali, P., Vadacca, S.: Exploratory search in multi-domain information spaces with liquid query. In: Proc. 20th Int. Conf. on World Wide Web, pp. 189–192. ACM, New York (2011)
8. Braga, D., Corcoglioniti, F., Grossniklaus, M., Vadacca, S.: Panta Rhei: Optimized and Ranked Data Processing over Heterogeneous Sources. In: Maglio, P.P., Weske, M., Yang, J., Fantinato, M. (eds.) ICSOC 2010. LNCS, vol. 6470, pp. 715–716. Springer, Heidelberg (2010)
9. Brin, S., Page, L.: The anatomy of a large-scale hypertextual web search engine. Computer Networks 30, 107–117 (1998)
10. Buneman, P., Chapman, A., Cheney, J., Vansummeren, S.: A Provenance Model for Manually Curated Data. In: Moreau, L., Foster, I. (eds.) IPAW 2006. LNCS, vol. 4145, pp. 162–170. Springer, Heidelberg (2006)
11. Cafarella, M.J., Halevy, A., Madhavan, J.: Structured Data on the Web. CACM 54(2), 72–79 (2011)
12. Campi, A., Ceri, S., Gottlob, G., Maesani, A., Ronchi, S.: Chapter 9: Service Marts. In: Ceri, S., Brambilla, M. (eds.) Search Computing. LNCS, vol. 5950, pp. 163–187. Springer, Heidelberg (2010)
13. Ceri, S.: Chapter 1: Search Computing. In: Ceri, S., Brambilla, M. (eds.) Search Computing. LNCS, vol. 5950, pp. 3–10. Springer, Heidelberg (2010)
14. Choudhury, G.S.: Case Study in Data Curation at Johns Hopkins University. Library Trends 57(2), 211–220 (2008)
15. Doorn, P., Tjalsma, H.: Introduction: archiving research data. Archival Science 7, 1–20 (2007)
16. Dopichaj, P.: Ranking-Verfahren für Web-Suchmaschinen. In: Lewandowski, D. (ed.) Handbuch Internet-Suchmaschinen. Nutzerorientierung in Wissenschaft und Praxis, pp. 101–115. AKA, Akad. Verl.-Ges., Heidelberg (2009)

17. Gray, J., Szalay, A.S., Thakar, A.R., Stoughton, C., van den Berg, J.: Online Scientific Data Curation, Publication, and Archiving. CoRR Computer Science Digital Library cs.DL/0208012 (2002)
18. Heidorn, P.B., Tobbo, H.R., Choudhury, G.S., Greer, C., Marciano, R.: Identifying best practices and skills for workforce development in data curation. Proc. American Society for Information Science and Technology 44(1), 1–3 (2007)
19. Hey, T., Trefethen, A.: The data deluge: An e-science perspective. In: Berman, F., Fox, G.C., Hey, A.J. (eds.) Grid Computing — Making the Global Infrastructure a Reality, pp. 809–824. Wiley (2003)
20. Inmon, W.: Building the Data Warehouse. Wiley Technology Publishing, Wiley (2005)
21. Kulikova, T., et al.: The embl nucleotide sequence database. Nucleic Acids Research 32(suppl. 1), 27–30 (2004)
22. Laudon, K., Traver, C.G.: E-commerce: business, technology, society, 9th edn. Pearson/Prentice Hall (2013)
23. Lee, C.A., Marciano, R., Hou, C.Y., Shah, C.: From harvesting to cultivating: transformation of a web collecting system into a robust curation environment. In: Proc. 9th ACM/IEEE-CS Joint Conference on Digital Libraries, pp. 423–424. ACM, New York (2009)
24. Levene, M.: An Introduction to Search Engines and Web Navigation, 2nd edn. Wiley (2010)
25. Lord, P., Macdonald, A., Lyon, L., Giaretta, D.: From data deluge to data curation. In: Proc. UK e-Science All Hands Meeting, pp. 371–375 (2006)
26. Meliou, A., Gatterbauer, W., Halpern, J.Y., Koch, C., Moore, K.F., Suciu, D.: Causality in databases. IEEE Data Eng. Bull. 33(3), 59–67 (2010)
27. Palmer, C.L., Allard, S., Marlino, M.: Data curation education in research centers. In: Proc. 2011 ACM iConference, pp. 738–740. ACM, New York (2011)
28. Ramírez, M.L.: Whose role is it anyway? a library practitioner's appraisal of the digital data deluge. ASIS&T Bulletin 37(5), 21–23 (2011)
29. Rusbridge, C., et al.: The digital curation centre: a vision for digital curation. In: Proc. 2005 IEEE Int. Symp. on Mass Storage Systems and Technology, pp. 31–41. IEEE Computer Society, Washington, DC (2005)
30. Sanderson, R., Harrison, J., Llewellyn, C.: A curated harvesting approach to establishing a multi-protocol online subject portal. In: Proc. 6th ACM/IEEE-CS Joint Conference on Digital Libraries, pp. 355–355. ACM, New York (2006)
31. Smith, P.L.: Where IR you?: Using "open access" to Extend the Reach and Richness of Faculty Research within a University. OCLC Systems & Services 24(3), 174–184 (2008)
32. Stahl, F., Schomm, F., Vossen, G.: Marketplaces for data: An initial survey. ERCIS Working Paper No. 12, Münster, Germany (2012)
33. Tan, W.C.: Provenance in Databases: Past, current, and future. IEEE Data Eng. Bull. 30(4), 3–12 (2007)
34. Van der Aalst, W., Van Hee, K.: Workflow Management: Models, Methods, and Systems. MIT Press (2004)
35. Vossen, G., Hagemann, S.: Unleashing Web 2.0: From Concepts to Creativity. Morgan Kaufmann Publishers (2007)
36. Witten, I., Frank, E., Hall, M.: Data Mining: Practical Machine Learning Tools and Techniques, 3rd edn. Morgan Kaufmann Publishers (2011)

Part II
Social Networks and e-Learning

Why Do People Stick to Play Social Network Sites? An Extension of Expectation-Confirmation Model with Perceived Interpersonal Values and Playfulness Perspectives

Ron Chuen Yeh[1], Yi-Chien Lin[2], Kuo-Hung Tseng[1], Pansy Chung[3], Shi-Jer Lou[4], and Yi-Cheng Chen[5]

[1] Graduate Institute of Business & Management
[2] Department of Applied Foreign Language
[3] Department of Information Management
Meiho University, 23 Pingquang Rd., Pingtung 91202, Taiwan
{x2051,x3179,x0003,x3046}@meiho.edu.tw
[4] Graduate Institute of Vocational and Technical Education
National Pingtung University of Science and Technology
1, Shuefu Road, Neipu, Pingtung 912, Taiwan
lou@mail.npust.edu.tw
[5] Department of Information Science & Management Systems
National Taitung University, 684, Sec. 1, Zhonghua Rd., Taitung 950, Taiwan
yc_bear@nttu.edu.tw

Abstract. The rapid progress of the Internet and communication technologies has changed our daily lives dramatically. The presence of the social networking sites (SNS) provides the users new types of communicating media. Understanding SNS users' needs, getting popular gathering, acquiring potential users and making users stick on the websites are critical for their sustainable operation. Based upon expectation-confirmation model (ECM), this study explores the factors influencing users' stickiness on SNS. The instrument for data collection was meticulously developed. The partial least squares technique was applied for validating the proposed research model and hypotheses. The findings of this study are expected to provide practical guidance to the SNS owners and serve as instrumental references to the research discipline to understand the causal effects of the related perceived factors that may influence users' satisfaction, continuance motivation and stickiness towards SNS. That will further enhance the business performance of the SNS operations.

Keywords: social networking site, stickiness, expatiation confirmation theory.

1 Introduction

With rapid increase in the world online population, the cyberspace is gradually becoming another massive virtual world outside of the physical living environment. Various types of social networking sites (SNS) and applications have emerged,

N.T. Nguyen et al. (Eds.): *Adv. Methods for Comput. Collective Intelligence*, SCI 457, pp. 37–46.
DOI: 10.1007/978-3-642-34300-1_4

including virtual communities, online communities, electronic communities, and Internet learning communities. SNS attract users of various social strata, age groups, information usages, and information needs to join them. In fact, communities are important organizational units on the group level, while SNS provide a convenient medium that can transcend spatial restrictions in the virtual world to Internet viewers who have similar interests in specific topics, so that users can engage in real-time information transmission and emotional interaction within their communities.

However, with rapid development of Internet technology and social networking sites, changes in user needs, and intense competition in e-commerce, SNS are facing pressures of management and operation. Thus, SNS operators must have in-depth understanding of user needs, be able to create various types of new applications to attract popularity, enhance user continuance motivation, and use key thinking and operational strategies to enhance user stickiness. These will be important issues that affect whether SNS will achieve sustainable operation. Observation and understanding of the actual usage behavior of SNS users show that factors that affect the continued usage and stickiness of SNS include the discrepancy between the expected functions and system performances, as well as personal perception factors, such as usage satisfaction, social interaction value with peers, friends, and relatives, and other psychological factors.

According to prior research on information system users' continuous intention, motivation, and actual usage, it is known that user decision-making regarding information systems would undergo a three-stage process, namely initial contact, influence of the experiences from the initial usage, and termination of use that overturns the initial idea or continuous usage of systems. This is the concept from Bhattacherjee's [1] "post-acceptance model of IS Continuance", based upon the extension of expectation confirmation model (ECM). In addition, regarding SNS users' perceived value, this study focuses on the perspective of perceived interpersonal value of SNS characteristics, along with the perceptive factor of user perceived playfulness at using SNS for integrated exploration. Thus, in order to truly understand the causal issues that affect the stickiness of SNS users to SNS, to gain an in-depth understanding of the needs of SNS users, and in turn, enhance user intention for continued usage and stickiness, this study utilizes the perspective of expectation confirmation theory as the theoretical basis, and applies stringent theoretical development procedures to construct a causal model, in hopes of providing an important reference for the management of SNS and related research issues to strengthen the operational performance of SNS.

2 Theoretical Development and Conceptual Model

Wellman [2] defined social networking as a kind of interpersonal interactive relationship network that provides socializing ability, information, a sense of belonging, and social identification supported by passion. Plant [3] proposed that social networking is a group of individuals or organizations that temporarily or permanently engage in interaction over electronic media to discuss common interests and issues. Catherine et al. [4] suggested that virtual social networks are a group of people who have common interests and thoughts who continuously use a location on the Internet or Internet tools for interaction, and spread information to form a group.

The expectation confirmation model (ECM), proposed by Oliver [5] in the marketing field, established the basic framework for researching the satisfaction of general consumers, and it has been widely applied to the evaluation of user satisfaction and post-sale behavior. Bhattacherjee [1] further extended and modified expectation confirmation theory, to explore the continued usage intention of e-bank users, adding the "perceived usefulness" of technology acceptance model in the research model, to replace the expected usage concept. This concept is set as post-use expectation to dispel the time changes of pre-use expectations. Research results showed that when users feel satisfied about information systems, and have a positive perception for the systems usage, it would enhance their continued usage intention. Positive confirmation of posterior expectation and the perceived usefulness for the system both positively influence the satisfaction of users. Confirmation of posterior expectation also affect user perceived usefulness, which means that if the performance of information systems is higher than user expectations, it would make them think that this system is useful. Another researcher also pointed out that, perceived usefulness, post-use expectation confirmation, and satisfaction from the prior usage are important factors influencing users' continued IS usage [6].

Additionally, the convenience and diversity of SNS lower the users' conversion costs, so that they can easily convert among different SNS. Thus, when website stickiness is seen as a key success factor for SNS management, how to attract user attention and continued usage, and further making users willing to spend more time on SNS has become a major challenge for SNS developers and operators. Stickiness refers to a website that attracts users' repeated use, and users who use this website with greater frequency, thus becoming continuous members. Besides, users' media habits are regarded as a sort of media behavior that demonstrates the people's preference to the media choice or content. The meaning of such media habits is in line with users' stickiness to SNSs. In several major studies, stickiness is treated as loyalty in the context of online game researches [10], where loyalty is defined as "the degree of willingness to re-play or re-participate". Adapted from concept of stickiness from online game playing, "stickiness" in this study is thus defined as the degree of users' willingness to return to and prolong their duration of each stay in SNS.

Gillespie et al. [7] used the time spent lingering on one SNS, website usage frequency, and the extent of depth at which the users use this SNS, to measure whether SNS have a high degree of stickiness. Allison et al. [8] concretized the standards for measuring SNS users, suggesting that stickiness refers to sustainability of Internet browsing, the number of visits, and the depth of browsing. The definition of stickiness by Maciag [9] is that users will continuously return to the website for browsing, and return to the website to browse for longer. Wu et al. [10] pointed out that psychological factors that affect game community users' satisfaction would also affect their continuance motivation, which further affects the stickiness of SNS significantly. In sum, there are indeed causal relationships among users' confirmation of expected performance of SNS (website systems), usage satisfaction, continuance motivation, and stickiness. User confirmation of SNS after usage would affect their level of satisfaction, and continuance motivation would affect the potency of SNS, and in turn, affect SNS stickiness. Thus, we proposes the following hypotheses:

H1: User continuance motivation has a significant effect on stickiness.
H2: User satisfaction has a significant effect on continuance motivation.
H3: User confirmation expectancy has a significant effect on satisfaction.

The SNS provides seamless integrated and convenient media, so that participants in SNS all over the world can transcend the restrictions of time and space for real-time information transmission and interaction of community emotions, thus becoming a social value interactive relationship network for modern people to engage in socializing, information transmission, emotional exchange, social identification, interpersonal support, and a sense of belonging. Zeithaml [11] proposed the conceptual view of perceived interpersonal value, suggesting that perceived interpersonal value is based on perceived quality of product and perceived price to motivate purchasing decisions, in other words, the customer's results of evaluating perceptions about the product or service. Value is the overall evaluation for the effectiveness of a product after consumers make a composite evaluation of the item purchased and the amount paid, and after considering the value of product efficacy. Sweeny and Soutar [12] used measurement of perceived interpersonal value to develop a multidimensional scale of PERVAL, to summarize the emotional, social, price, and perceived interpersonal value of quality. Among these, social perception refers to the perception for objects in society, and this primarily refers to the measurement and perception of interpersonal relationships. It involves perceptions regarding the individual or group features in the social environment, not only for external features such as personal expressions, language, and posture, but also involves core value considerations of interpersonal relationships, intrinsic motivation, intentions, views, beliefs, and character traits. Social perception is the process by which people compose and explain information about others. Thus, the perceived interpersonal value provided by SNS to users should affect to some degree their usage satisfaction and continuance motivation regarding SNS. Thus, this paper proposes the following hypotheses:

H4: Perceived interpersonal value has a significant effect on continuance motivation.
H5: Perceived interpersonal value has a significant effect on satisfaction.

Davis et al. [13] proposed that the concept of playfulness is an important basis to explore the intrinsic and extrinsic motivations of human behavior, which uses the perspective of human behavioral motivation to observe whether human usage of information technology is affected by related intrinsic and extrinsic motivation factors. Among them, extrinsic motivation focuses on the actions done, and the purpose is to achieve certain objectives and return; intrinsic motivation is a sense of perceived playfulness and satisfaction when engaging in an action, which further strengthens the shortcoming of technology acceptance model. Intrinsic motivation measures the perceived usefulness and perceived ease-of-use of information technology, but lacks consideration of internal psychological motivation and factors. "Intrinsic behavior" is defined as the performance of behavior as a result of pleasure and liking to perform it, regardless of any form of reward upon performance, while "Extrinsic motivation", on the other hand, refers to the performance of a behavior in order to obtain certain rewards or achieve certain goals attached to the performance of the behavior. In communities of entertainment technology, the significance of perceived usefulness will decrease in comparison to in problem-solving technology communities. In this study, "perceived playfulness" is defined as "the extent to which the user perceives that his or her attention is focused on the interaction with the specific entertainment application, and is curious during the interaction, and finds the interaction intrinsically enjoyable or interesting". Teo et al. [14] pointed out that for Internet users, they use the Internet primarily to enhance their work efficacy, followed by feeling perceived playfulness and ease of use from the Internet. Hsu and Lu [15] explored the loyalty of online gaming community users, and found that Perceived playfulness and social norms significantly affect user satisfaction and loyalty of gaming communities. Sweeter and Wyeth [16] suggested that making game community users feel playful is the most important goal of all computer games. When gamers do not feel enjoyment, then they would not play the game. According to the above studies, the nature of Internet technology application completely differs from the traditional single-machine system. Evaluation of users participating in online games not only should consider the effect of extrinsic motivation, but also the intrinsic motivation factors, regardless of online games or Internet technology. For SNS users, Perceived playfulness should significantly affect their satisfaction and continuance motivation in using SNS. Thus, we proposes the following hypotheses:

H6: Perceived playfulness has a significant effect on continuance motivation.
H7: Perceived playfulness has a significant effect on satisfaction.

In short, this study uses the theoretical basis of expectation confirmation theory, and applies the above theoretical development procedure to construct the causal conceptual framework, as shown in Fig. 1.

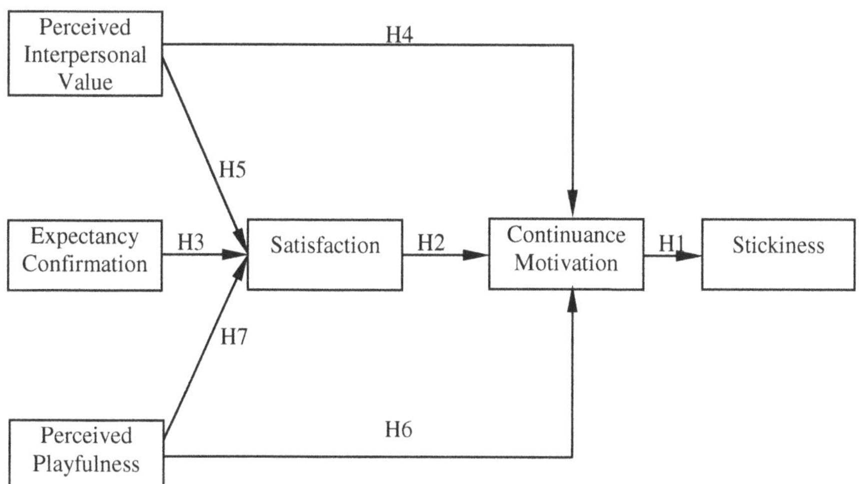

Fig. 1. Research model

3 Research Methodology

At the outset, we developed the major constructs of our conceptual framework and the associated measures. A number of prior relevant studies were reviewed to ensure that a comprehensive list of measures was included. For constructing a deliberate questionnaire to achieve the research goal, initial design and subsequent refinement of the instrument was done by researchers via several rounds of in-depth personal interviews with the experienced social network users and academic experts of universities/colleges in Taiwan using the questionnaire to structure their questions.

This process was continued until no further modifications to the questionnaire were necessary. Feedback from the in-depth personal interviews served as the basis for refining the experimental scales of the survey instrument. The researchers tested and revised the questionnaire several times before starting to gather the large-scale survey data. Finally, the questionnaire contained two major parts including a portion for the respondent's basic data portion and another for the responses to this study constructs. The basic data portion requested social network users to give their own demographic characteristics. The second part contained 31 questionnaire items relating to the six research constructs. Data were collected using a 5 point Likert-type scale.

The empirical data were collected using a questionnaire survey administered from January to March, 2012. Overall, we sent out 500 questionnaires and 273 completed questionnaires were returned. Ten responses were considered incomplete and had to be discarded. This left 263 valid responses for the statistical analysis, and a valid response rate of 52.6 % of the initial sample. Using a t-test and Chi-square tests to compare the responding and non-responding healthcare organizations' type, number of beds, and geographic location, the results signified no significant differences ($p > 0.05$). In addition, the potential non-response bias was also assessed by comparing the early versus late respondents that were weighed on several demographic characteristics. The results indicated that there are no statistically significant differences in demographics between the early and late respondents. These results suggest that non-response bias was not a serious concern.

The empirical data collected were analyzed using the partial least squares (PLS) method, which is particularly suitable for identifying the variance and validating the causal relationships between latent variables comprising complex theoretical and measurement models [18]. The hypotheses, proposed for the predictive and nomological structure of the research model. The PLS method allows the validations of the measurement model and the estimation of the structural model. The questionnaire administered in the large-scale questionnaire survey included items worded with proper negation and a shuffle of the items to reduce monotony of questions measuring the same construct. The statistical analysis strategy involved a two-phase approach in which the psychometric properties of all scales were first assessed through confirmatory factor analysis (CFA) and the structural relationships were then validated by the bootstrap analysis.

4 Analysis and Results

All of the constructs in the conceptual framework were modeled as reflective and were measured using multiple indicators. The measurement model relating the

IT-healthcare skill/knowledge and attitudes-to-develop-collaboration scale items to their latent constructs was analyzed by PLS-Graph 3.0. The assessment of item loadings, reliability, convergent validity, and discriminant validity was performed for the latent constructs through a confirmatory factor analysis (CFA). Reflective items should be uni-dimensional in their representation of the latent variable, and therefore correlated with each other. Item loadings should be above 0.707, showing that more than half of the variance is captured by the constructs. As shown in Table 1, all items of the instrument have significant loadings higher than the recommended value of 0.707.

Table 1. Descriptions and confirmatory factor loadings of scale items

Major Construct	Items	Factor Loading	Mean	AVE	Composite Reliability
Perceived Interpersonal Value (PIV)	5	0.81-0.88	3.50-3.70	0.70	0.92
Perceived Playfulness (PP)	5	0.73-0.85	3.60-3.83	0.64	0.90
Expectancy Confirmation (EC)	5	0.75-0.80	3.67-3.84	0.60	0.88
Satisfaction (SAT)	5	0.72-0.83	3.80-3.92	0.63	0.90
Continuance Motivation (CM)	6	0.74-0.86	3.44-3.67	0.60	0.90
Stickiness (STK)	5	0.82-0.91	3.40-3.71	0.77	0.94

All constructs in the measurement model exhibit good internal consistency as evidenced by their composite reliability scores. The composite reliability coefficients of all constructs in the proposed conceptual framework (Fig. 1) are more than adequate, ranging from 0.87 to 0.93. To assess discriminant validity, (1) indicators should load more strongly on their corresponding construct than on other constructs in the model and (2) the square root of the average variance extracted (AVE) should be larger than the inter-construct correlations [18]. The percent of variance captured by a construct is given by its average variance extracted (AVE). To show discriminant validity, each construct square root of the AVE has to be larger than its correlation with other factors. As the results shown in the Table 2, all constructs meet this requirement. Finally, the values for reliability are all above the suggested minimum of 0.7 [19]. Thus, all constructs display adequate reliability and discriminant validity. All constructs share more variance with their indicators than with other constructs. Thus, the convergent and discriminant validity of all constructs in the proposed conceptual framework can be assured.

Table 2. Inter-correlation and discriminant validity among major constructs

Major Construct	PIV	PP	EC	SAT	CM	STK
Perceived Interpersonal Value (PIV)	0.77					
Perceived Playfulness (PP)	0.76	0.84				
Expectancy Confirmation (EC)	0.66	0.61	0.80			
Satisfaction (SAT)	0.60	0.44	0.51	0.77		
Continuance Motivation (CM)	0.60	0.50	0.54	0.74	0.80	
Stickiness (STK)	0.67	0.62	0.75	0.40	0.45	0.88

*Diagonal elements are the square roots of average variance extracted (AVE).

The path coefficients and explained variances for the conceptual model in this study are shown in Fig. 2. T-statistics and standard errors were generated by applying the bootstrapping procedure with 200 samples. All of the constructs in this study were modeled as reflective and were measured using multiple indicators, rather than summated scales. A test of the structural model was used to assess the nomological network of the research model. As can be seen from Fig. 2, hypotheses, H1, H2, and H3, effectively drawn from CM to STK, SAT to CM and EC to SAT, are supported by the significant path coefficients. That is, the expectancy-confirmation theory (ECT) as proposed by Oliver [5] was also confirmed in this study context. On the other hand, with the significant path coefficients, the analysis results also provide support for the Hypotheses, H4 and H5, drawn from PIV to SAT and CM. In addition, the path from PP to SAT and CM are also significant. The direct effects from all of the major constructs, perceived interpersonal value (PIV), perceived playfulness (PP), expectancy confirmation (EC), and satisfaction (SAT), through continuance motivation (CM), totally account for 35% of the variance explained in stickiness (STK). The constructs of PIV, PP, and SAT together explain 55% of the variance explained in the construct of users' continuance motivation (CM). Besides, The constructs of PIV, EC, and SAT together explain 59% of the variance explained in the construct of users' satisfaction (SAT) on SNS usage. An F test is further applied to test the significance of the effect size of the overall model. Both of the two dependent variables are significant. Therefore, the model has strong explanatory power for the proposed theoretical model.

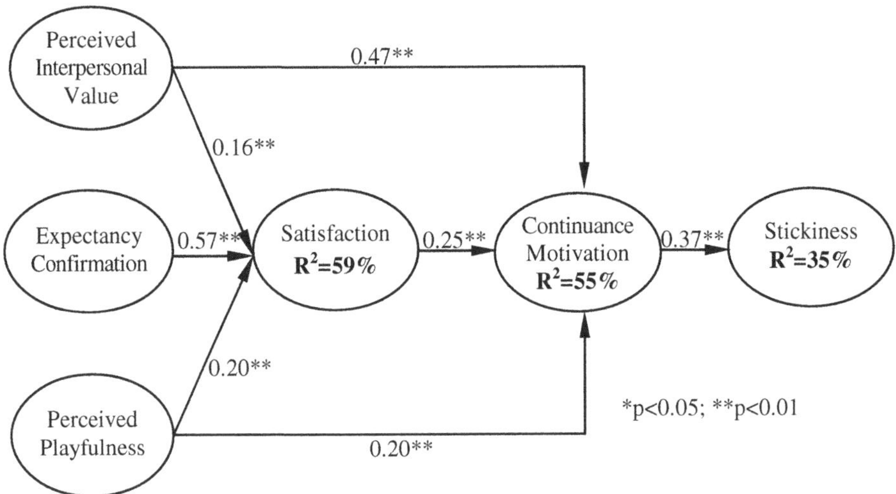

Fig. 2. PLS analysis results

5 Discussions and Conclusion

This study used the perspective of expectation-confirmation theory (ECT) as the theoretical basis, to engage in an in-depth exploration of the factors affecting SNS user stickiness. A careful procedure for theoretical development was used to construct a causal model for SNS stickiness. The partial least squares method was used to verify and evaluate the constructed conceptual framework and research hypotheses. Research results proved that the proposed antecedents (i.e., perceived interpersonal value, perceived playfulness, expectancy confirmation, and satisfaction, and user continuance motivation) have significant positive effects on users' stickiness of SNS, and also has a very high predictive explanatory power. In other words, if the users have higher continuance motivation, they may stick more around the SNS environment. In addition, from the perspective of expectation confirmation theory, regarding user confirmation of expected performance, empirical results showed that confirmation of expected performance significantly produces a positive effect on satisfaction; satisfaction would further cause a significant positive effect on the motivation of continued usage of the SNS, which also means that when the confirmation of initial expectations and the functional performance provided by SNS are higher, the user satisfaction would be higher, and there would be further strengthening of motivation to continue usage. For perceived interpersonal value, the empirical results showed that perceived interpersonal value has a significant positive effect on continuance motivation. This means that in website perceived interpersonal value, perceived utility has a high influence on behavioral intention. If users' perceived interpersonal value is high, then the user continuance motivation would also be high.

For Perceived playfulness, analytical results showed that Perceived playfulness has significant positive effect on continuance motivation. In other words, greater perceived playfulness obtained by users in community websites would lead to higher user satisfaction. This conclusion is consistent with Hsu and Lu [15] who explored online gaming community loyalty, and found that most users seek intrinsic playfulness and satisfaction from the gaming communities. Furthermore, Perceived playfulness also has a significant positive effect on satisfaction. This result is consistent with Sweeter and Wyeth [16], who explored computer usage behavior, and suggested that a very important usage purpose for users is to gain playfulness from Internet communications technology.

Regarding the research limitations of this study, since some of the retrieved questionnaires were conducted online, these conditions could not be fully grasped, and may have resulted in error of the analytical results. In addition, this study only focused on SNS users as research subjects, so it cannot represent the opinions of all SNS users. This study primarily focused on the SNS in Taiwan, so it could only represent the current conditions of SNS in Taiwan, and could not apply to usage conditions of SNS users in other countries. However, the findings can serve as references to SNS operators and researchers to effectively understand important perception factors of SNS users, and provide important research references and practical management of relationships among the major constructs of this study.

Acknowledgments. This research was supported by the National Science Council of Taiwan, under operating grant NSC-100-2511-S-276-003 and was partially supported by operating grant NSC-100-2511-S-143-004.

References

1. Bhattacherjee, A.: Understanding information systems continuance: an expectation-confirmation model. MIS Quarterly 25(3), 351–370 (2001)
2. Wellman, B.: For a social network analysis of computer networks. In: SIGCPR/SIGMIS Conference on Computer Personal Research, vol. 1(1), pp. 1–11 (1996)
3. Plant, R.: Online communities. Technology in Society 26(1), 51–65 (2004)
4. Catherine, R.M., Gefen, D., Arinze, B.: Some antecedents and effects of trust in virtual communities. Journal of SIS 11(3), 271–295 (2002)
5. Oliver, R.L.: A cognitive model of the antecedents and consequences of satisfaction. Journal of Marketing Research 17(4), 460–469 (1980)
6. Kim, S.S., Malhotra, N.K.: A longitudinal model of continued IS use: An integrative view of four mechanisms underlying post-adoption phenomena. Management Science 51(5), 741–755 (2005)
7. Gillespie, A., Krishna, M., Oliver, C., Olsen, K., Thiel, M.: Using stickiness to build and maximize web site value: Owen Graduate school of Management (1999)
8. Allison, N.C., Bagozzi, R.P., Warshaw, P.R.: Extrinsic and intrinsic motivation to use computers. Journal of Applied Social Psychology 22(14), 1111–1132 (1999)
9. Maciag, G.A.: Web portals user in, drive away business. National Underwriter Property and Casualty-Risk and Benefit Management 11(1), 1–9 (2000)
10. Wu, J.H., Wang, S.C., Tsai, H.H.: Fall in love with on-line game. Computers in Human Behavior 26(3), 1271–1295 (2010)
11. Zeithaml, V.A.: Consumer perception of price, quality and value: a means-end model and synthesis of evidence. Journal of Marketing 52(1), 2–22 (1988)
12. Sweeney, J.C., Soutar, G.: Consumer perceived interpersonal value: the development of multiple item scale. Journal of Retailing 77(2), 203–220 (2001)
13. Davis, F.D., Bagozzi, R.P., Warshaw, P.R.: Extrinsic and intrinsic motivation to use computers. Journal of Applied Social Psychology 22(14), 1111–1132 (1992)
14. Teo, T.S.H., Lim, V.K.G., Lai, R.Y.C.: Intrinsic and extrinsic motivation in internet usage. Omega 27(1), 25–37 (1999)
15. Hsu, C.L., Lu, H.P.: Consumer behavior in online game communities. Computers in Human Behavior 23(3), 1642–1659 (2005)
16. Sweetser, P., Wyeth, P.: GameFlow: a model for evaluating user enjoyment in games. Computers in Entertainment 3(3), 7–16 (2005)
17. Nunnally, J.: Psychometric Theory. McGraw-Hill, New York (1978)
18. Fornell, C., Larcker, D.F.: Structural equation models with unobservable variables and measurement error. Journal of Marketing Research 18(1), 39–50 (1981)
19. Chin, W.W.: Issues and opinion on structural equation modeling. MIS Quarterly 22(1), 7–16 (1998)

Collaborative Problem Solving in Emergency Situations: Lessons Learned from a Rescue Mission*

Radosław Nielek[1], Aleksander Wawer[2], and Adam Wierzbicki[1]

[1] Polish-Japanese Institute of Information Technology, ul. Koszykowa 86, 02-008 Warszawa, Poland
{nielek,adamw}@pjwstk.edu.pl

[2] Institute of Computer Science Polish Academy of Sciences, ul. Jana Kazimierza 5, 01-237 Warszawa, Poland
axw@ipipan.waw.pl

Abstract. The focus of this paper is on time-constrained collaborative problem solving using common web-based communication systems. The rescue action of a missing Polish kite surfer conducted with help of the kiteforum.pl community created a unique opportunity to investigate how people organize such actions in the Internet and how they collaborate under the enormous pressure of time. Quantitative and qualitative analysis of phenomena that occurred during the people's interaction, has been done with help of natural language processing techniques. A list of recommendations for the designers of collaborative problem solving systems and people involved in such action has been proposed.

Keywords: collaborative problem solving, phpBB, emergency, rescue.

1 Introduction

Teamwork, or more generally the approach of splitting a big task into smaller ones and assigning it to people, is probably as old as communities themselves, but the development of the Internet has substantially changed this process, both quantitatively and qualitatively. Even the biggest infrastructural projects conducted before the Internet era (among them the Egyptian pyramids) amassed at most tens of thousands of workers. In comparison, Wikipedia, which is the most famous crowdsourcing project in the Internet, has, only for the English version, almost 17 million editors[1] and they have created 4 million articles[2] that are comparable to the Encyclopaedia Britannica in the matter of quality [3].

The way people collaborate is shaped by many factors, among them: technology design [9], social presence [11], level of involvement, education of participants

* Research supported by the grant "Reconcile: Robust Online Credibility Evaluation of Web Content" from Switzerland through the Swiss Contribution to the enlarged European Union.

[1] `http://en.wikipedia.org/wiki/Wikipedia:Wikipedians`

[2] `http://en.wikipedia.org/wiki/Wikipedia:Size_of_Wikipedia`

N.T. Nguyen et al. (Eds.): *Adv. Methods for Comput. Collective Intelligence*, SCI 457, pp. 47–58.
DOI 10.1007/978-3-642-34300-1_5

and the pressure of time. Most of laboratory experiments lack the pressure of time and strong emotions. Therefore, the rescue action of the missing Polish kite surfer has created a unique opportunity to look at how people organize such actions on the Internet and how they collaborate. The SOS signal sent by Jan Lisewski has started a serious, community driven rescue action coordinated on the biggest Polish kite surfing forum `http://kitesurfing.pl` (KF). Users on the KF were: *brainstorming for the best solution, exploring social networks, disseminating information, calculating the most probable drift and collecting money for private charter a helicopter/boat.* As it is shown in the next chapters the efficiency of the collaboration and coordination was strongly influenced by the technology and the design of the phpBB based forum. Although the rescue action was successful, it is evident that many aspects suffered because of the limitation of technology.

The main objective of this paper is to study the way people have collaborated in organizing a successful rescue action with help of the Internet forum as an intermediary.The rest of the paper is structured as follows: the second chapter describes the expedition and the data set used in this paper, qualitative phenomena that appeared during cooperation is described in the third chapter, followed by the fourth chapter focused on quantitative analysis with help of natural language processing tools. The last chapter concludes the paper.

2 Related Work

Related work concerns the study of diverse Web2.0 or crowdsourcing systems that have been used for collaboration under time pressure. Among these, the Wikipedia is still the most important and most well-researched. A question that stands out in research on crowdsourcing systems is the issue of motivations for participation. This issue is especially important for time-constrained collaborative work, which requires a fast mobilization of participants.

Although recent reports indicate that there is a constant decrease in the number of active editors on the Wikipedia [1] and a substantial gender bias exists among editors [2] the Wikipedia is an impressive achievement, considering that Wikipedia editors do not receive financial gratification. If not money, then what motivates people to spend their free time in such crowdsourcing systems?

Heng-Li and Cheng-Yu show in their publication [4] that recognition and reputation are not primary motivations of wikipedians. More important are the need to feel self-confident and efficacious. According to the NYU Students Survey conducted by Stacey Kuznetsov [7] the most important driving force behind people's commitment is reciprocity. Heavy users of Wikipedia are more willing to do some editing (e.g. correct errors and omissions) probably because they assume that other people do the same and, thus, they will benefit from the higher quality of Wikipedia articles in the future. People active in free software/open software projects point out a slightly different motivation. The will to learn new skills and share knowledge are the two most commonly mentioned reasons to join open source projects according to the survey conducted by Berlecon Research[3].

Wikipedia is just one, although the most prominent, examples of successful collaborative problem solving systems on the Internet. Crowdsourcing has been used by Facebook to translate the web site [6]. People have been asked to choose the right word mapping and then everyone could vote for the most relevant translation. The creators of alternate reality games (ARG) present a less utilitarian approach - entertainment. ARGs are based on the interaction between players and characters that have been created and are actively controlled by game designers. Usually, ARGs are designed to blur the border between the real world and the game and to encourage cooperation between participants (some of them even require it because of the level of complexity, among them one of the first and most famous game "the Beast" designed by Microsoft to promote the Steven Spielberg movie Artificial [5]. See [8] for an in-depth study of alternates reality games and identification of design concepts, which support collaboration and involvement.

3 The Expedition and the Rescue Mission

3.1 The Red Sea Crossing

In the late-July of 2011, Jan Lisewski, a 42 years old famous Polish kite surfer, has crossed the Baltic Sea with a kite. He passed over 90 miles starting from Polish coast and finishing in Sweden without any assistance. It was the world's first successful attempt to crossing the Baltic Sea on a board with a kite. Almost half a year later, encouraged by the success of the first expedition, Jan Lisewski decided to take the next challenge: crossing from El Gouna in Egypt to Duba in Saudi Arabia (ca. 130 miles). The general idea was quite similar to the first passage except the location: no boat assistance, no bidirectional communication (neither satellite phone nor VHF radio[4]), practically no support teams in Egypt and Saudi Arabia. The whole project was intended to be low-cost.

The success of the project was strongly dependent on favourable and stable weather. A kite surfer can sail close-hauled (the dead angle is strongly dependent on the type of board and kite but can reach around 100 degrees) but cannot sail without wind (if wind is weaker than three on the Beaufort scale). Although the speed record on a kite board is over 55 knots, the average speed during crossinging is smaller because of tacking, weaker wind and waving. The Baltic Sea crossing has taken 12 hours what means that the average speed was ca. 8.5 knots.

In contrast to the Baltic Sea crossing, which has been a great success (although there were also some technical problems like flooded GPS or lack of connectivity),

---

[3] `http://flossproject.org/report/`

[4] There is contradictory information about the full list of equipment but there is no evidence of bidirectional communication during the rescue action.

the attempt to cross the Red Sea has failed and required a serious rescue action. After ca. 10 hours, 50 kilometres from Saudi Arabian coast, Jan Lisewski sent the 911 HELP signal with a SPOT2 device[5] (Satellite Personal Tracker)[6].

3.2 Dataset

The data used in this paper comes from the biggest Polish kite forum: Kiteforum.pl which has been founded in 2003 and is the oldest, and at the same time the biggest, Polish community focused on kite surfing. Currently, this forum contains almost 300 thousands posts in 38,000 threads and over 10,000 registered users. Technically, this web site is implemented using the phpBB[7], a very popular, open-source, flat-forum bulletin board software. KF seems to be purely community based, and it is not associated with a company (or foundation). Nor are there any attempts on KF to monetize the web audience. Only registered users can publish posts (but anyone can read even without login) and each post is accompanied by information about author (including the date of first registration and *the number of published posts*) and date of publication (with accuracy of minutes).

The thread, which is the main object of the study described in this paper, refers to the Jan Lisewski attempt to cross the Read Sea with a kite. The thread is composed of 1519 posts that have been published within 3 months from the mid-December 2011 to the beginning of March 2012 (will be denoted in the further part of this paper as "Red Sea crossing"). For comparison, the thread related to the crossing of the Baltic Sea (composed of ca. 450 posts) was also analysed (denoted as "Baltic crossing"). All post that have appeared in these two threads have been downloaded and parsed. In addition, some statistics have been calculated for the whole forum.

4 Qualitative Effects of Collaborative Work on KF Forum

4.1 Interaction with Mass Media

Rapid provisioning of reliable, interesting, full of emotion, eye-catching and unique information is the mission of the vast majority of media. Therefore, it is not surprising that the rescue action has raised their interest. In the beginning information published by media were strongly based on posts from the

[5] SPOT is not a professional device and is dedicated mostly to hobbyists. In the basic version it uses a one-directional communication and allows sending by satellite three signals: SOS, HELP and Check-in/OK, all bundled with exact position obtained from GPS. The main difference between HELP and SOS signals is that the first one notifies only friends/families (email addresses and mobile phones given *a priori* by spot owner) and the second one also local (according to location taken from GPS) emergency response centre.

[6] `http://www.findmespot.eu/en/`

[7] `http://www.phpbb.com/`

KF. Because journalists have not always credited the source (or have not done it clearly enough) for users on the KF forum it appeared as an external acknowledgment that mistakenly increased the credibility of information published on KF. Users have posted information from media and de facto created a bi-directional feedback, which could mislead people organizing the rescue. This shows the journalists' responsibility in emergency situations and the importance of source credibility management (which includes trustworthiness evaluation and exact source denotation). Later on, users on the KF forum noticed the journalists' practice, which finally changed the attitude towards journalists to negative (one of the users has even threatened legal action against media[8]).

Of course, the role of media cannot be seen one-sidedly. Viewership of national TV stations and number of visitors on the most popular web sites were a huge opportunity to mobilize public opinion and created pressure on policy makers. People directly involved in organizing the rescue action put a lot of effort to bring media on board and give the issue publicity with help of TV and radio stations. KF users have used their own social network to alarm particular broadcastings. Media also create a unique opportunity to reach people with a rare knowledge and valuable social connection but neither media nor rescue team noticed and used this chance (despite the fact that at least in few cases such a need existed, e.g.: *"If there is anyone who dive know anything about www.daneuroe.org please call me."*,*"URGENT!!! what is the ID number of the SPOT device"*, *"Does anyone have any contact in Saudi Arabia"*).

Although at the beginning of the rescue action media mostly cited the KF posts, after some time (ca. 1-2 hours) the national broadcasters started to gather their own information. Some high profile sources were, and probably will be, out of reach for KF community and are relatively easy accessible for traditional journalists, e.g. Foreign Minister, Polish Consul in Saudi Arabia or famous experts. Interviews with Radoslaw Sikorski (Polish Foreign Minister) and Igor Kaczmarczyk (Polish Consul in Riyadh) have delivered a rich material for analysis on KF but could be even more valuable if journalists would be equipped with precise information about the expedition and present status of rescue action (creating and maintaining a list of questions should be a high priority task for KF community).

The news about successful rescue of Jan Lisewski was accessible at the same time for the best journalists and the whole society because was firstly twitted by Radoslaw Sikorski (Polish Foreign Minister). Shortly after he twitts this information it was republished on KF.

4.2 Backup Communication Channel

One of the important functions of the KF during the rescue action was a backup communication channel. Most communication was carried out through mobile phones and KF was at the beginning used to exchange and broadcast phone numbers (e.g.: Kowis: *"Gregor, please, call me. I have two person speaking Arabic*

[8] *"cited all without credits. I expect a legal action later on."* posted by flash on 2012-03-04 01:32:00.

which can translate to English. My phone number is ..."). Later on users have started posting asks for calls either because some numbers were busy or out of range (e.g. *"Kargul, Pinio, pick up the phone!!!"*, *"Zielwandam, call me if you can. My phone number is ..."*). Widespread distribution of phone numbers of people organizing a rescue action (and phone number of SAR in Poland also freely available in the Internet) have caused that a lot of people started to call those numbers just to gather some information. As a consequence many phone lines have been blocked making communication almost impossible. Therefore, an appeal on the KF has appeared asking for stop calling and blocking lines. Gregorg, one of the person directly involved into the rescue action posted: *"Please do not call Janek. You are blocking the telephone line."*

4.3 Problems with Performance and Bandwidth

As already has been mentioned, the KF is a non-commercial project and, therefore, it is equipped only with a limited computational power to keep it low-cost. As the dissemination of the information about the missing kite surfer was speeding up (with substantial help of national TV stations), more and more people wanted to read first-hand information (and maybe also help) and visited KF web site. The easy to predict consequence was a slowdown of the web site. People have encountered problems with browsing the forum and posting. Although users have noticed problems and even posted about it (*"the kiteforum works very slow"*) no serious remedies have been implemented (except requests for logout). Visitors who could not gather information from the KF forum moved to other web sites related to rescue action, e.g. findmespot.com where tracking information from the SPOT device was published and caused a quasi-denial of service attack. It virtually cut off the rescue team from information about exact position of the castaway (*"People not involved directly into rescue action please log out and leave the findmespot.com web site because it is overloaded."*).

Limiting the access for people from outside of the community will decrease chances for finding crucial social connections (jaroo posted: *"My friend from the maritime academy gave me an URL address"*) and knowledge (flash posted: *"I'm not interested in kitesurfing but I've knowledge about sea rescue"*). 46 newcomers[9] (15% of all users taking a part in discussion) wrote 179 posts, which accounts for 11% of the total *but all except one were posted after the first SOS signal.* In comparison, the Baltic Sea crossing has attracted only 5 new users (ca. 3% of all active users), who have posted 8 times (total number of posts is 465).

4.4 Lack of Updated Information

A single web page of the phpBB contains (at least in the configuration used by the KF) 15 posts. If this number is exceeded, content is paged and users can move between previous and next posts by clicking on arrows at the top (or

[9] Not every new account can be attributed to a newcomer. Eska, a member of the core coordination team, has created separate accounts for this purpose (login: Rescue).

bottom) of the web page. Users have published during the rescue period on average 22 posts per hour (detailed information can be found on fig. 2). The rapid flow of posts made following the situation a little bit difficult. Informative posts were pushed up by short comments and new visitors had to click backward to find the most relevant information. Many users instead of doing it preferred to repeatedly asking about the up-to-date information (Stani21:*"Do you have any new information about location?"*,Bastek_kite: *"Any new information?"*, Solution: *"Gregor, do you have any information?"*etc.).

Next to the design of the phpBB bulletin another reason causing the flood of questions was the very basic dilemma: if there is no information it could be either because of lack of new facts or because no one shares it on the forum. Everyone involved into the rescue action[10] wanted to be as up-to-date as possible. The core members of the rescue team had not noticed this issue but after 30 hours and the explicit request made by Ksiaze they have created a continuously updated whiteboard[11] with all contacts. It was only a moderate success because of problems with effective redirecting all people looking for information to this web page.

4.5 Credibility Problems

Similarly to other open knowledge communities, the credibility of disseminated information must be verified by the participants. The KF forum participants reacted strongly to the posting of incorrect information. In one case, a new users published one post that claimed that the missing kite surfer was already rescued. As expected, this resulted in a heated discussion. When the post turned out to be wrong, the user was banned from the forum. However, such a strong reaction creates the concern about reducing voluntary participation. A more appropriate reaction would be through a trust management or reputation system for source credibility evaluation.

5 Quantitative Analysis

5.1 People's Involvement

The idea of crossing the Red Sea with a kite was preceded by the successful crossing of the Baltic Sea. Both expeditions were extensively commented on the kiteforum.pl. Similar subject but drastically different emotions make the comparison between these two threads extremely interesting. The threads vary widely in the matter of people's involvement, dynamic of discussion, mentioned topics and behaviour patterns. Topics are discussed in-depth in the next chapter; therefore, in this section special focus will be put on purely quantitative effects. As is shown in Table 1 the Red Sea crossing has attracted more attention than the Baltic crossing, both in the matter of users and the length of comments.

[10] It is worth to notice that for many crowdsourcing systems there is no procedure for accepting new team member. Everyone can join freely. Being a member of a team is actually a personal mental state.

[11] `http://www.kiteforum.pl/forum/viewtopic.php?f=27&t=44642`

Table 1. Basic statistic for different parts of the Red Sea crossing thread

	Red Sea crossing	Baltic crossing
Avg. length of comments (in characters)	184	165
No. of posts	1510	465
No. of users	312	178
Posts per user	4.83	2.61
No. of new users	46	5
Timespan	Eleven weeks	Six weeks

The substantial difference in the posts per user ratio between the Red Sea crossing thread and the Baltic crossing is caused by the hyper activity of the minority of users. 20% of the most active users generated 67.2% and 55.7% posts, respectively. Gini index values are respectively 0.60 and 0.48 and show that the distribution of the activity of users in the Red Sea crossing threads has a much longer tail.

The Red Sea crossing thread can be divided into four separate periods following the most important events. *Preparation* was started by the first comment in this thread and is followed by *crossing* triggered by the information about the start of Jan Lisewski. The first intercepted SOS signal indicates the end of *crossing* and the beginning of *rescue* period. Confirmed information about finding the kite surfer finished the rescue period and started the last one - congratulations. Table 2 shows basic statistics for all four periods.

Table 2. Basic statistic for different parts of the Red Sea crossing thread

	Preparation	crossing	Rescue	Congratulations
Avg. length of comments (in characters)	104.33	112.75	176.44	256.91
No. of posts	117	60	1020	313
Posts per hour	0.06	5.45	22.6	11.2
No. of users	70	34	203	168
Posts per user	1.67	1.76	5.02	1.86

The user activity smoothly follows the dynamics of events even if there are no dramatic emotions. During the crossing people have posted almost 100 times more often than during the preparation but the length of comments were very similar. Except the rescue period users usually post on average less than two comments. Even if they have a lot of thoughts and observations to share they choose to make comments longer, rather than publish many posts. The rescue period is presented separately on Figure 1.

5.2 Sentiment in Posts

This section describes the sentiment analysis tools and algorithms applied to forum posts. In short, sentiment computation method is a lexicon-based one

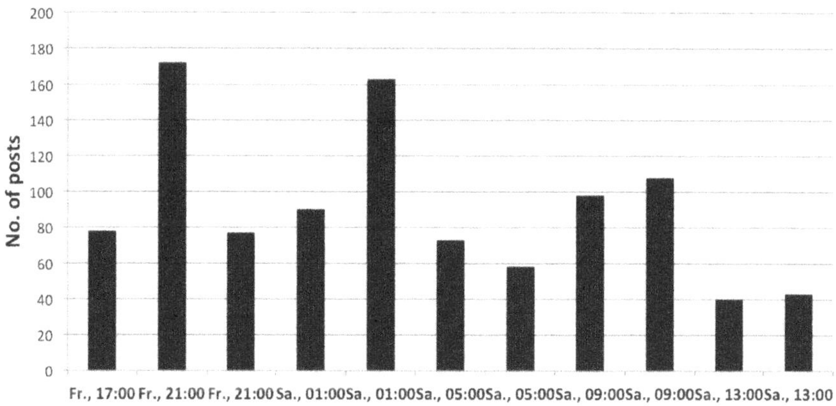

Fig. 1. Number of posts for the rescue period in two-hour intervals

extended by a shallow parser. An version of the system used in this paper has been described in [10]. Figure 2 presents aggregated results of sentiment analysis. Sentiment of each message has been computed as positive (sum of positive word scores in a message) minus negative (sum of negative word scores) divided by the number of words in a message.

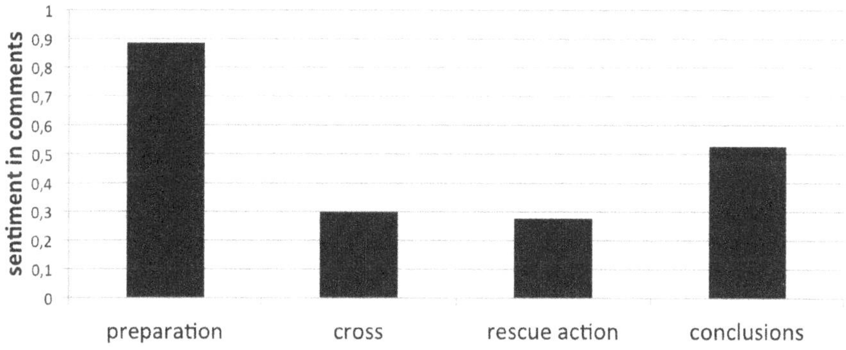

Fig. 2. Average level of posts' sentiment in four periods

The results show a strong change in emotions after the preparation stage. Positive emotions appear again in the conclusions stage. The negative emotions in the crossing stage can be explained by the fact that despite of promises given by team preparing the Red See Crossing, live tracking was available only by a limited time (at the beginning). Therefore, the initial optimism in the comments has been later on displaced by concerns. So, posts published during crossing period were less positive than would be expected.

5.3 Posts Topic Modeling

This section describes the results of automatic topics identification of the discussions in the two threads of the KF forum related to the Baltic Sea and Red Sea crossings. The main hypothesis related to this investigation has been that the two threads concern a variety of diverse topics, and that the thread organization of the KF forum is not fine-grained enough to allow such a usage (which resembles sub-threads).

The topic modeling technique we have used is the well-known Latent Semantic Indexing (LSI). The algorithm involves two steps: construction of a term-document matrix and application of singular value decomposition (SVD) to obtain singular value and singular vectors matrices. One popular claim of the method is its ability to capture conceptual content of a corpus of texts by establishing associations between terms that tend to co-occur. Singular vectors can be seen as representations of latent topic variables and their contents approximated by terms with the highest weights.

The discussion of results of topic modeling analyses on both datasets, Baltic and Red Sea, is mostly qualitative. It focuses on interpretability and usefulness of obtained topic representations. Quantitative parameters such as energy spectrum covered (LSI) are of considerably less importance if topic representations are not meaningful. Each topic is approximated by a set of its distinctive terms with their associated weights (the meaning of weights depends on an algorithm).

The results have been presented for 10 most important topics (in terms of singular values). Limiting to 10 factors discards 61.980% of energy spectrum in the case of Red Sea data and 58.209% of energy spectrum for Baltic data. The LSI technique, originally not designed for topic modeling, performs good. Both Baltic crossing and Red Sea crossing data generate quite distinctive and mostly interpretable topic representations. This is especially evident in capturing English language in two separate topics in Red Sea and one topic in Baltic data, not disclosed.

In Red Sea crossing data, five topics are generic discussions of the trip which include mentions of Saudi Arabia, coast guard, surfer's name and surfing alone. Two presented topics are special and more interesting. Topic #6 is focused on organizing help using the satellite signal, while topic #7 on reading gps positions from the receiver and time-position speculations. Selected two topic representations for Red Sea are as follows (English translations in parentheses):

- #6 : -0.250*(device) + 0.237*(be able) + -0.213*lisewski + 0.208*(help) + -0.169*(sea) + 0.157*(satellite) + -0.139*(signal) + -0.132*(number)
- #7 : 0.251*(be able) + -0.242*gps + -0.206*(clock) + -0.196*(receiver) + -0.159*(ground) + -0.132*(position) + -0.128*(use) + 0.128*(arab) + -0.127*(readings) + -0.121*(satellite)

Baltic data topics are focused on concepts like keeping fingers crossed, gsm range and words of joy such as congratulations, achievement. Topic #0 is the one mostly focused with gsm network range. Topic #2 is devoted to wave discussions:

breaking waves, wave crests, but also kite. Two selected Baltic data topics are as follows:

- #0 : 0.300*(wave) + 0.262*(transmitter) + 0.163*(range) + 0.153*(longitude) + 0.140*gsm + 0.134*(to sail) + 0.122*(Baltic) + 0.116*(sea)
- #2 : -0.434*(wave) + -0.197*(break) + -0.178*(crest) + -0.178*(move) + 0.170*(transmitter) + -0.144*(Baltic) + -0.128*(kite)

6 Conclusions

Each rescue action is different, therefore, any generalizations should by made very carefully. There is no single recipe for making collaboration of a bunch of anonymous people efficient but even if it was the pressure of time, emotion and limited resources could effectively prevent its implementation. A list of recommendation that come from the analysis of Jan Lisewski case are presented below:

- People prefer to use tools they know and stay within the existing communities; there is no easy way redirect traffic and, thus, it is better to ask people for help in places where they feel comfortable; although same of dedicated web sites (communities) were created to help Jan Lisewski the most active still remains the kiteforum.pl.
- Over time, even the most dramatic events evoke less emotion; sustaining commitment and mobilizing people is crucial, especially since no one knows how long the rescue mission will take.
- More people usually means a greater chance of success; therefore it is important to encourage people to join the community and share their thoughts (for the kiteforum.pl the number of visitors easily surpassed few thousands but only a 312 users have posted anything).
- Everyone who wants help should be confronted with a list of tasks where he/she can pick up that what best fits his/her skills and resources; volunteers in a crowdsourcing system should not be forced to ask for task allocation.
- Users should be instructed that discussion, even full of guessing and presumptions, is good, but should be supplemented with information about author's level of knowledge, experience and education because it will protect people from implementing untested assumptions.
- Few simple templates for repeatable (periodical) messages will eliminate the omissions (e.g. every post about signal from the SPOT device should contain type of signal, position and exact time).
- Information is indispensibly connected with the source (it will help judge credibility of information).
- Translation and information dissemination were two tasks which were carried particularly efficient by the KF community.

The rescue action of Jan Lisewski, and particularly cooperation of ordinary people in the Internet, creates a unique opportunity to investigate how the Internet technology influence the efficiency of such tasks. Further research should be extended to a survey of people involved in the rescue action and an in-depth analysis of interaction between different communities and web sites.

References

1. Angwin, J., Fowler, G.A.: Volunteers log off as wikipedia ages (September 27, 2009)
2. Collier, B., Bear, J.: Conflict, criticism, or confidence: an empirical examination of the gender gap in wikipedia contributions (2012)
3. Giles, J.: Internet encyclopaedias go head to head. Nature 438(7070), 900–901 (2005)
4. Heng-Li, Y., Cheng-Yu, L.: Motivations of wikipedia content contributors. Computers in Human Behavior 26(6), 1377–1383 (2010)
5. Kim, J.Y., Allen, J.P., Lee, E.: Alternate reality gaming. Commun. ACM 51(2), 36–42 (2008)
6. Kirkpatrick, D.: The Facebook Effect: The Inside Story of the Company That Is Connecting the World. Simon & Schuster (2010)
7. Kuznetsov, S.: Motivations of contributors to wikipedia. ACM SIGCAS Computers and Society 36(2) (2006)
8. McGonigal, J.E.: This Might Be a Game: Ubiquitous Play and Performance at the Turn of the Twenty-First Century. Ph.D. thesis (2006)
9. Miller, K.M., Dick, G.N.: Computer-based collaboration in student work: does a preference for using technology affect performance? (2005)
10. Wawer, A.: Mining co-occurrence matrices for so-pmi paradigm word candidates. In: Proceedings of the Student Research Workshop at the 13th Conference of the European Chapter of the Association for Computational Linguistics, pp. 74–80. Association for Computational Linguistics, Avignon (2012), http://www.aclweb.org/anthology/E12-3009
11. Weinel, M., Bannert, M., Zumbach, J., Hoppe, H.U., Malzahn, N.: A closer look on social presence as a causing factor in computer-mediated collaboration. Comput. Hum. Behav. 27(1), 513–521 (2011)

Biometric and Intelligent Student Progress Assessment System

Artūras Kaklauskas*, Edmundas Kazimieras Zavadskas, Mark Seniut, Andrej Vlasenko, Gintaris Kaklauskas, Algirdas Juozapaitis, Agne Matuliauskaite, Gabrielius Kaklauskas, Lina Zemeckyte, Ieva Jackute, Jurga Naimaviciene, and Justas Cerkauskas

Vilnius Gediminas Technical University, Sauletekio av. 11, Vilnius, LT-10223, Lithuania
arturas.kaklauskas@vgtu.lt

Abstract. A number of methodologies (Big Five Factors and Five Factor Model, intelligence quotient tests, self-assessment) and strategies for web-based formative assessment are used in an effort to predict a student's academic motivation, achievements and performance. These methodologies, biometric voice analysis technologies and 13 years of authors' experience in distance learning were used in development of the Biometric and Intelligent Student Progress Assessment System for psychological assessment of student progress. Also the BISPA system was developed in consideration of worldwide research results involving the interrelation between a person's knowledge, self-assessment and voice stress along with instances of available decision support, recommender and intelligent tutoring systems.

Keywords: e-Learning, Voice Stress Analysis, Intelligent System, e-Self-Assessment, e-Examination, Historical Information, Reliability of Results.

1 Introduction

All parties involved in distance learning would like to know how well the students have assimilated the study materials being taught. The analysis and assessment of the knowledge students have acquired over a semester are an integral part of the independent studies process at the most advanced universities worldwide. A formal test or exam during the semester would cause needless stress for students.

Various tensions arise between an examiner and the person being examined during an exam. These tensions are caused by different perceptions about the degree of preparedness for the exam; such perceptions spring from the self-assessment of an individual taking the exam. In an effort to decrease such tension, restore an educational sphere and teach collegial team work, tolerance and the goal of a better and more objective appraisal of knowledge and abilities, the authors have developed the Biometric and Intelligent Student Progress Assessment (BISPA) System.

The aforementioned methodologies (Big Five Factors and Five Factor Model, intelligence quotient tests, self-assessment), biometric voice analysis technologies and

* Corresponding author.

N.T. Nguyen et al. (Eds.): *Adv. Methods for Comput. Collective Intelligence*, SCI 457, pp. 59–69.
DOI: 10.1007/978-3-642-34300-1_6

13 years of authors' experience in distance learning were used in development of the the BISPA system. These methodologies are very helpful and significant, while the results of their practical application are often, though not always, reliable. Therefore, the analysis how the Big Five Factors, intelligence quotient tests and self-assessment helped to deal with the reliability issues of self-assessment in practical applications was particularly significant in the development of the BISPA system.

The structure of this paper is as follows. Following this introduction, Section 2 provides a description of the BISPA system. Section 3 follows with a case study. Finally, the concluding remarks appear in Section 4.

2 Biometric and Intelligent Student Progress Assessment (BISPA) System, the BISPA System Architecture, Components, Used Methodology and Algorithms

Our research rests on two hypotheses: (1) a well-prepared self-assessment process may enable rather reliable forecasting of a student's academic achievements; (2) micro-tremor frequencies depend on the student's stress and, therefore, of the level of his/her readiness for the exam. The major purpose of our experimental design was to develop an empirical model to approximate an unknown relationship (the interrelation between a person's knowledge, self-assessment and voice stress) up to a given level of accuracy. This empirical model be likely algebraically not easy to simulate and ill-structured.

The BISPA system was developed in consideration of worldwide research results involving the interrelation between a person's knowledge, self-assessment and voice stress along with instances of available subsystems of a decision support system [1], recommender systems [2] and intelligent tutoring systems [3]. The results obtained from testing the BISPA system were in concord with results obtained worldwide from the research mentioned herein.

Fundamental components of decision support system (DSS) architecture are the databases and database management system, the models and models management system and the user interface. This article describes all these fundamental components, which are a composite part of the BISPA system.

Recommender systems are software tools and techniques that recommend items of information likely to be of interest to the user or that provide suggestions for items useful to a user. The Recommendations Providing Model of the BISPA system is similar to traditional recommender system components.

Traditionally Intelligent Tutoring Systems (ITS) consist of four different subsystem modules: the expert (domain) module, student module, tutor module and interface module. All these modules are components of the BISPA system's ITS.

The innovations involved in the BISPA system, developed by the authors of this article, are several. One is that the voice stress analysis technologies available across the world are integrated with the known decision support, recommender and intelligent tutoring systems. Another aspect is that the research on the BISPA system is superior to other research due to innovative models of the Model-base, also developed by these authors, which permit testing and a more detailed analysis of the knowledge attained by a student.

The BISPA system comprises of the following components (see Figure 1): the Database Management Subsystem and Databases, the Equipment Subsystem, the Model-base Management Subsystem and the Model Bases with User Interface. The architecture (see Figure 1) and components of BISPA system are briefly analyzed below.

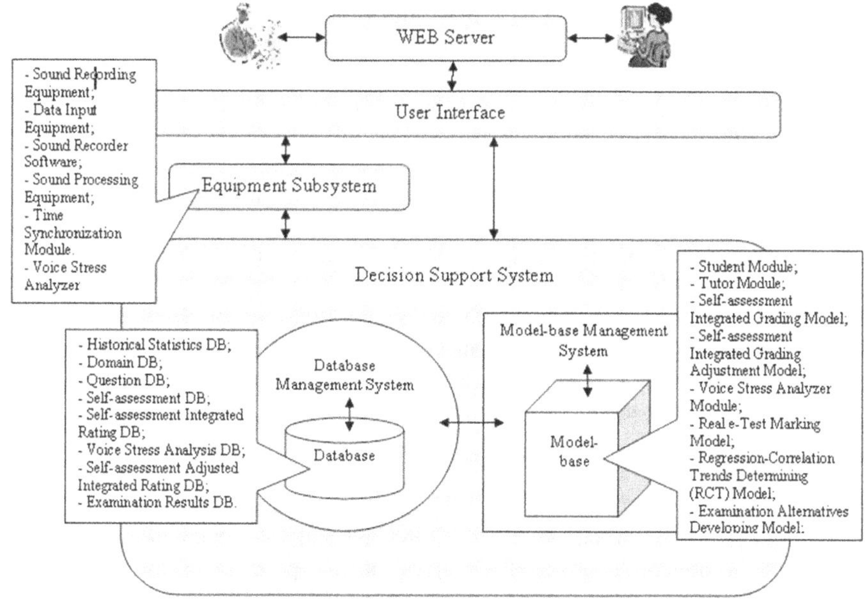

Fig. 1. The architecture of BISPA system

The *Equipment Subsystem* consists of sound recording equipment, data input equipment, sound recorder software, sound processing equipment and a time synchronization module and Voice Stress Analyser.

The *Database* contains the developed Historical Statistics Database, Domain Database, Question Database, Voice Stress Analysis DB, Self-assessment DB, Self-assessment Integrated Rating DB, Self-assessment Adjusted Integrated Rating DB and Examination results DB.

The *Historical Statistics Database* accumulates historical statistical data collected from the Self-assessment Integrated Grading Model, Self-assessment Integrated Grading Adjustment Model and real e-test (test questions; test results with information about correct/incorrect answers, time spent on each question, and the number of times a student has changed an answer to each test question; voice stress data; correlation between emotional stress; correct answers data; and such).

The *Domain Database* contains learning materials: information and knowledge related to specific subjects. Over the first three semesters of master's courses, students take seven core modules and five optional modules. Students may choose one elective from 21 modules in *Real Estate Management* and 17 modules in *Construction Economics* (both Master's degree study programmes), and must also pass five examinations on optional modules. During the fourth semester, master students write their final theses. Once they have registered, students mark in electronic questionnaires the sections of the elective modules they want to study . The *Domain* Database also offers to students study materials based on the repetitive keywords in different optional modules. A mixed approach is also possible and available. The received information is combined to make action plans, "mini curricula" that guide a learner/student.

The *Question Database* accumulates the following information: questions sorted by modules, possible answers to each question, and correctness assessments of possible answer versions. The *Self-assessment DB*, *Self-assessment Integrated Rating DB*, *Voice Stress Analysis DB*, *Self-assessment Adjusted Integrated Rating DB* and *Examination Results DB* accumulate following information from corresponding modules (see Figure 1).

The *Model-base* consists of nine models (see Figure 1). The *Student Model* stores data that is specific to each individual student. The *Tutor Model* formulates questions of various difficulties, specifies sources for additional studies and helps to select literature and multimedia.

Students pass two tests (e-psychological and real). After an e-psychological test the students are passing a real exam electronically and compare the received subjective and objective rating marks amongst themselves. The description of e-psychological and real tests in parallel with modules is follows.

e-Psychological Test

During an e-Psychological Test questionnaire is used as the method of data collection. Each question in the questionnaires was formulated and their evaluation procedure was set enabling students to assess themselves on a 10-point scale during the self-assessment. It facilitated estimation of their self-rated forecasted exam marks. Initially students answer 16 questions both verbally and electronically; these questions are not directly related with the learning contents. Next 16 criteria used as the basis for the e-self-assessment are named: What mark are you expecting for the exam? How many days have you spent studying? What was your average for the last semester? In your opinion, is the subject you've been studying essential and necessary for your future profession? Were the studies of this course interesting? Did the time you spent studying this discipline correspond with the benefit you received? What average do you expect for this semester? What was your average for your undergraduate studies? What percent of the exam material have you assimilated? Did you attend all lectures and practice sessions? How many days did you need to prepare for this exam to get the expected mark? How many days, in average, did you need to prepare for past exams to get the expected mark? What is you accuracy, as a percentage, at guessing your exam mark? If this subject was an elective, and you were able to choose, would you have chosen this or some other subject? Did the instructor present the material understandably and clearly? Was the subject you studied personally interesting and important to you?

Based on the regression model, the Self-assessment Integrated Grading Model sets three formulas to calculate grades: using two criteria: "What grade are you expecting for the exam?" and "What was your average for the last semester?" (equation 1); and using all 16 criteria (equations 2 and 3; all questions/criteria are listed in Section 2).

In the first case, the data collected by examining 151 students was used. The calculations produced a linear equation so that both questions would best explain the actual e-test grade:

$$P_1 = 2{,}505 + 0{,}261 \cdot k_1 + 0{,}417 \cdot k_3, \quad (1)$$

where k_1 is a student's answer to the question "What grade are you expecting for the exam?"; and k_3 is the student's answer to the question "What was your average for the last semester?".

This equation best explains the grade produced by answers to the two above criteria and makes it possible to predict the performance of other students.

In the second case, with the data collected by examining 208 students, the following equation for self-assessment grades was produced:

$$P_2 = 4{,}354 + 0{,}016 \cdot k_1 - 0{,}042 \cdot k_2 + 0{,}304 \cdot k_3 - 0{,}030 \cdot k_4 - 0{,}083 \cdot k_5 + 0{,}071 \cdot k_6 - 0{,}010 \cdot k_7 + 0{,}125 \cdot k_8 + 0{,}124 \cdot k_9 - 0{,}136 \cdot k_{10} + 0{,}020 \cdot k_{11} + 0{,}016 \cdot k_{12} - 0{,}034 \cdot k_{13} - 0{,}010 \cdot k_{14} - 0{,}063 \cdot k_{15} - 0{,}014 \cdot k_{16}. \quad (2)$$

This equation best explains the grade produced by answers to all 16 questions and makes it possible to predict the performance of other students.

In the third case, calculations of self-assessment grades are based on the following equation:

$$P_3 = k_1 q_1 + k_2 q_2 + \ldots + k_i q_i + \ldots + k_{16} q_{16}, \quad (3)$$

where k_i is the answer to a question; q_i is the answer's weight.

The equation 3 must satisfy the following condition:

$$q_1 + q_2 + \ldots + q_i + \ldots + q_{16} = 1. \quad (4)$$

Hence, the equation to calculate self-assessment grades in the third case is as follows:

$$P_3 = 0{,}4 \cdot k_1 + 0 \cdot k_2 + 0{,}1 \cdot k_3 + 0{,}01 \cdot k_4 + 0{,}03 \cdot k_5 + 0{,}04 \cdot k_6 + 0{,}05 \cdot k_7 + 0{,}08 \cdot k_8 + 0{,}16 \cdot k_9 + 0{,}02 \cdot k_{10} + 0 \cdot k_{11} + 0 \cdot k_{12} + 0 \cdot k_{13} + 0{,}05 \cdot k_{14} + 0{,}03 \cdot k_{15} + 0{,}03 \cdot k_{16}. \quad (5)$$

During self-analysis tests, some students were noticed experimenting with the BISPA system by giving made up answers unrelated to their exam. Such misleading answers were eliminated with the help of Voice Stress Analyser. Any answer with microtremor frequencies above 11 Hz was eliminated from the analysis.

Afterwards the students complete an electronic questionnaire which serves as the basis for establishing the level of a student's self-assessment (high, average, low). By employing a Self-assessment Integrated Grading Adjustment Model, the BISPA system evaluates the micro-tremors and digital information submitted by a student and writes a subjective rating of a psychological assessment. After this the students are able to take a real exam electronically and compare the received subjective and objective rating marks amongst themselves.

Correlation between Emotional Stress and Examination Mark

Our research also rests on the second hypothesis "Micro-tremor frequencies depend on the student's stress and, therefore, of the level of his/her readiness for the exam". Figure 2 shows the relation between real mark given students during exam and students ID number (dark line of the trend) and the average microtremor frequency (determined by Voice Stress Analyzer Module) of the answers to test questions and students ID number (bright line of the trend). The x-axis shows the students ID number who had passed the exam. The right side of the y-axis shows the correct answers (by percent). The left side of the y-axis shows the average microtremor frequency of each student. In addition Figure 2 shows two correlating curves obtained during the research: the direct relationship between exam marks and students ID numbers (dark columns); the direct relationship between average microtremor frequency and ID number questions (dark broken line). The charts shown in Figure 2 lead to the following conclusions: a correlation between the level of emotional stress during the examination and the marks scored in the actual examination was determined for students who took part in the experiment; considerably higher stress levels (microtremor frequencies) were recorded for students who scored less than 6 points in the examination.

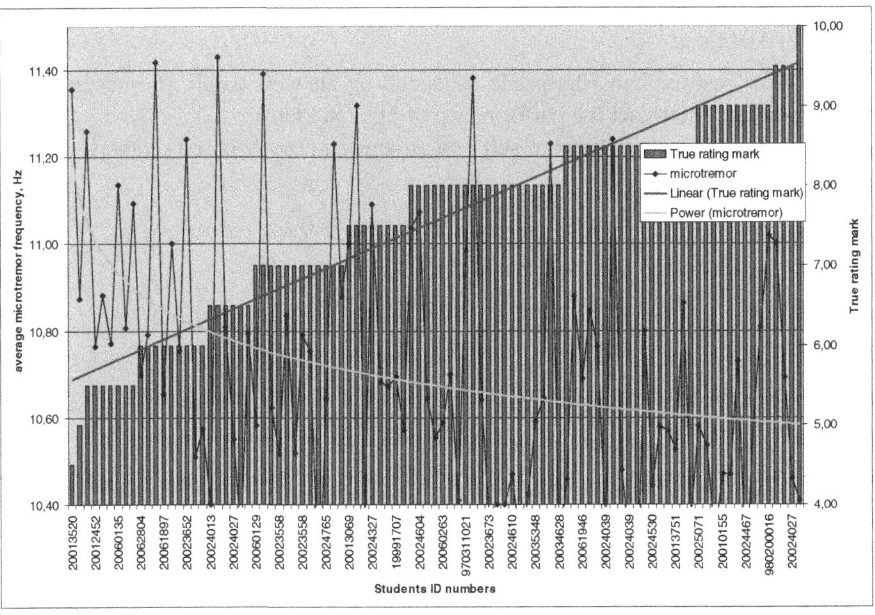

Fig. 2. Correlation between emotional stress and examination mark in the e-test: x-axis – student Ids; y-axis – examination marks of students on a ten-point scale (on the right) and the average microtremor frequency (Hz) in a student's voice (on the left)

Real e-Test

The *Real e-test Marking Model* gives several options as answers from which students, taking an actual test, are to select the best answers to multiple-choice questions. An incorrect answer gives 0 points and a correct answer gives 1 point; intermediate answers are scored between 0 and 1. A question's difficulty is determined taking into account the results of previous tests taken by other students.

The *Regression-correlation Trends Determining (RCT) Model* was developed by the authors as an aid to collect, tabulate and plot numerical information, to analyse and interpret numerical data and to make informed decisions. The RCT model makes it possible to discover interesting patterns that show how students using the system behave and to store student interactions and feedbacks in the BISPA system; thereby past memory experiences are retained, and new ways of teaching may be derived. The RCT model provides information on testing processes in matrixes and in graphical form, including information about correct/incorrect answers, time spent on each question, the number of times a student has changed an answer to a test question, and the like. Complex parameters are also available, besides the correctness of an answer they also assess the time required to answer the question along with hesitations to chose an option. The acquired knowledge is likely to get a different score once an answer has been evaluated using a complex parameter. The RCT model, which is based on statistical information from the Self-assessment Integrated Grading Model, Self-assessment Integrated Grading Adjustment Model and the real e-test, shows various regression-correlation dependencies between different parameters and data. Some of the dependencies are presented and explained in the Case Study.

Such integrated information on the testing process, along with the Question Database, makes it possible to create tests in a nonrandom manner, to customise them by choosing the number of questions, their difficulty and the proportion of questions from different subjects for each student individually. Such customisation is the function of the *Examination Alternatives Developing Model*. All test results are stored in the Examination Results Database. Students may use the statistics available in the RCT model to see the difficulty of a question and the average grade of the entire group; students then see their performance in relation to their group before and after their studies. Once the data about the difficulty of questions has been stored, it is possible to give easier questions first, gradually moving on to more difficult ones. The subjects covered by questions may also be selected similarly, moving from an easier subject to a more difficult one, later repeating the most difficult ones. A base of questions, therefore, makes it possible to customise test questions for each student, rather than select them in a random manner; customised tests will differ by the number of questions, their difficulty and the proportion of questions from different modules. Furthermore, easier questions may be given at the beginning, and then the test would proceed on to the more difficult questions. Similarly, the subjects taught can be selected starting with easier subjects and moving on to more complex ones; the subjects that have not yet been mastered can be repeated.

The *Recommendations Providing (RP) Model* collects historical information about a student's responses, provides feedback that helps to determine the strengths and weaknesses of the student's knowledge and then provides various recommendations for further education. The RP model explains why one or another answer is incorrect and offers certain additional literature and multimedia to clarify the questions answered incorrectly. Likewise, the RP model—since its basis is integrated information on the testing process—can show areas of improvement to the instructor in charge of the module. If, for example, more than 200 students spend, on average, more than 25% of their time answering the test questions in the *Real Estate Market Analysis* section (as compared to the rest of the modules), and their scores for this exam section are more than 2 points below the module's average, the RP model recommends the instructor to add more material and more explicit explanations for the more difficult areas, and so on.

The system has a student-friendly *interface* to facilitate the use of the teaching services. It includes a guide on the system's basic functions, a comprehensive window-based menu and other relevant information.

3 Case Study

The research involved a group of some 200 volunteer students (at Vilnius Gediminas Technical University, Lithuania) who took a real computer-based examination and had to select (tick) the correct answer on a computer display; they were also asked to say the selected answer aloud. During the exam, students had to mark and say the correct answers aloud to 20 questions within 10 minutes.

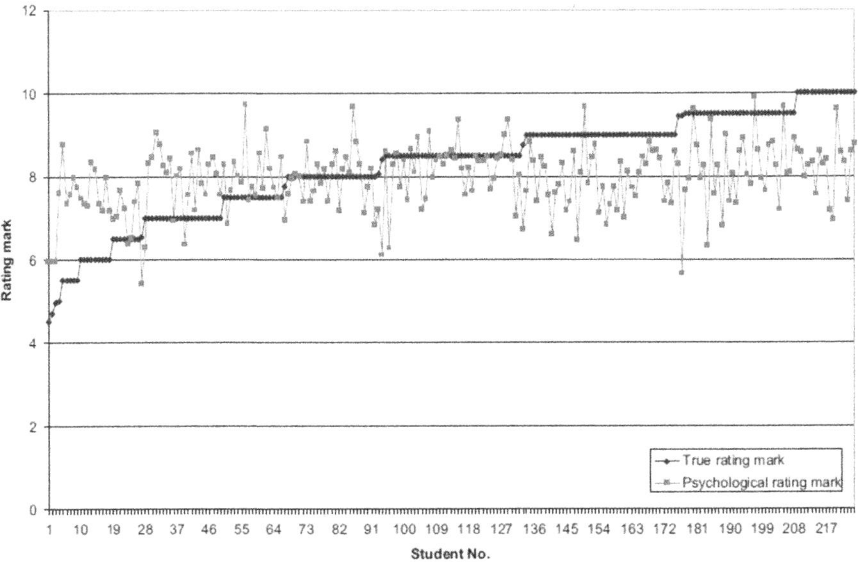

Fig. 3. Comparison of student grades after the e-psychological test done before the e-test, and the grades given after the e-test. Legend: y-axis shows student grades on a ten-point scale; x-axis shows student IDs; "grades after the e-psychological test" shows student grades given by the BISPA system after the e-psychological test done before the e-test ; "grades of the actual test" shows actual student grades given by the Intelligent Testing System after the e-test; "linear (grades of the psychological test)" shows the regression-correlation linear trend that describes student grades given by the BISPA system after the e-psychological test done before the e-test ; "linear (grades of the actual test)" shows the regression-correlation linear trend that which describes the actual student grades given after the actual e-test

Now, with the help of the BISPA system, students' knowledge can be assessed automatically, rather than by an exam, on the basis of student psychological tests, accumulated historical voice stress data, determined regression equation and a special algorithm. The BISPA system makes automatic assessment of a student's knowledge before an exam, taking into account the student's verbal/oral answers and questionnaire answers. The grade determined by the psychological test is also verified using the student's answers in the online self-assessment questionnaire. For example, when a teacher/lecturer gives a student questions such as "How many days did you study?" or "Was this course interesting to learn?" or "Was the material provided by the instructor understandable and clear?" before an exam, the BISPA system can give a precise grade assessing the student. Figure 3 compares the grades given to students after the e-psychological test done before the exam and their actual exam grades (using the BISPA system). The regression-correlation curves in Figure 3 show the interrelation between the grades given after the e-psychological test and the grades given after the actual e-examination. The marks based on their self-evaluation is statistically significant (with a significance of α=10%).

The testing shows that students with higher exam grades had poorer psychological test results than those whose grades were lower. It leads to a conclusion that the students with better actual grades assess their knowledge as average on the psychological test. Whereas the students with lower actual grades often overestimate their abilities. This is quite clear from Figure 3. Such dependencies were also noticed by Sung *et al.* [4], Papinczak *et al.* [5], etc. Sung *et al.* [4] believe that low- and high-achieving students tend respectively to over- and underestimate the quality of their own work in self-assessments. Papinczak *et al.* [5] believes that students tend to overestimate their competence, especially lower-performing students, while young or highly capable students are more likely to understate their work.

4 Discussion and Conclusions

In the future, the authors of this paper intend to integrate the BISPA system with Student Progress Assessment with the Help of an Intelligent Pupil Analysis System [6], [7], the Web-based Biometric Computer Mouse Advisory System [8], Recommended Biometric Stress Management System [9] and mobile technologies, which will permit a more detailed analysis of the knowledge attained by a student. Currently the BISPA system is under adaptation to mobile phones HTC Titan, Samsung i19100 Galaxy S2 and iPhone 4S 16GB.

Such integration is not a very new area. For example, Wang [10] believes that other, affect-related signals, such as skin conductance, blood volume pulse, eye blink rates, pupil size and such, could also be incorporated into eye-tracking systems. In more naturally occurring settings, pupillary responses could be compared with other subjective or psychophysical measures, such as skin conductance, heart rate variability, blinks, subjective ratings and the like. By combining the results from various measures, researchers are more likely to identify the unique cause that could explain all of them [10].

Other researchers in such an area of integration worked in the following manners. For example, Kahneman *et al.* [11] analyzed pupillary, heart rate and skin resistance changes during a mental task. Kahneman and Peavler [12] tested papillary responses together with skin conductance and heart rate. Lin *et al.* [13] and Lin *et al.* [14] compared pupillary responses and heart rate variability (as well as subjective techniques) to assess the user cost of playing a computer game.

Furthermore inaudible voice frequencies shall be analyzed in our research. It has been proven that different individuals have different top limits of an audible boundary. Such limits are frequently close to 15 or 16 kHz in young adults and below 13 kHz in the elderly, whereas some individuals can perceive air vibrations of 20 kHz as sound. Thus the studies by these authors will attempt to analyze a cut-off frequency of a fairly high 26 kHz to fully exclude contamination by audible sound components. Furthermore students will also be asked to answer questions during an oral examination with their eyes closed for an analysis of inaudible frequencies to eliminate any effects of visual input. These results would shed a great deal of light on the viability of these physiological channels and their applications to future technologies.

References

1. Uraikul, V., Chan, C.W.: Paitoon Tontiwachwuthikul. Artificial intelligence for monitoring and supervisory control of process systems. Engineering Applications of Artificial Intelligence 20(2), 115–131 (2007)
2. Blanco-Fernández, Y., López-Nores, M., Pazos-Arias, J.J., García-Duque, J.: An improvement for semantics-based recommender systems grounded on attaching temporal information to ontologies and user profiles. Engineering Applications of Artificial Intelligence 24(8), 1385–1397 (2011)
3. Barros, H., Silva, A., Costa, E., Bittencourt, I.I., Holanda, O., Sales, L.: Steps, techniques, and technologies for the development of intelligent applications based on Semantic Web Services: A case study in e-learning systems. Engineering Applications of Artificial Intelligence 24(8), 1355–1367 (2011)
4. Sung, Y., Chang, K., Chang, T., Yu, W.: How many heads are better than one? The reliability and validity of teenagers' self- and peer assessments. Journal of Adolescence 33(1), 135–145 (2009)
5. Papinczak, T., Young, L., Groves, M., Haynes, M.: An analysis of peer, self, and tutor assessment in problem-based learning tutorials. Medical Teacher 29, 122–132 (2007)
6. Kaklauskas, A., Zavadskas, E.K., Babenskas, E., Seniut, M., Vlasenko, A., Plakys, V.: Intelligent Library and Tutoring System for Brita in the PuBs Project. In: Luo, Y. (ed.) CDVE 2007. LNCS, vol. 4674, pp. 157–166. Springer, Heidelberg (2007)
7. Zavadskas, E.K., Kaklauskas, A.: Development and integration of intelligent, voice stress analysis and IRIS recognition technologies in construction. In: 24th International Symposium on Automation & Robotics in Construction (ISARC 2007), Kochi, Kerala, India, September 19-21, pp. 467–472 (2007)
8. Kaklauskas, A., Zavadskas, E.K., Seniut, M., Dzemyda, G., Stankevic, V., Simkevičius, C., Stankevic, T., Paliskiene, R., Matuliauskaite, A., Kildiene, S., Bartkiene, L., Ivanikovas, S., Gribniak, V.: Web-based Biometric Computer Mouse Advisory System to Analyze a User's Emotions and Work Productivity. Engineering Applications of Artificial Intelligence 24(6), 928–945 (2011)
9. Kaklauskas, A., Zavadskas, E.K., Pruskus, V., Vlasenko, A., Bartkiene, L., Paliskiene, R., Zemeckyte, L., Gerstein, V., Dzemyda, G., Tamulevicius, G.: Recommended Biometric Stress Management System. Expert Systems with Applications 38, 14011–14025 (2011)
10. Wang, J.T.: Pupil Dilation and Eye-Tracking. In: Schulte-Mecklenbeck, M., Kuhberger, A., Ranyard, R. (eds.) A Handbook of Process Tracing Methods for Decision Research: A Critical Review and User's Guide. Psychology Press (2010)

11. Kahneman, D., Tursky, B., Shapiro, D., Crider, A.: Pupillary, heart rate, and skin resistance changes during a mental task. Journal of Experimental Psychology 79, 164–167 (1969)
12. Kahneman, D., Peavler, W.S.: Incentive effects and pupillary changes in association learning. Journal of Experimental Psychology 79, 312–318 (1969)
13. Lin, T., Imamiya, A., Hu, W., Omata, M.: Combined User Physical, Physiological and Subjective Measures for Assessing User Cost. In: Stephanidis, C., Pieper, M. (eds.) ERCIM Ws UI4ALL 2006. LNCS, vol. 4397, pp. 304–316. Springer, Heidelberg (2007)
14. Lin, T., Imamiya, A., Mao, X.: Using multiple data sources to get closer insights into user cost and task performance. Interacting with Computers 20(3), 364–374 (2008)

Scaffolding Opportunity in Problem Solving – The Perspective of Weak-Tie

Min-Huei Lin[1] and Ching-Fan Chen[2]

[1] Aletheia University, Tamsui Dist . 251 New Taipei City, Taiwan, R.O.C.
Ph.D. Student, Graduate Institute of Information and Computer Education, NTNU
au4052@au.edu.tw
[2] Tamkang University, Tamsui Dist .
251 New Taipei City, Taiwan, R.O.C.
cfchen@mail.tku.edu.tw

Abstract. The importance of students' problem solving ability has been stressed and the goal of teaching database analysis, design and applications is that students can transfer what they learned in class to solve ill-structured problem in real life. Therefore teachers need to provide students a meaningful learning context for solving problems of the authentic context. But novices are deficient in problem solving owing to lack of domain-specific prior knowledge and metacognitive knowledge, so teachers have to perceive the students' difficulties during problem solving from the point of view of 5W1H learning issues and provide dynamic scaffolding within students' ZPDs. The aim of this study is to propose the double-spiral model to monitor and analyze students' problem solving processes based on IDEAL by HCCS, and teachers play the role of weak-tie to link clusters of students through identifying the weak-tie emerged from students' data. Based on the selected weak-tie, teachers can design scaffolding strategies within students ZPDs to support their problem solving.

Keywords: scaffolding, zone of proximal development, problem solving, weak-tie.

1 Introduction

The goal of teaching database analysis and design along with database applications is to prepare students to gather information and to solve problems. Database applications are served as tools to help solving problems and making sound decision. According to Voss and Post (1988), learners represent problems by examining the concepts and relations within a problem, isolating the major factors and the accompanying constraints, and recognizing divergent perspectives on the problem. However, in real life, problems are typically complicated and ill-structured. Yet many prior researches have pointed to students' deficiencies in problem solving, for instance, a failure to apply knowledge from one context to another, especially when solving ill-structured problems. Students' difficulties in problem solving have been attributed to both limited domain and metacognitive knowledge [5]. Only the scaffolding built in the ZPD is effective, and the ZPD and scaffoldings are of dynamic properties.

N.T. Nguyen et al. (Eds.): *Adv. Methods for Comput. Collective Intelligence*, SCI 457, pp. 71–81.
DOI: 10.1007/978-3-642-34300-1_7

This study employed information technology, which is named as Human-Centered Computing System that is based on experts' recognition and concept of textual data, to collect the textual data related to the phase of identifying problem in the ill-structured problem-solving, and to use the experts' implicit knowledge (value-driven) to activate information technology so as to integrate data (data-driven) with a view to showing the ZPDs created by the problem solvers. And In this study, the researcher investigates the problem solving learning of database modeling in department of MIS in Aletheia University from the learning issues of 5W1H, that is 'Why, What, When, Where, Whom, How'. Teachers play the role of weak-tie recognizer and identify the emerged weak-tie to design scaffolding strategies to support students performing ill-structured problem solving in database analysis, design and applications.

2 Literature Review

2.1 Problem-Solving and Learning

Problem solving refers to student's efforts to achieve a goal for which they do not have an automatic solution, and becomes more important when teachers move away from lockstep, highly regimented instruction. Problems have an initial state - the problem solver's current state or level of knowledge, and a goal - what the problem solver is attempting to attain. Bransford and Stein (1984) formulated a way to solve problems known as IDEAL, in which the steps of problem solving included identify the problem (I: write problem in solver's words), define and represent the problem (D: list all plans considered), explore possible strategies (E: which strategy is used and why), act on the strategies (A: show all work to arrive at solution), and look back and evaluate the effects of solver's activities (L: does the solution make sense, is it reasonable?) [1]. The links between learning and problem solving suggest that students can learn heuristics and strategies and become better problem solvers, and it is best to integrate problem solving with academic content. According to a contemporary information processing view, problem solving involves the acquisition, retention, and use of production systems. These productions are forms of procedural knowledge that include declarative knowledge and the conditions under which these forms are applicable [9].

Generally, problems can be categorized as being either well-structured or ill-structured. Using a database to solve ill-structured problems adds another dimension to problem solving, since the problem lacks a clear problem statement and clear solutions. To solve an ill-structured problem, students need to represent the problem, determine what information to retrieve or update from the database, select appropriate procedures to operate the information, perform the procedures, and decide if the results generated by the database are accurate [2]. Yet many prior researches have pointed to students' deficiencies in problem solving, for instance, a failure to apply knowledge from one context to another, especially when solving ill-structured problems. Students' difficulties in problem solving have been attributed to both limited domain and metacognitive knowledge [5].

This study aims to investigate the conceptual and operational framework for scaffolding the ill-structured problem solving process in database design and applications. The IDEAL is employed to guide the problem solving process, and the information technology is used to collect and analyze the students' mental representations in each phase of problem solving. Externalized supports during problem solving may be accomplished by recognizing the weak-tie among the cognition of teachers and students.

2.2 ZPD and Scaffolding

The concept of the zone of proximal development (ZPD) has been defined as "the distance between the actual developmental level as determined by independent problem solving and the level of potential development as determined through problem solving under teachers guidance or in collaboration with more capable peers" [9], as shown in figure 1. Scaffolding instruction theory is originated from the constructivism, it emphasized that one's knowledge is constructed by individual, and the one's constructive process is through the interaction with others in social systems. The meaning of scaffolding extends to instruction, it stands that adults or teachers provide temporary scaffolding or support form to assist learners to develop their learning ability. When learners have developed advanced capabilities, the scaffolding will be withdrawn [8].

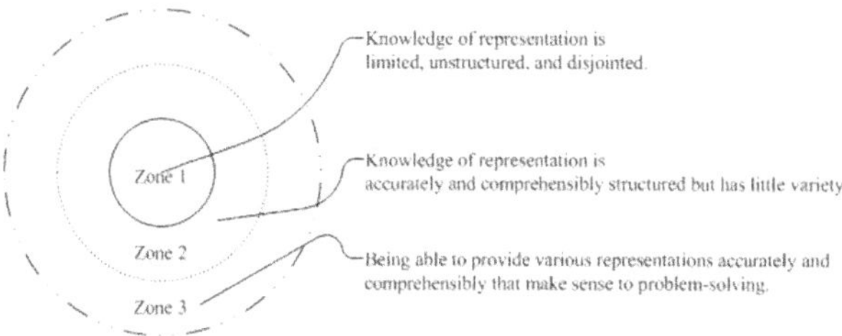

Fig. 1. Zone of proximal development

During the processes of problem-solving based on the IDEAL, the key point of implementing scaffolding instruction is whenever the teachers or capable peers provide assistance they can really discover the zone of proximal development of learners, they just can provide appropriate assistance at the right moment. Therefore this study tries to make use of the information technology to gather and analyze the data about students performing problem solving to design and implement dynamic scaffoldings within the proper ZPD.

3 Methodology

To investigate the ZPDs of students and implement the effective scaffolding during the phases of problem solving, this study adopts the four-step decision making process to solve problems, in which are intelligence, design, choice, and review activities, proposed by Simon (1955). Among these four steps, the intelligence activity which searches the environment for conditions calling for decision is the most essential one. If the experts can observe the environment and identify the problem clearly, it would be easier to get good result [10]. Therefore, this study employs the Human-centered Computing System in the intelligence activity to recognize the problems of students' problem-solving. Subsequently, teachers identify the target clusters and the weak-tie emerged among them, and they design the alternatives according to the weak-tie among the ZPDs of students (design activity). Lastly, an appropriate one will be selected after analyzing the different clusters (choice activity) to assist students to engage in ill-structured problem solving. Such a research model named double-spiral model, it means that there are a series of interactions between human and the computing system, is proposed as follows (figure 2).

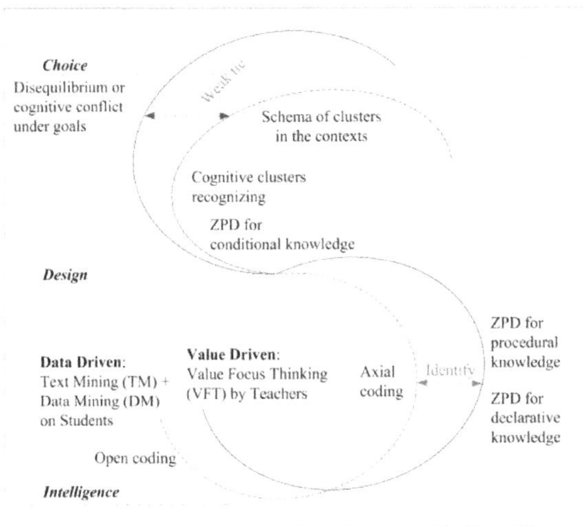

Fig. 2. Double-spiral research model

The solid "S" is a value-driven process, it means the instruction designed and conducted by a teacher, namely the trajectory of a teacher's instruction about the tacit knowledge and value of the teacher in the domain-specific ill-structured problem solving. The dotted "S" is a data-driven process, in that the actually development levels of students are analyzed by data mining and text mining the self-reporting and self-monitoring data, that is to say, it records the reactions and progresses of individuals or clusters of students' problem solving under the instructional guidance. The red section of the dotted "S" represents the actual developmental level of students that they really practice, internalize, and perform.

3.1 Intelligence for Design and Choice

In this study, the problem representations of all students are rated according to some aspects about the data flow and control flow of information system in advance. These ratings include the completeness and depth of declarative knowledge and procedural knowledge that map students into four quadrants (clusters) as shown in figure 3. The ratings for the declarative knowledge are divided into two categories: fundamental and elaborative cognition (factual knowledge and conceptual knowledge), the ones for procedural knowledge include the interactions among users, human-computer interfaces, and database.

After that, the problem representation of every cluster of students in an individual quadrant is analyzed through the intelligence activity – Human-Centered Computing System [4].

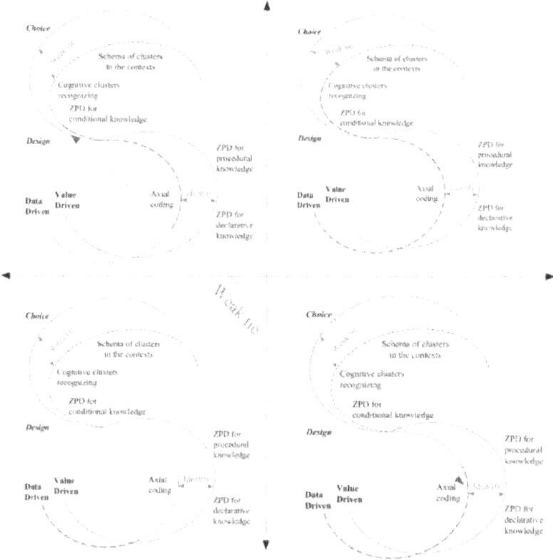

Fig. 3. Double-spiral analyses for four quadrants

3.2 Human-Centered Computing System for Recognizing the ZPDs

Gick (1986) states that representation in problem solving among experts in a field is schema-driven, the thinkers will attempt to identify critical information from memory, drawing on extremely large amounts of information. When constructing relevant problem representations, knowledgeable subjects will retrieve existing knowledge, rather than integrating complex information into working memory that is externally presented. In novices, an incomplete or biased knowledge base and lack of metacognitive knowledge will hinder their ability to problem solve and indirectly create extraneous load on the working memory [2].

The Human-centered computing system has been constructed by employing grounded theory and text mining techniques, it can be used to assist teachers to collect and analyze the self- reporting and monitoring problem representing articles of students to obtain the current problem-solving schema that they have built in their long-term memory. When their ZPDs have been visualized and clustered, teachers may have chance to find the weak-tie between the teacher and clusters of students, or the weak-tie between clusters of students, and then design the effective scaffolding within the ZPD of the specific cluster. The detailed process is listed below.

***Step 1*: Data preprocess**
1-1H [1]) Teachers define the key words and concept terms according to content knowledge and cognitive activities.
1-1C [2]) Teachers specify a time period and course unit, and then retrieve the required students' learning data from the database of the learning management system.
1-2H) Teachers recognize the students learning data according to their domain knowledge, and then eliminate meaningless words, tag words, and append the concept labels with these words.

***Step 2*: Terms co-occurrence analysis (open coding)**
The system processes the open-coding analysis, which contributes to the visualization by using data association analysis. The visual image helps the teacher to focus on the extractable various knowledge structures which are obtained from the data analysis.
2-1C) The association value of two terms can be computed as formula 1. It is based on their co-occurrence in the same sentence.

$$assoc(W_i, W_j) = \sum_{s \in D} \min\left(|W_i|_s, |W_j|_s\right) \quad (1)$$

where W_i and W_j are the ith word and the jth word; s denotes a sentence and is a set of words; D is a set of sentences and includes all the students learning data. $|W_i|_s$ and $|W_j|_s$ denote the frequency of words W_i and W_j occurred in the sentence s.

2-2C) The result of co-occurrence analysis will be visualized to co-occurrence association graph.
2-1H) The co-occurrence association graph can help teachers to recognize the concepts and categories inside it, and stimulate teachers preliminary understand the association of students' cognitive clusters appeared from learning data.

***Step 3*: To develop various cognitive clusters (Axial coding)**
Based on the characteristics of "knowledge", "importance" and "urgency", teachers extract the data to create ZPDs of four quadrants from learning data. The process is illustrated as follows.
3-1H) Teachers need to decide what categories of knowledge ($w_{knowledge}$) are.
3-1C) Process the categories of knowledge from the first to the last one.
3-2C) Categories of knowledge ($w_{knowledge}$) are used to sift out the meaningful sentences. This variable is used to confirm the research topics and to remove irrelevant sentences with a view to narrowing down the data range and to sift out valid sentences.
3-3C) To integrate all the valid sentences related to knowledge terms, and create integrated association diagrams, then go to 3-1C.

***Step 4*: To decide the weak-tie based on value focus thinking**
4-1H) Based on their domain knowledge, teachers confirm the key factors and identify weak-tie to design scaffolding strategies.

[1] 'H' means that actions in such a step are performed by human.
[2] 'C' means that the actions in such a step are performed by computer.

4 Case Study

In this study, students are asked to solve the ill-structured problem in the context of designing an on-line signing up system for school's athletic game. Students start the task from the problem representation. After the students represent their mental representations with the textual format, the human-centered computing system is used to analyze these data subsequently.

4.1 The Context for the Ill-Structured Problem Solving

The present study integrates the ideal of constructivism and cognition load to design the procedure of ill-structured problem solving under the process of IDEAL [6][2][7]. The first step in problem solving is to form a mental representation of the problem, and then the solver performs operations to reduce the discrepancy between the beginning and goal states. If problem solvers incorrectly represent the problem by not considering all aspects or by add too many constraints, it is unlikely to identify a correct solution path. Therefore, problem-solving learning programs typically devote a lot of time to the representation phase. And the present study focuses on the problem representation.

Table 1. Instruction design for ill-structured problem solving in the database design and applications based on IDEAL

Steps	Contents
Identify	Students write the problem representations in their own words, define the mini world of the information system, and clarify the entities, attributes, relationships, and functional procedures for the context of the problem.
Define	Students list all plans they consider. Serving the flowcharts of the information system and the database diagram as the tools for students to elaborate their thoughts and enhance their thinking transparency.
Explore	Students propose which strategy is used and explain why is used. Students relate the flowchart with database diagram to identify the interaction process between control flow and data flow.
Act	Students show all work to arrive at solution. In this phase, students design the SQL statements to operate data and generate some solutions for the problem.
Look back and evaluate	Students make sure that does the solution make sense, is it reasonable? In this phase, students evaluate the outcomes according to their expected goal.

4.2 Data Resource

The experimental data was collected from seventy-two students, who have experiences in learning programming, concepts of information management, management mathematic, in the course of database management and application in March, 2012. They have built the basis for programming languages, algorithms, flowchart and entity-relationship model and diagrams.

4.3 Human-Centered Computing Phase: Extract the ZPDs of Clusters in Various Quadrants

Ratings. According to the rubrics of the rating system, the problem representations of all students are rated, and all students are dispatched to one of the four quadrants (figure 4). The first quadrant means that students are of complete procedural and declarative knowledge, students in the second quadrant are of complete procedural knowledge but incomplete declarative knowledge, students in the fourth quadrant are of complete declarative knowledge but incomplete procedural knowledge, and both declarative knowledge and procedural knowledge of students in the third quadrant are incomplete. As shown in figure 4, that most of the students fell into the third quadrant reminded the teacher of the much more supports these students need. The few students are in the first quadrant and their mental models are most close to the teacher's. However the mental models of students in the second and fourth quadrants are close to the ones which students in the first quadrant have. So it is necessary to investigate the problem representations of students in each quadrant through HCCS to visualize the mental representations in each quadrant. Through these association graphs, the weak-tie between various quadrants may be emerged, and the opportunity to implement dynamic scaffolding may be accomplished.

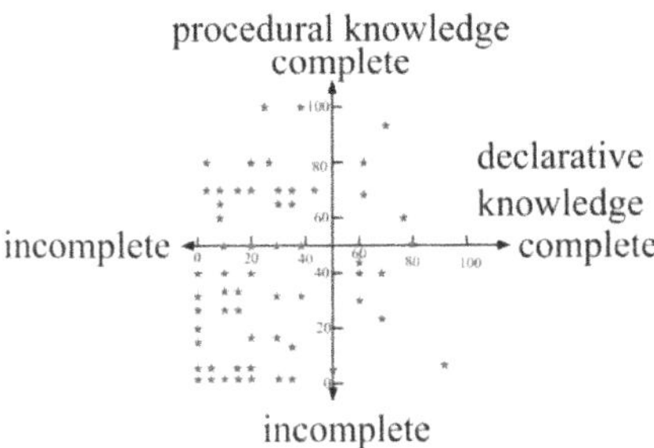

Fig. 4. Distribution of students in four quadrants

Data analysis within-quadrant and between-quadrants. The ZPD of every quadrant is visualized by the HCCS and teachers may detect and recognize the weak-tie between various quadrants to design the scaffolding strategy. According to the step 1 and 2 described in section 3.2, after the data preprocess and open coding are completed, some key aspects of problem solving are recognized and used for the axial coding, that they are declarative and procedural knowledge about single contests, team contests, and sign up for participation (step 3). For example, the declarative knowledge of single contests in the four quadrants is shown in figure 5, and the declarative knowledge of team contests in the four quadrants is shown in figure 6.

As shown in figure 5, the mental model of the fourth quadrant is similar to the second quadrants but is more incomplete and disconnected than the ones in the second quadrant. Therefore when entering the step 4 in section 3.2, the weak-tie used to link the second and fourth quadrants may be built. The peer scaffolding may be established by recommending the students in the second quadrant to the ones in the fourth quadrants. Through the observation and communication of similar models, students may have higher self-efficacy and perform problem representation better.

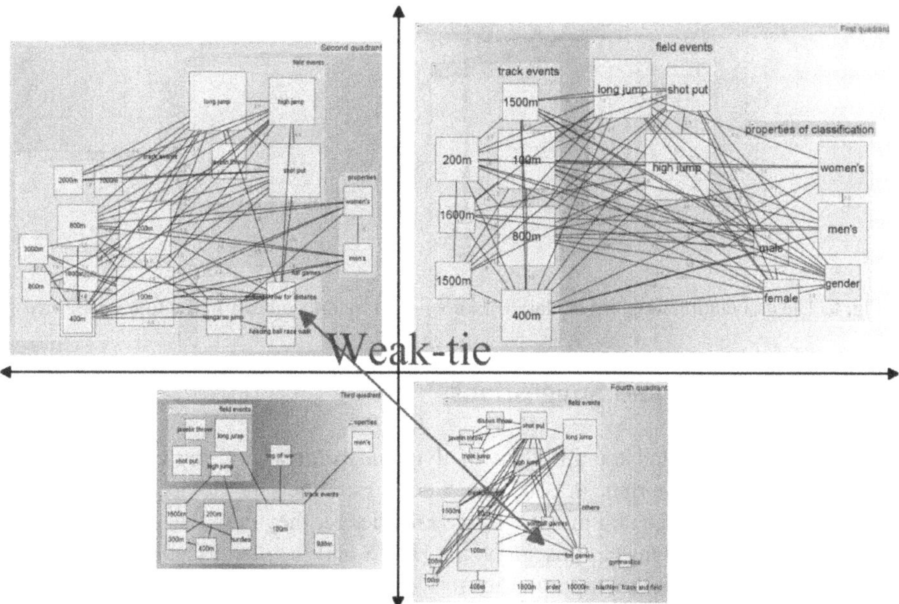

Fig. 5. The association graph of declarative knowledge - personal contests in four quadrants

In figure 6, the links among concepts are loosely in the second quadrant, but the mental model of the second quadrant is similar to the one in the first quadrant. So the weak-tie across the second and first quadrants can be built, then the peer scaffolding can be designed based on the weak-tie. The breadth and depth of knowledge in the fourth quadrant are more insufficient than the ones in the second quadrant, but there exist some concepts that may be seen as the weak-tie between the second and fourth quadrants. Therefore the peer scaffolding may be established by recommending the students in the second quadrant to the ones in the fourth quadrants, while the students in the two quadrants interact with each other and tutor peers in their misconceptions.

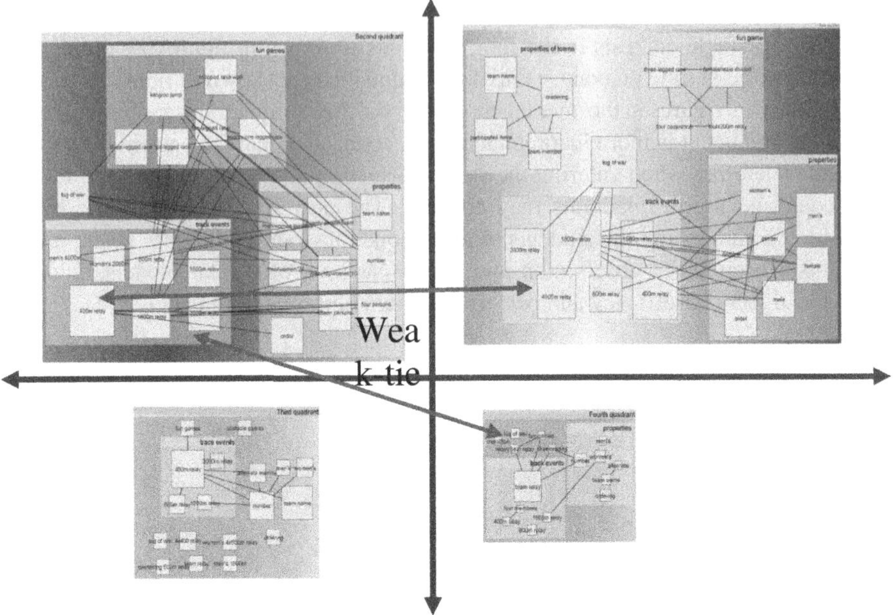

Fig. 6. The association graph of declarative knowledge - team contests in the four quadrants

5 Conclusions and Suggestions

Both the scaffolding and the ZPD are of dynamic natures, and the competence of problem solving is important in educational settings. When the novices lack the domain-specific knowledge and metacognitive knowledge, they may feel difficult in ill-structured problem solving. In this study, the HCCS and the identification of weak-tie indeed provide teachers the opportunities to identify the ZPDs of students and implement effective scaffoldings in the ZPD. After students develop more capabilities, the scaffoldings will be withdrawn. Teachers also play the role of weak-tie to link the two clusters of students, and use the emerged weak-tie from students' data to design the scaffolding strategies that support students performing problem representations well. The effectiveness of scaffolding strategies will be evaluated and the transfer of students' knowledge to authentic problem solving also be observed in the future study.

References

1. Bransford, J., Stein, B.: The IDEAL problem solver. W. H. Freeman, New York (1984)
2. Chen, C.: Teaching problem solving and database skills that transfer. Journal of Business Research 63(2), 175–181 (2010)
3. Chen, C.H.: Promoting college students' knowledge acquisition and ill-structured problem solving: Web-based integration and procedure prompts. Computers and Education 55(1), 292–303 (2010)

4. Hong, C.F.: Qualitative Chance Discovery – Extracting competitive advantages. Information Sciences 179, 1570–1583 (2009)
5. Ge, X., Land, S.M.: Scaffolding students' problem-solving processes in an ill-structured task using question prompts and peer interactions. Educational Technology Research and Development 51(1), 21–38 (2003)
6. Jonassen, D.H.: Learning to solve problems: an instructional design guide. Pfeiffer, San Francisco (2004)
7. Kalyuga, S., Chandler, P., Tuovinen, J.E., Sweller, J.: When problem solving is superior to studying worked examples. J. Educ. Psychol. 93(3), 579–588 (2001)
8. Wood, C., Bruner, J.S., Ross, G.: The role of tutoring in problem solving. Journal of Child Psychology and Psychiatry 17, 89–100 (1976)
9. Schunk, D.H.: Learning theories: An educational perspective, 6th edn. (2011)
10. Simon, H.A.: A Behavioral Model of Rational Choice. Quarterly Journal of Economics 79, 99–188 (1955)

A Fast Algorithm for Predicting Topics of Scientific Papers Based on Co-authorship Graph Model

Nhut Truong Hoang, Phuc Do, and Hoang Nguyen Le

University of Information Technology, Ho Chi Minh,
VNU-HCM
nhut.truonghoang@niit.edu.vn,
phucdo@uit.edu.vn,
nlhoang1203@gmail.com

Abstract. This paper focuses on the problem of predicting the topic of a paper based on the co-authorship graph. Co-authorship graph is an undirected graph in which paper is represented by a node and two nodes are linked together by a link if they share at least one common author. The approach of link-based object classification (LBC) is based on the assumption that papers in the same neighbourhoods of the co-authorship graph tend to have same topic, and the predicted topic for one node in the graph depends on the topics of the another nodes that linked to it. In order to solve LBC, we have a traditional relaxation labeling to be proposed by Hoche, S., and Flach. Based on this algorithm, we propose an improvement of this algorithm. Our proposed algorithm has the processing speed faster than the traditional one. We test the performance of the proposed algorithm with the ILPnet2 database and compare the experimental result with the traditional algorithm.

Keywords: co-authorship, topic, relaxation labeling, social network.

1 Introduction

Nowadays, the number of research papers about networked data such as social network data, the hyperlinks … is increasing rapidly. A social network contains nodes (actors) and links (or ties). Since the amount of data on social network is increasing rapidly, the number of research papers about data mining on social network also increases rapidly. According to the traditional methods of machine learning and data mining, we limit the scope of research to the set of identical objects that are mutually independent. In link mining, the scope of research is the set of un-identical objects that are linked together. Link mining contains several problems such as link analysis, web mining and graph mining [4]. The purpose of LBC is to predict the final attribute of object based on its attributes, links and the attributes of another objects that are linked to it [1][4][5]. The problem of LBC may be stated as follows:

N.T. Nguyen et al. (Eds.): *Adv. Methods for Comput. Collective Intelligence*, SCI 457, pp. 83–91.
DOI: 10.1007/978-3-642-34300-1_8

• **Input:** Given a network with a set of nodes, a set of undirected edges and a subset of "unknown-attribute" nodes

• **Output:** Predict the attributes of the "unknown-attribute" nodes

We use LBCto predict topics of scientific papers in a co-authorship networks where nodes are papers with known-attribute and unknown-attribute. The LBC can be used in the following networks:

- Collaboration network where nodes are authors and links are the collaboration between authors.
- Co-authorship network where nodes are papers and links are co-authors.
- Co-citation network where links are co-citing or citing.

We can use LBC to discover the community structure, abnormal patterns and identify the roles of the objects in social network. This paper uses the collective classification algorithm based on relaxation labeling to predict topics of papers [2]. We propose an improvement of this algorithm in order to increase the speed of the traditional algorithm. We use the ILPnet2 database [3] to test the performance of our improved algorithm and compare the experimental result with the traditional algorithm. The rest of this paper is organized as follows: Section 2: Predicting topics of papers from the neighborhood; Section 3: An improved algorithm; Section 4: Experimental Result and Section 5: Conclusion and future works.

2 Predicting Topics of Papers from the Neighborhood

As mentioned above, depending on the problem we have different views of the data. Based on the bibliographic data, we can make different types of graphs. In collaboration graphs, the authors are nodes, and edges are the scientific collaboration. In co-citation graphs, the papers are nodes, and edges represent one paper cites another paper or it is cited by another one. In co-authorship graphs, nodes represent scientific papers and undirected edges between two nodes express two corresponding papers having common authors. Here, we also determine the numbers of authors who are the co-authors of two papers; this number will be the weights of the edge that links two corresponding papers. Giving a set of scientific papers in which some papers already have topics, we can predict the topics of the remaining papers based on the set of known topics and the linking structure of the co- authorship graph. An example of LBC is shown in figure 1.

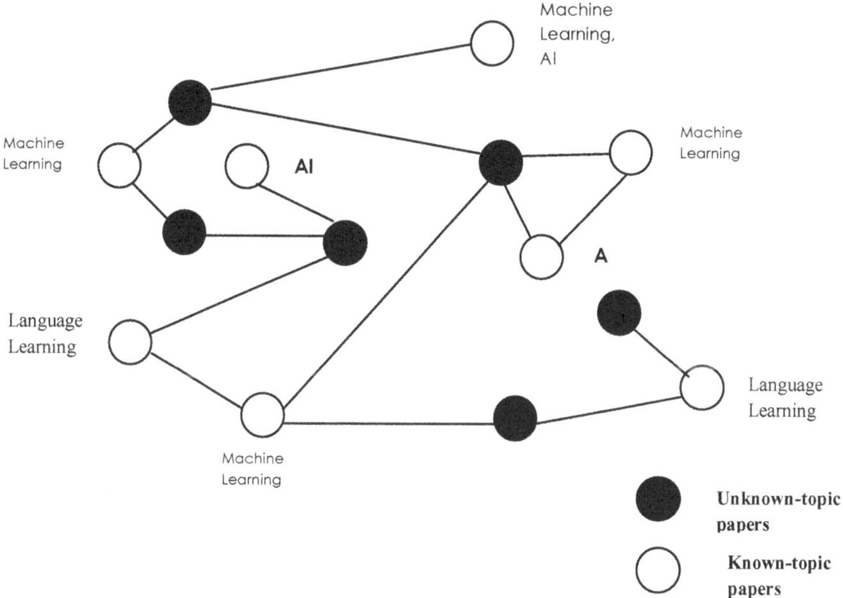

Fig. 1. Co-Author graph of scientific papers

As mentioned above, the LBC model will be used to predict the topics of the papers. LBC model not only uses the "local" attributes but also uses the "external" attributes such as attributes of linked objects. The connections between objects will influence the result of the problem. Figure 2 is an example of predicting the unknown-topic papers.

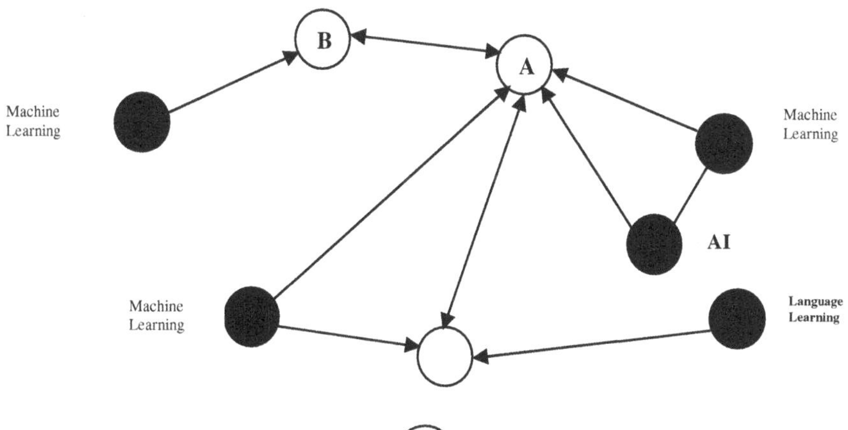

Fig. 2. Predicting the papers ◯ based on their neighborhoods

Finally, we use a relaxation labeling based on collective classification algorithm [2] to solve the problem of predicting linked objects. Given a graph G = (V,E) in which V is the set of nodes (papers) and E is the set of undirected edges. Let V = {v_1, v_2, ... v_n} and E = {(e_i,e_j)} be the undirected weighted edges between $v_i,v_j \in V$. Let V = $K \cup U$, $K \cap U = \emptyset$. K is the set of known-topic papers and U is the set of unknown-topic papers. The algorithm will find the correct topics of the unknown nodes in U from a set L of limited topics where L = {L_1, L_2, ... L_n} by using the nodes in K and the link structure of graph G. The main idea of algorithm is as follows:

Initial model: Find the initial probabilities of observing all nodes in U according to each label in L_i by the following formula:

$$P_{ik}(1) := (n_k + p_m) / |K| + m$$

where

|K| denotes the cardinality of set K
m = 1 and p = 0.5
n_k denotes the number of times that the topic l_k in K

The Collective inference model is as follows:

Let $P(t)(y_i=l_k|N_i)$ be the probability in the t-th relaxation labeling iteration of observing topic l_k for paper y_i based on y_i's neighbor papers called Ni.

After calculating the initial prediction probability $P_{ik}(1)$ for each paper $v_i \in U$ and each topic L_i without using any structural information of a given graph G, we iteratively update the prediction probability based on existing probability of the neighbor papers that are adjacent to paper v_i.

For each paper $v_i \in U$ and each topic L_i, we update prediction probability at the time t+1 by using the formula: $P_{ik}(t+1)=P(t+1)(y_i=l_k|N_i)$ and the existing probability $P_{jk}(t)=P(t)(y_j=l_k|N_j)$. We use the following equation for updating:

$$P_{ik}(t+1)=(1/\Sigma_{nj\in N_i}W_{ij})\ \Sigma_{nj\in N_i}W_{ij}P_{jk}(t) \qquad (1)$$

Where W_{ij} is the weight of the edges between two papers v_i and v_j.

In the undirected graph, equation (1) is recursive. In order to decrease the influence between linked nodes, equation (1) will be repeated 100 times. According to [2], since the relaxation labeling strategy does not necessarily guarantee convergence, Susanne Hoche and Peter Flach have adjusted the equation (1) to equation (2) as below:

$$P_{ik}(t+1)=\beta_{(t)}.\ (1/\Sigma_{nj\in N_i}W_{ij})\ \Sigma_{nj\in N_i}W_{ij}P_{jk}(t) + (1-\beta_{(t)}).\ P_{ik}(t) \qquad (2)$$

Where β determines the balance between the influence of neighbor papers N_i and v_i itself with estimates t = 1,2, ... 100. The initial value β is assigned by 1. After each iteration, β is adjusted as $\beta_{(t+1)} = 0.99\beta_{(t)}$

This equation shows that the initial influence of the prediction paper v_i on its neighbor papers is high but gradually decreases for each iteration.

3 An Improved Algorithm

According to equation (2), we have:

$$P_{ik}(t+1)=\beta_{(t)}.\ (1/\Sigma_{nj\in\ Ni}W_{ij})\ \Sigma_{nj\in\ Ni}W_{ij}P_{jk}(t)\ +(1-\ \beta_{(t)}).\ P_{ik}(t)$$

The initial value β is assigned by 1. After each iteration, β is adjusted by $\beta_{(t+1)} = 0.99\beta_{(t)}$ Where t = 1, 2, … N so that:

$$|P_{ik\,(t+1)}-\ P_{ik(t)}\,| > 0.001 \quad (3)$$

The variation of P corresponding to time T of equation (3) is shown in Figure 3. With the collective classification algorithm based on relaxation labeling (RL), we propose that after 100 times of iteration, the value of prediction probability is almost unchanged. In our improved algorithm, there is no need to repeat 100 times; we just need to do iteration until the value of prediction probability of predicted node in the previous time and in the current time have nearly the same value. This improvement gives the same predicting results as the original RL algorithm; but our proposed algorithm is executed two times faster than the original RL algorithm.

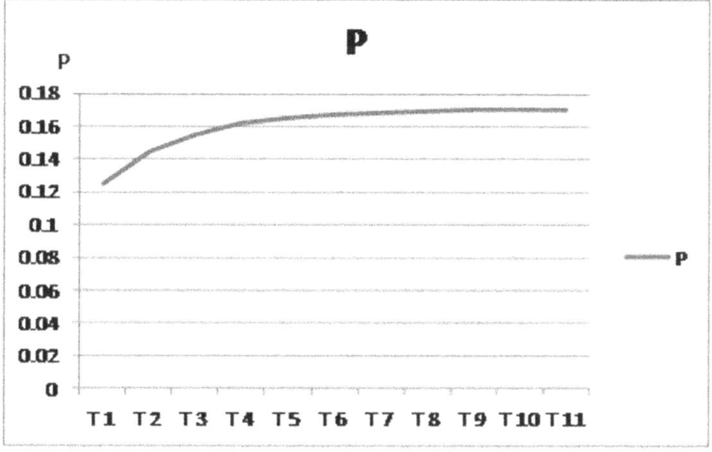

Fig. 3. Variation of P corresponding to time T

4 Experimental Results

a- Setup experimental data

ILPnet2 is a popular web database, it contains the information of papers in ILP database from year 1970 to year 2003 [3]. This database is called ILP Publications database. This database is used by Susanne Hoche and Peter Flach [2] to test the original algorithm mentioned in this paper. Therefore, this database is highly reliable.

This database contains 1,112 scientific papers to be written by 761 authors with 31 different topics. There are 822 "unknown-topic" papers and there are only 290 papers has topics belonging to 31 different scientific topics. Each author has average 1.46 papers; and each topic has average 35.9 papers. The distribution of number of authors and number of papers is shown in Table 1.

Table 1. Distribution of number of authors and number of papers

Number of Authors	Number of Papers	Percentag e %
1	363	32.6
2	423	38.0
3	203	18.3
4	84	7.6
5	22	2.0
6	15	1.3
7	1	0.1
8	1	0.1
Total	**1112**	**100**

b- Testing the improved algorithm

We use some measurements as follows:

- TP(True positive): the number of correct predicted papers in Positive sample.
- FP(False positive): the number of incorrect predicted papers in Negative sample.
- TN(True negative): the number of correct predicted papers in Negative sample.
- FN(False negative): the number of incorrect predicted papers in Positive sample.
- TPRate=TP/P where P is positive sample
- FPRate=FP/N where N is negative sample

  Accuracy=(TP+TN)/(P+N) ; Precision=TP/(TP+FP)

Table 2 and 3 are the comparison of measurements of relaxation algorithm and our improved algorithm. We use two test cases as follows:

- **Test case #1:** We select P= (1/2) x number of papers in the **"ILP_Application"** topic and N=(1/2) x numbers of papers in the topic different from **"ILP_Application"** topic.
- **Test case #2:** Select P=(1/2) x Number of papers having **Language_Learning**" topic and N=(1/2)x number of Papers in the topics different from **"Language_Learning"** topic.

Based on the testing results, we discover that the prediction results of two algorithms are the same.

Table 2. Comparison of TPRate, FPRate, Accuracy and Precision of papers in the "ILP_Application" topic with test case #1

Random Selection of papers	TPRate	FPRate	Accuracy	Precision
1	0.75	0.37	0.65	0.32
2	0.75	0.29	0.72	0.38
Average	0.75	0.33	0.69	0.35

Table 3. Comparison of FPRate, Accuracy and Precision of papers in the "Language_Learning" topic with test case #2

Random Selection of papers	TPRate	FPRate	Accuracy	Precision
1	0.75	0.19	0.80	0.6
2	1	0.19	0.86	0.67
Average	0.86	0.19	0.83	0.64

Table 4. Comparison of processing speed between two algorithms

Number of Samples	Susanne Hoche and Peter Flach (second)	Improved algorithm (second)
16	1	1
22	2	1
88	7	5
724	2969	1088

Evaluation

Analyzing the experimental result, we recognize that paper in the "Language_Learning" topic has more connections to linked papers with the same topics (Positive Degree) than the number of papers in the "ILP_Application" topic. It means that there are many papers with different topics in the neighborhood of the paper in the "ILP_Application" topic. Let graph G_T denote the set of papers in the topic T. We define Positive Degree is the average degree of papers in the same "chosen" topic, or the number of nodes in the same topic that are linked to the "chosen" node. The values of TPRate, FPRate, Accuracy and Precision of the papers in the "Language_Learning" topic will be greater than the papers in the "ILP_Application" topic or the average degree (Positive Degree) of the "Language_Learning" topic is higher than the average degree of the "ILP_Application" topic.

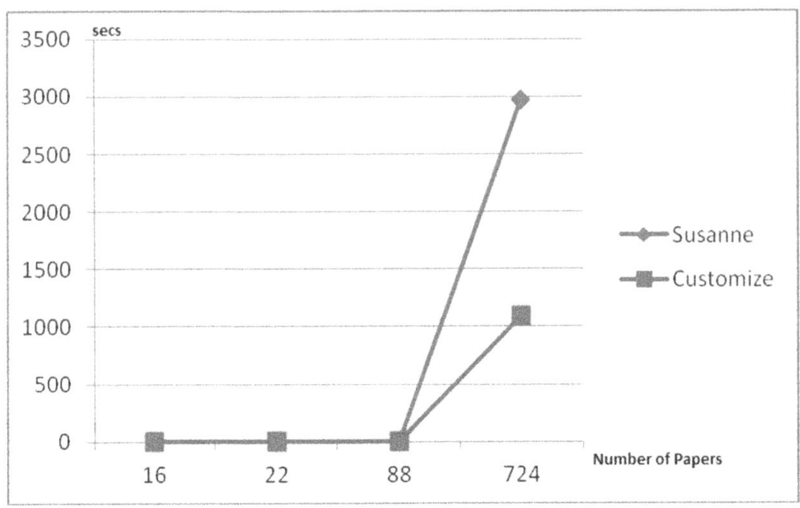

Fig. 4. Comparison of the processing speed between two algorithms

5 Conclusions and Future Works

We proposed an improving collective classification algorithm based on relaxation labeling to be used by Susanne Hoche and Peter Flach. We studied and implemented two algorithms: the collective classification algorithm based on relaxation labeling and our improved algorithm. These algorithms both have the assumption that the papers and its neighborhoods in the co-authorship graph tend to have the same topic. Therefore the task of LBC will depend on nodes that have direct links to it. The testing result on the database ILPnet2 shows that our proposed algorithm is effective when the density of the connection of the same topic nodes is high; and otherwise if the neighbor nodes have many different topics, the rate of success will very low. About the processing speed, our improved algorithm has the processing speed two times faster than the algorithm of Susanne when the number of papers reach approximately 1000 papers. We consider that our model still has some limits when the neighborhood of a chosen paper has many papers in different topics. In the future, according to the suggestion of Susanne Hoche and Peter Flach [2] about this limitation, the solution is to use the classification technique of papers or modeling the nodes' influence on the prediction for a linked node. Moreover, when considering node's influence, we have to use the positional attributes or structural attributes in the co-authorship graph; it means that we have to use a variety of node centrality measures. The model above just only uses a local attribute (AuthorID) to build the co-authorship graph; we can use more attributes to build graphs such as co-author (AuthorID) and co-topic (KeyworkID) to make more types of different graphs. Besides, we will do more experiments to update the target probabilistic function as mentioned in the formula (3) based on the characteristics of the problem instead of using an absolute value to be proposed subjectively.

References

[1] Getoor, L., Friedman, N., Koller, D., Taskar, B.: Learning probabilistic models of link structure. Journal of Machine Learning Research 3, 679–707 (2003)

[2] Hoche, S., Flach, P.: Predicting Topics of Scientific Papers from Co-Authorship Graphs: a Case Study. In: Proceedings of the 2006 UK Workshop on Computational Intelligence (UKCI 2006), pp. 215–222 (September 2006)

[3] The ILPnet2 (online retrieved), http://www.cs.bris.ac.uk/~ILPnet2/Tools/Reports/

[4] Lu, Q., Getoor, L.: Link-based classification. In: Proceedings of International Conference on Machine Learning (2003)

[5] Macskassy, S., Provost, F.: A simple relational classifier. In: Workshop on Multi-Relational Data Mining, pp. 64–77 (2003)

Part III

Agent and Multiagent Systems

Cooperating and Sharing Knowledge in an Intelligent Multi-Agent System Based on Distributed Knowledge Bases for Solving Problems Automatically

Nguyen Tran Minh Khue and Nhon Van Do

University of Information Technology, HCMC.
Km 20, Hanoi highway, Thuduc District, Hochiminh City, Vietnam
minhkhuevn@yahoo.com,
nhondv@uit.edu.vn

Abstract. In this paper, we present cooperating and sharing knowledge between agents in an Intelligent Multi-Agent System for solving problems automatically. This system is built on three distributed knowledge bases. We also illustrate developing the system based on JADE with three fields of Mathematics: plane geometry, 2D analytic geometry, and algebra. Finally, we show the effects of the Intelligent Multi-Agent System.

Keywords: Multi-Agent System (MAS), cooperating, sharing knowledge, solving problem, distributed knowledge bases.

1 Introduction

Multi-Agent Systems (MAS) can support E-Learning and high-ranking demand such as solving problems automatically. Knowledge needs to be shared in many different fields and different places to deal with most complex issues. The advantage of MAS is combining intelligence with coordination of agents in the system. So, it is useful to develop an Intelligent MAS to solve problems automatically. Nowadays, some MAS have been built on JADE, JACK, AGLETS, JAT Lite, ... such as works in [5], [6], [9-12], [15], [20]. Some MAS have been built in Java, .NET environments such as works in [4], [7], [8], [13], [14], [16-19]. They concern issues such as enhancing comprehension of learners, increasing effects of studying documents, improving the interaction between teachers and students in distance learning, building ontologies of subjects, coordinating between the model of distance learning and the model of knowledge management, ... However, they are not suitable to solve issues such as solving problems automatically in Mathematics. The main reasons are that ontologies or knowledge bases in these systems are simple. Some ontologies have forms of trees, some knowledge bases consist of simple components whereas dealing with issues about Mathematics needs more complex knowledge bases and stronger inference machineries. Besides, Maple, Matlab, Mathematica ... only support some single functions about Mathematics. In reality, there are many questions that we have to infer results from many steps. These softwares can not deal with issues relating to

N.T. Nguyen et al. (Eds.): *Adv. Methods for Comput. Collective Intelligence*, SCI 457, pp. 95–105.
DOI: 10.1007/978-3-642-34300-1_9

many fields. So, we propose here a method to deal with them. In previous paper such as in [1], we presented a MAS model to solve problems automatically in two fields: plane geometry and 3D analytic geometry. In [2], we built a model of a MAS system, architectures of agents in the system, and proposed a method to test the system. In this paper, we focus on a method of cooperating and sharing knowledge between agents in an Intelligent MAS based on JADE to solve problems automatically in three fields: algebra, plane geometry, and 2D analytic geometry. When increasing the number of fields, amount of agents will increase. So, the process for finding results will be more complex. The aim of our research is building a system in many fields. It can simulate a complex, nature, and reasonable organization of human. Developing a method of cooperating and sharing knowledge in three fields is a step to expand the system.

2 A Model of an Intelligent Multi-Agent System

Agents reside and are executed in *Places*. There are four kinds of Agents : *User Agent, Intermediary Agent, Knowledge Agent* and *Notice Agent*. Among these kinds, User Agent and Knowledge Agent are kinds of *Intelligent Agents*, other Agents are kinds of *Mobile Agents*. A *Knowledge Base* stores knowledge relating to a field. *Storage* stores facts about states of local environment in Place and facts about states of global environment in the system. In the model, we organize four Places. *Place 1* is used to interact with users. UAgent is a *User Agent*. IAAgent, IBAgent and ICAgent are *Intermediary Agents*. *Place 2* is used to deal with knowledge in the field A. IAAgent is an Intermediary Agent. AAgent is a Knowledge Agent in field A. NAgent is a Notice Agent of Place 2. Knowledge Base A stores the knowledge in field A. *Place 3* and *Place 4* are similar to Place 2, but they are for knowledge in field B or knowledge in field C [2]. Three knowledge bases of Place 2, Place 3, and Place 4 are COKB [3]. *Maple* is a Mathematics soft-ware integrated into the system. It supports Knowledge Agents for solving some simple and single functions. Here are reasons to develop the system based on JADE: JADE is an open source software, therefore we can easily deploy an Intelligent Multi-Agent System based on it. It is suitable for integrating with Maple, XML, MathML, MathJax, Java libraries and building distributed systems. Besides, JADE supports actions: creating, cloning, suspending, resuming, killing agents [1]...

3 An Activity Model for Solving Problems Automatically

Fig. 1 shows functions of four kinds of agents and the relationship between agents. Users input into the system with mathematical symbols through User Agent. This agent converts symbols into MathML (Mathematical Markup Language) [2], divides the problem into facts and objects. Then it classifies these facts and objects according to field A, field B or field C and inform to Intermediary Agent. After agents coordinate to find results, User Agent shows answers with mathematical symbols. Intermediary Agents check goals. If the set of goals is in the set of facts, Intermediary Agents inform

[1] `http://jade.tilab.com/`

[2] `http://www.w3.org/TR/MathML2/`

results to User Agent. If the set of goals is not in the set of facts, Intermediary Agents inform to Notice Agents. Notice Agents inform objects, facts to Intermediary Agents in other Places. Intermediary Agents inform objects, facts and request demand to Knowledge Agents. Knowledge Agents have two options. In the first way, they use COKB and existing objects, facts to create new objects, facts and inform them to Intermediary Agents. In the other way, they create conditional proofs to be proven by existing objects, facts. In the process to find answers, if there is not any new object or fact created, Intermediary Agents check goals and the set of the goals are not in the set of the facts; Intermediary Agents have to change the method of classification according to fields, because knowledge can be in two or three fields. After using all methods of classification according to fields, if there is not still any new object or fact created, the loop is stopped. In this case, the system can not find answers. Engineers insert, delete, update knowledge in COKB and they can list, read, search, view, select, load knowledge from COKB.

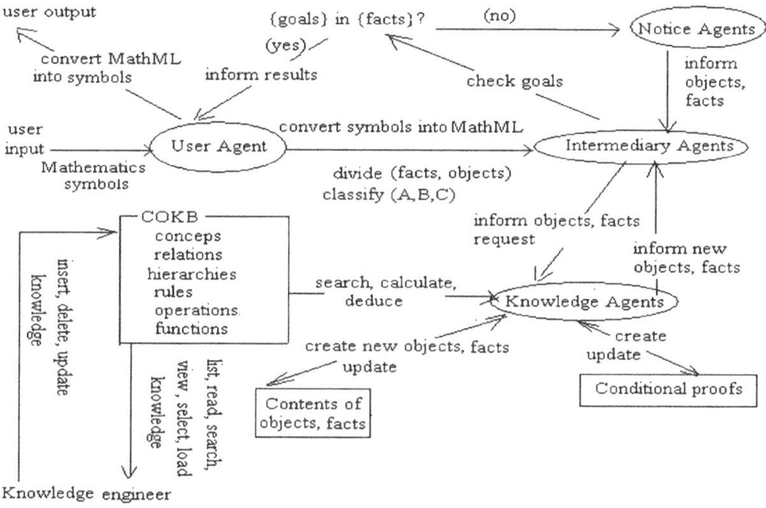

Fig. 1. An activity model for solving problems automatically

4 The Architecture of the System

Components to develop Place 1, Place 2, Place 3 and Place 4 include: COKB, Maple 13, XML, MathJax, JDK 7, Suim, MozSwing, MathML, Apache Tomcat 7. Users input questions by using a Java Server Page through a web browser (IE, Firefox, …). The web page includes Suim. Suim is a software to input mathematical symbols[3]. It can create MathML codes from symbols inputted. MozSwing is a software that supports inputting MathML codes. It is Java Swing combined with XUL (XML User

[3] http://suim.jp/

Interface Language). XUL is a part of Mozilla, it supports building applications that can run on different operating systems and environments connect with Internet or not. We can easily adjust texts, graphics or interfaces of applications using XUL[4]. After receiving MathML codes, MozSwing can show mathematical symbols created from these codes. MathJax uses CSS, JavaScript library and font Unicode for typesetting well and building interfaces of mathematical applications[5]. Web server Apache Tomcat receives information about the users and the questions, then, it stores them into files. The User Profiles Management System monitors these files and transfers MathML codes to Inference System. Maple is a software supporting calculation with numbers and symbols. It includes many packages for algebra, geometry, analytic, and arithmetic ... We can create new libraries for Maple. We combine Java with Maple by using OpenMaple[6]. Inference System uses knowledge in COKB to search, calculate, deduce knowledge and uses OpenMaple to calculate functions. Then new objects and facts are created. We use the method of forward chaining combine with backward chaining. Weakness of forward chaining is causing many new objects and facts that are redundant. Weakness of backward chaining is causing many conditional proofs and some of them have never been proven by existing facts. Therefore, we combine two methods to reduce amount of steps in finding answers. Engineers input/output knowledge in COKB through Knowledge Update System.

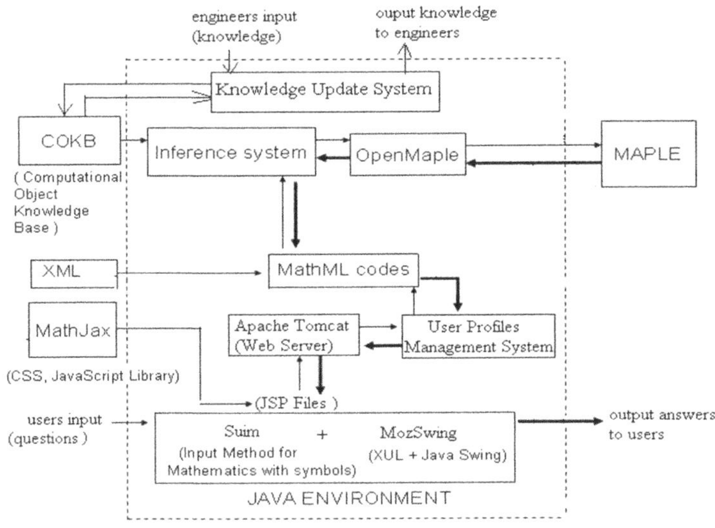

Fig. 2. The architecture of the system

[4] https://developer.mozilla.org/en/XULrunner,
https://developer.mozilla.org/En/XUL

[5] http://www.mathjax.org/

[6] http://www.maplesoft.com/

5 Implement the System

The system use JADE platform. Fig 3 shows the structure of Intelligent Multi-Agent System based on JADE in a LAN.

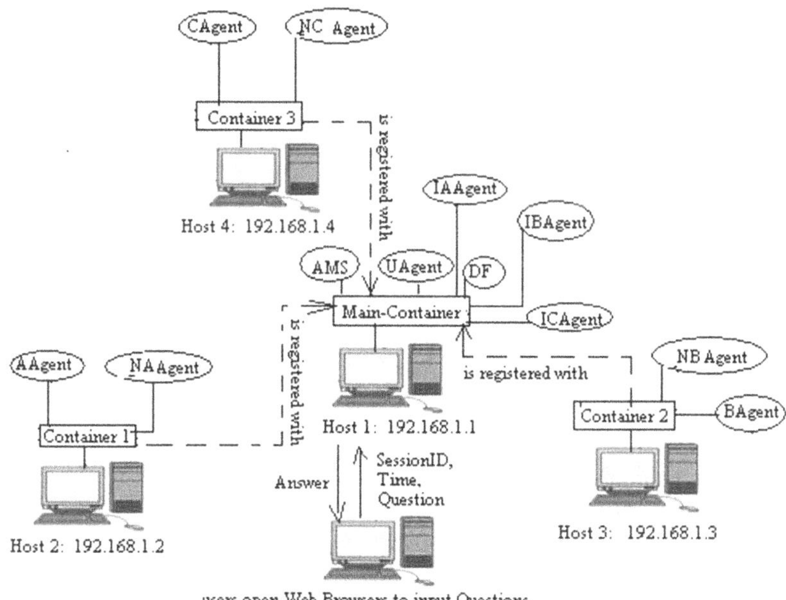

Fig. 3. The structure of Intelligent Multi-Agent System based on JADE in a LAN

The system includes 4 hosts. Host 1 (192.168.1.1) is illustrated Place 1, Host 1 (192.168.1.2) is illustrated Place 2, Host 3 (192.168.1.3) is illustrated Place 3, and Host 4 (192.168.1.4) is illustrated Place 4. Main-Container is built in Host 1. Container 1 is built in Host 2. Container 2 is built in Host 3. Container 3 is built in Host 4. Container 1, Container 2, and Container 3 are registered with Main-Container. Main-Container has default agents such as AMS and DF. Users open Web Browsers to input questions. The system stores Session ID of each user, Time (when the user submits question), content of the question. After the system finds the result, it will show the answer to the user.

6 An Example

Here is an example about solving a problem relating to 3 fields of knowledge: plane geometry (field A), 2D analytic geometry (field B), and algebra (field C). In coordinate plane Oxy sets A[1,3], B[4,2], D $\in$ Ox, DA=DB. Find coordinate of point D and perimeter of OAB. [21]. Input: Oxy: coordinate_plane; A: point; B: point; D: point; A[1,3]; B[4,2]; in (D,Ox); equal (DB,DA); ? coordinate (D); ? perimeter (OAB)

The process of finding answers as following: The first goal of the system is to find coordinate of point D. The system searches a concept about the coordinate of a point in 2D analytic geometry Knowledge Base. Then, the system finds xD and yD. The system use knowledge about 2D analytic geometry to calculate yD. We have: D $\in$ Ox => yD = 0

In the following process, the system use knowledge about 2D analytic geometry to calculate the module of DA and DB.

DA=DB => $\sqrt{(xA-xD)^2+(yA-yD)^2} = \sqrt{(xB-xD)^2+(yB-yD)^2}$

=> $\sqrt{(1-xD)^2+(3-0)^2} = \sqrt{(4-xD)^2+(2-0)^2}$

Then, the system use knowledge about algebra to find value of xD from following equation: $\sqrt{(1-xD)^2+(3-0)^2} = \sqrt{(4-xD)^2+(2-0)^2}$

=> xD = $\frac{5}{3}$ (use Maple to find solutions). The system use knowledge about plane geometry to calculate perimeter of triangle OAB. perimeter (OAB) = OA + AB + OB . After calculating modules of OA, AB, OB, the system receives results: perimeter (OAB) = $2\sqrt{10} + 2\sqrt{5}$

Output: coordinate (D) = [$\frac{5}{3}$, 0]; perimeter (OAB) = $2\sqrt{10} + 2\sqrt{5}$

The structures of a question and an answer:

The structure of a question: fact_group_1; fact_group_2; fact_group_3; question_group. We can change the order of fact_group_2 and fact_group_3, or change the order of facts which are between two semicolons in fact_group_1, fact_group_2, and fact_group_3. That doesn't change the question. The structure of the fact_group_1: object_name: concept_name; …; object_name: concept_name. Some names of concepts: coordinate_plane, point, segment, triangle, circle, … [3]. An example about fact_group_1: Oxy: coordinate_plane; A: point; B: point ; D: point ; F: point; W: point. Name_Set is a set of characters in object names of fact_group_1. Example: Name_Set = {O,x,y,A,B,D,F,W}. The structure of the fact_group_2: point_name [valueX,valueY]; …; point_name[valueX,valueY]

point_name is in Name_Set; valueX: value of abscissa of a point whose name is point_name; valueY: value of ordinate of a point whose name is point_name. An example about fact_group_2: A[1,3]; B[4,2]

The structure of the fact_group_3: relation_name (object_name_i; object_name_j); relation_name (object_name_o; object_name_p); …; function_name (object_name_g) = value; function_name (object_name_h) = value, …

Characters in object_name_i, object_name_j, object_name_o, object_name_p, object_name_g, object_name_h are in Name_Set.

Some names of relations: in, equal, parallel, perpendicular, ... Some names of functions: area, perimeter, coordinate, … [3]

An example about fact_group_3: in (D,Ox); equal (DB,DA); area (BFW) = 75

The structure of the question_group: ? function_name_1 (object_name); …; ? function_name_K (object_name); ? relation_name_1(object_name_l; object_name_t); …; ? relation_name_H (object_name_e; object_name_r)

The structure of an answer: function_name_1 (object_name) = value; …; function_name_K (object_name) = value; relation_name_1(object_name_l; object_name_t) = true/false; …; relation_name_H (object_name_e; object_name_r)= true/false

Characters in object name, object_name_l, object_name_t, object_name_e, object_name_r are in Name_Set. K: the number of questions about calculating value of objects; H: the number of questions about defining the relation between objects; M=K+H: the number of questions.

An example about the question_group: ? coordinate (D); ? perimeter (OAB); ? equal (AW, DF). An example about the answer: coordinate (D) = $[\frac{5}{3}, 0]$; perimeter (OAB) = $2\sqrt{10} + 2\sqrt{5}$; equal (AW,DF) = true

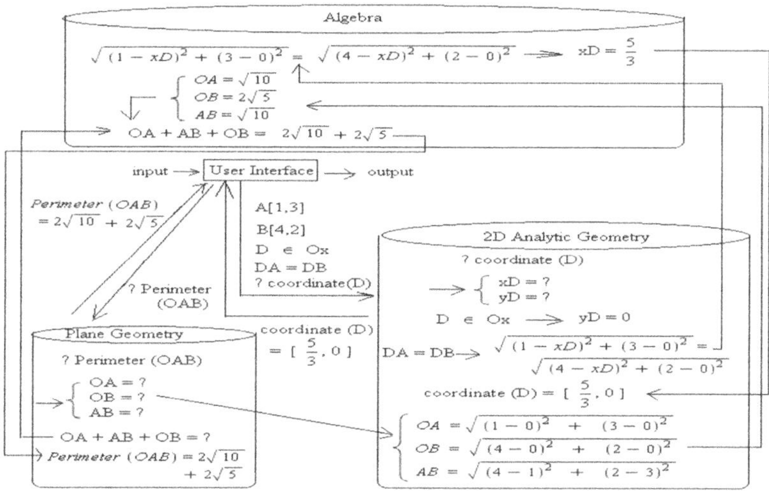

Fig. 4. An example of finding answers

In the above example, the process of finding answers happens in three distributed knowledge bases: Algebra, Plane Geometry, and 2D Analytic Geometry. Each knowledge base consists of 6 components: Concepts, Hierarchy, Relation, Operators, Functions, and Rules. When a user inputs a question, the system transfers its facts to knowledge bases. After the system finds answers, it transfers the answers to users. Fig 5 shows the process of accessing components of knowledge bases and deducing knowledge to find results of the question in the above example. A component of a knowledge base is relating to other components in this knowledge base, and a component of a knowledge base is relating to components in other knowledge bases.

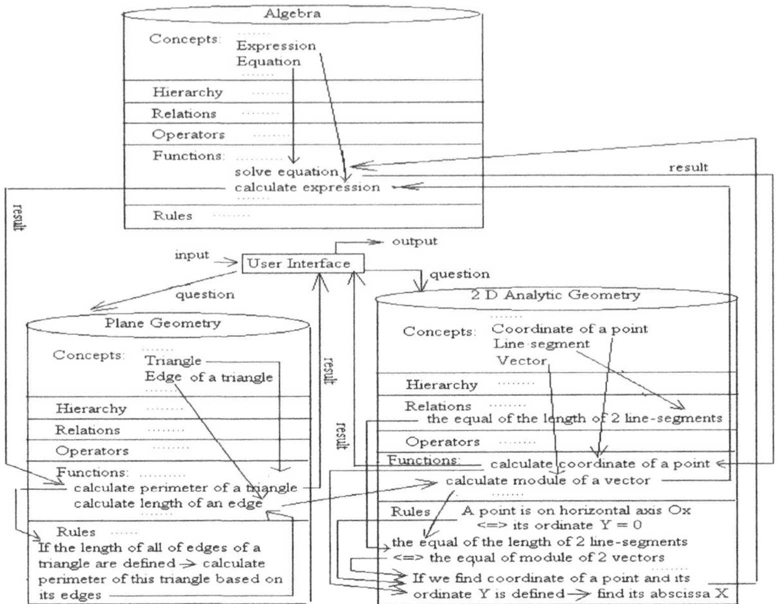

Fig. 5. The process of finding answers in three distributed knowledge bases

7 The Effects of the System

Here are the effects of the intelligent Multi-Agent system based on distributed knowledge bases of different fields. Firstly, the time to solve a problem in Multi-Agent paradigm is less than the time to solve the same problem in Client-Server paradigm. Secondly, when the total size of knowledge bases increases, the time to solve a problem in Client-Server paradigm increases much. However, the time to solve the same problem in Multi-Agent paradigm increases little when the total size of knowledge bases increases.

❖ In Multi-Agent paradigm: Let T the time to solve a problem of an agent.

- If the agent is Mobile Agent then $T = t_{proccess} + t_{migration}$ (1)

$t_{proccess}$: the time to execute in a Place, $t_{migration}$: the time to migrate in a network

Factors which can influence $t_{proccess}$ include the size of the agent, the size and the complication of the question, the size of knowledge base (the number of concepts, hierarchies, relations, functions, operations, rules). Factors which can influence $t_{migration}$ include the size of the agent and the bandwidth of the network [20].

Extending (1), we have: $$T = t_{agent} + \sum_{i=1}^{n} (t_{proccess}(i) + t_{agent}(i, i+1)) + t_{n+1} \qquad (2)$$

t_{agent}: the time to create agent and migrate to the first Place; $t_{proccess}$ (i) : the time to execute in Place (i); t_{agent} (i, i+1): the time to migrate from Place (i) to Place (i+1); t_{n+1} : the time to migrate from the last Place and kill itself

- If the agent is Intelligent Agent then T = $t_{proccess}$ (3)

$t_{proccess}$: the time to execute in a Place

Factors which can influence $t_{proccess}$ include the size of the agent, the size and the complication of the question, the size of knowledge base. $\forall$ x $\in$ {UAgent, AAgent, BAgent, CAgent, IAAgent, IBAgent, ICAgent, NAAgent, NBAgent, NCAgent}. Let t_{begin} (x) the begin-time to solve a problem of the agent x and t_{end} (x) the end-time to solve a problem of the agent x. We have: $t_{end}(x) = t_{begin}(x) + T(x)$ (4). Let D_{MA} the time to solve a problem of the system in Multi-Agent paradigm. $D_{MA} = \max\{t_{end}(x)\} - \min\{t_{begin}(x)\}$ (5). Let Average $_{MA}$ the average-time to solve a problem of the system in Multi-Agent paradigm. $\text{Average}_{MA} = (1/M) * \sum_{j=1}^{M} D_{MA(j)}$ (6) with M: the number of problems; j $\in$ {problem 1, problem 2, … , problem M}; $D_{MA(j)}$ the time to solve the problem j of the system in Multi-Agent paradigm.

❖ In Client-Server paradigm: Let D_{CS} the time to solve a problem of the system.

$D_{CS} = \sum_{i=1}^{n} (t_{process}(i) + t_{tranmission}(i,0)) + \sum_{i=1}^{n} t_{client}(i)$ (7) with $t_{process}$ (i) : the time to process in server at the ith time; $t_{tranmission}$ (i,0): the time for server to respond the result to client at the ith time; t_{client} (i): the time to process in client at the ith time.

Let Average $_{CS}$ the average-time to solve a problem of the system: $\text{Average}_{CS} = (1/M) * \sum_{j=1}^{M} D_{CS(j)}$ (8) with M : the number of problems; j $\in$ {problem 1, problem 2, …, problem M}; $D_{CS(j)}$ the time to solve the problem j of the system in Client-Server paradigm.

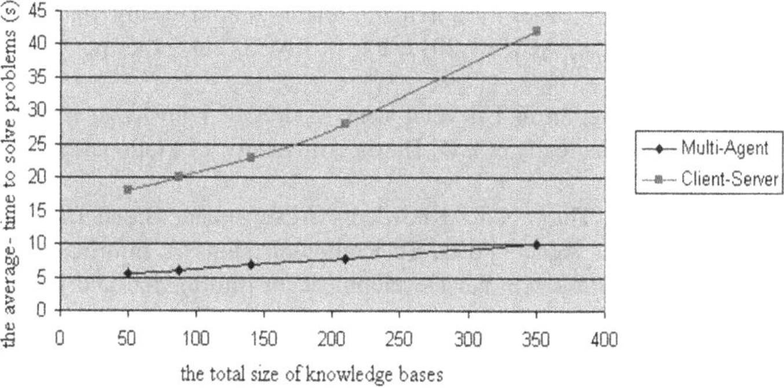

Fig. 6. The chart about the average-time to solve problems

Fig 6 shows the results of our experiment with M=100 problems relating to plane geometry, 2D analytic geometry, and algebra in Multi-Agent paradigm and Client-Server paradigm. The average-time to solve problems in Multi-Agent paradigm is less than 18.94 s.

8 Conclusions

In this paper, we proposed models, methods, techniques and technologies about developing an Intelligent MAS based on JADE to deal with knowledge relating to many different fields. Besides, we showed integrating methods of artificial intelligence with software engineering and distributed technologies. Finally, we illustrated building a system to solve problems automatically in three fields: algebra, plane geometry, and 2D analytic geometry. The system has some advantages. Firstly, the system has flexibility and intelligence. It can solve complex issues based on distributed knowledge bases that are able to be updated. It supports effectively for organizing distributed knowledge bases to solve complicated problems. Secondly, the system in Multi-Agent Paradigm can work when it is run in the network, whose connection is not continuous. In few cases, it stops working when the time the connection is interrupted is more than the time of life-cycle of agent. In all cases, the system in Client-Server Paradigm can not work in the network, whose connection is not continuous [2]. Thirdly, we can save resources because agents are able to suspend when the system don't need them. Besides, we can use resources from many hosts in the network. In the future, we will research a general model of the MAS in many fields, and experiment the system with WAN.

References

1. Van Nhon, D., Khue, N.T.M.: Building a model of Multi-Agent systems and its application in E-Learning. Posts, Telecommunications & Information Technology Journal, Special Issue: Research and Development on Information and Communications Technology (18), 100–107 (October 2007)
2. Khue, N.T.M., Van Do, N.: Building a Model of an Intelligent Multi-Agent System Based on Distributed Knowledge Bases for Solving Problems Automatically. In: Pan, J.-S., Chen, S.-M., Nguyen, N.T. (eds.) ACIIDS 2012, Part I. LNCS, vol. 7196, pp. 21–32. Springer, Heidelberg (2012)
3. Van Nhon, D.: Construction and development models of knowledge representation for solving problems automatically. Ph.D Thesis, University of National Science, Ho Chi Minh City (2001)
4. Thuc, H.M., Hai, N.T., Thuy, N.T.: Knowledge Management Based Distance Learning System using Intelligent Agent. Posts, Telecommunications & Information Technology Journal, Special Issue: Research and Development on Information and Communications Technology (16), 59–69 (April 2006)
5. Barcelos, C.F., Gluz, J.C., Vicari, R.M.: An Agent-based Federate Learning Object Search service. Interdisciplinary Journal of E-Learning and Learning Objects 7 (2011)

6. El Kamoun, N., Bousmah, M., Aqqal, A., El Morocco, J.: Virtual Environment Online for the Project based Learning session. Cyber Journals: Multidisciplinary Journals in Science and Technology, Journal of Selected Areas in Software Engineering, JSSE (January 2011)
7. Mikic-Fonte, F.A., Burguillo-Rial, J.C., Llamas-Nistal, M., Fernández-Hermida, D.: A BDI-based Intelligent Tutoring Module for the e-Learning Platform INES. In: 40th ASEE/IEEE Frontiers in Education Conference, Washington, DC, October 27-30 (2010)
8. Wang, M., Ran, W., Jia, H., Liao, J., Sugumaran, V.: Ontology-Based Intelligent Agents in Workplace eLearning. In: Americas Conference on Information Systems, AMCIS (2009)
9. Hunyadi, D., Pah, I., Chiribuca, D.: A Global Model for Virtual Educational System. WSEAS Transactions on Information Science and Applications 6(3) (March 2009)
10. Henry, L., Sancaranrayanan, S.: Intelligent Agent based Mobile Learning System. International Journal of Computer Information Systems and Industrial Management Applications (IJCISIM) 2, 306–319 (2010) ISSN: 2150-7988
11. Kazar, O., Bahi, N.: Agent based approach for E-Learning. MASAUM Journal of Computing 1(2) (September 2009)
12. AlZahrani, S.S.: Regionally Distributed Architecture for Dynamic e-Learning Environment, RDADeLE (2010)
13. Repka, V., Grebenyuk, V., Kliushnyk, K.: Pro-Active Multi-Agent System in Virtual Education. In: Multi-Agent Systems - Modeling, Control, Programming, Simulations and Applications (2010)
14. De la Prieta, F., Gil, A.B.: A Multi-agent System that Searches for Learning Objects in Heterogeneous Repositories. In: Demazeau, Y., Dignum, F., Corchado, J.M., Bajo, J., Corchuelo, R., Corchado, E., Fernández-Riverola, F., Julián, V.J., Pawlewski, P., Campbell, A. (eds.) Trends in PAAMS. AISC, vol. 71, pp. 355–362. Springer, Heidelberg (2010)
15. El-Bakry, H.M., Mastorakis, N.: Greece Realization of E-University for Distance Learning. WSEAS Transactions on Computers 8(1) (January 2009)
16. El Bouhdidi, J., Ghailani, M., Abdoun, O., Fennan, A.: A New Approach based on a Multi-ontologies and Multi-Agent System to Generate Customized Learning Paths in an E-Learning Platform. International Journal of Computer Applications (0975 – 8887) 12(1) (December 2010)
17. Bentivoglio, C.A., Bonura, D., Cannella, V., Carletti, S., Pipitone, A., Pirrone, R., Rossi, P.G., Russo, G.: Intelligent Agents supporting user interactions within self regulated learning processes. Journal of e-Learning and Knowledge Society 6(2), 27–36 (2010)
18. Mahdi, H., Attia, S.S.: Developing and Implementing a Multi-Agent system for Collaborative E-learning (2010)
19. Masrom, S., Rahman, A.S.A.: An Adaptation of Agent-Based Computer-Assisted Assessment into E-Learning Environment. International Journal of Education and Information Technologies 3(3) (2009)
20. Quah, J.T.S., Chen, Y.M., Leow, W.C.H.: Networking E-Learning hosts using Mobile Agents. In: Intelligent Agents for Data Mining and Information Retrieval (2004)
21. Geometry Textbook 10, VN, p. 45. Educational Publisher (2010)

Multi-agent Platform for Security Level Evaluation of Information and Communication Services

Grzegorz Kołaczek

Institute of Informatics, Faculty of Computer Science and Management,
Wroclaw University of Technology,
Wybrzeże Wyspiańskiego 27, 50-370 Wrocław, Poland
Grzegorz.Kolaczek@pwr.wroc.pl

Abstract. The paper presents an original multi-agent approach to security level evaluation in service oriented systems for telecommunication sector, especially for providers of communication and information services. The platform is built from the following elements: user interface which has been used to select and configure the mode of operation of the analytical algorithms implemented within the platform; inter-modules interfaces which are used to supply the algorithms with appropriate datasets and to provide the user the results of analysis; analytical services which are responsible for security level evaluation and finally data repositories. The platform provides the security evaluation functionality as a web service according to Security as a Service (SaaS) model. This way of providing security assumes that the analytical methods are implemented independently from the evaluated objects and can be provided to any requesting subject as web services. The paper presents the architecture of the platform and the provided functionalities. Finally, the example scenario of the security level evaluation has been demonstrated and discussed. The presented platform for security level evaluation is an integral part of the comprehensive solution called PlaTel which has been designed for management and execution of information and communication services.

1 Introduction

The telecommunication market has always been 'service oriented'. This means that from early beginning changes in the telecom sector have been motivated by changes in an offered set of services (service-driven). To provide an end user high level quality services, telecom operator must for each service: design the architecture, define the cooperation between the existing components with the application of communication protocols, standardize the content and steering messages exchange. This way of providing new services is relevant to International Telecommunication Union (ITU) recommendation. ITU prepared the three step procedure for modeling telecommunication systems which includes: the description of provided for end user functionality, the architecture definition, the interdependencies and protocols definition [1,2,3].

The changes in telecom sector driven by changing technology and users expectation started in eighties. Then appeared the idea of *Intelligent Network* (IN) which was

N.T. Nguyen et al. (Eds.): *Adv. Methods for Comput. Collective Intelligence*, SCI 457, pp. 107–116.
DOI: 10.1007/978-3-642-34300-1_10

the first approach to programmable and reconfigurable telecommunication systems. The idea of IN has been extended by International Telecommunication Union - Telecommunication Standardization Sector (ITU-T) to iterative approach of Intelligent Network Conceptual Model (INCS). Although the growing number of implementations and wildly accepted standards the IN and INCS have not reached the expected level of interest. This motivated the next phase of changes in telecommunication systems in the nineties (Figure 1). Then several new application programming interfaces (API) for telecommunication systems have been developed, e.g. Parlay, 3rd Generation Partnership Project (3GPP), Open Service Architecture or Java API for Integrated Networks (JAIN). Finally, the simplified version of Parlay specification (Parlay/OSA – Parlay X) has been approved as the standard API for telecommunication services. The main assumption of Parlay X are as follows: service intelligence is separated from signalization, service intelligence can be moved outside the telecommunication network, the connection between a service logic and telecommunication network is defined by API. Additionally Parlay X uses the elements known from classic web systems such as Extensible Markup Language (XML), Universal Description, Discovery and Integration registry (UDDI), Web Services Description Language (WSDL) and Simple Object Access Protocol (SOAP) [4].

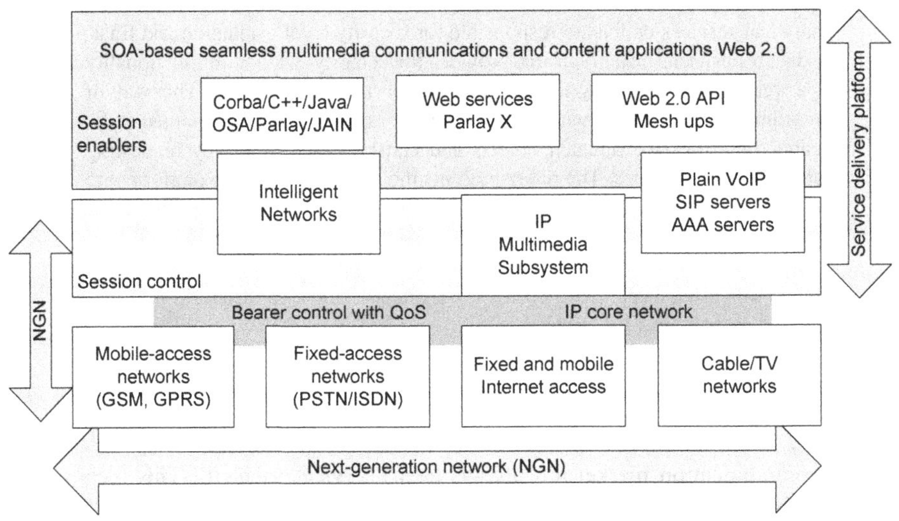

Fig. 1. Layered architecture of SOA-based ICT systems

Next section presents general overview of PlaTel – universal platform for management and execution of information and communication services. The third section describes the architecture and functionalities of security level evaluation module of PlaTel. This section is summarized by an example of practical application of the proposed framework in the task of security level evaluation in telecommunication systems. The last part of the paper contains conclusions and the further work.

2 Universal Platform for Management and Execution of Information and Communication Services

The universal platform for management and execution of information and communication services – PlaTel – has been developed to integrate several layers of abstraction on the way of providing intelligent communication and computational services. PlaTel offers complementary support in business processes management and execution for all type of users (e.g. telecomm services providers, end users or telecomm operators).

PlaTel has been developed in accordance with Service Oriented Architecture (SOA) paradigm. This means that PlaTel is a set of loosely coupled services constitute a set of basic functionalities provided as separate modules. There are a few basic modules which are responsible for: business process and organization modeling (PlaTel-O and PlaTel-P), service composition (PlaTel-U), mapping of communication-based requirements (PlaTel-K), execution environment management (Platel-R), security evaluation (PlaTel-W). Each of the modules may act independently using its own execution environment and input data. However, the maximum synergy can be achieved when all modules interact with each other. Combining several well know as well as originally developed methods of knowledge processing makes PlaTel an original collective intelligence tool for management and execution of information and communication services. In such case, the entire business process can be supported by PlaTel from the stage of the business process definition through the services composition, mapping them to the communication and computational resources and executing, till the evaluation of the final effect in terms of security and quality of services[5].

2.1 Security Problems and Security Level Evaluation of Communication and Information Services

PlaTel-W is a type of security monitor for the service-oriented systems which evaluates the level of security on the basis of detection of the anomalous patterns in network traffic or other characteristic values of the services available in the system [6]. In this paper we consider a service-oriented data exchange system. The generic system's architecture that is monitored with PlaTel-W has been depicted in Figure 2. It consists of services that have been composed (PlaTel-U) requested by user and which are intended to implemented a particular business process (PlaTel-P). After composition, the services are mapped on set of available services using the computational and communication requirements (PlaTel-K). Finally, services are executed (PlaTel-R). The execution of the services is monitored and all relevant data are collected in PlaTel repositories. Data from repositories are available for all PlaTel modules and allows for constant composition, mapping, execution, etc. improvement. PlaTel-W uses collected data to evaluate security level of the executed services and specially to find some anomalies in service execution which may be related to security breaches.

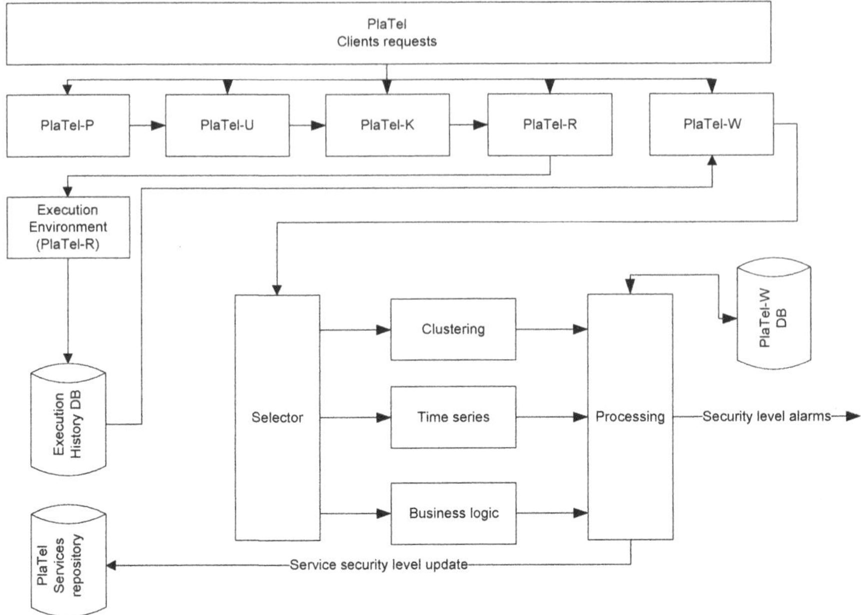

Fig. 2. Architecture of PlaTel-W

2.2 Functionality

The main components of PlaTel-W are responsible for analyzing the data provided by monitoring modules implemented by services execution environment (PlaTel-R). The analysis performed by PlaTel-W focuses on detection of anomalous events in service execution logs. There are three types of analysis performed which uses outlier detection, time series analysis and business logic abuses approach.

Outlier detection

The aim of outlier detection may be defined as the search for an observation which deviates so much from the other observations as to arouse suspicions that it was generated by a different mechanism [10]. This follows the general intuition which can be described by two following statements:

– Normal data object follow a "generating mechanism", e.g. some given statistical process
– Abnormal objects deviate from this generating mechanism

Outlier detection methods can be divided between univariate methods and multivariate methods that usually form most of the current body of research. Another fundamental taxonomy of outlier detection methods is between parametric (statistical) methods and nonparametric methods that are model-free. PlaTel-W implements a class of outlier detection methods which is founded on clustering techniques

(Expectation Maximization based method and Distance-Based Otliers Detection), where a cluster of small sizes or small densities, below the given threshold, can be considered as clustered outliers .

Time series analysis

The other method applied in PlaTel-W uses time series analysis approach to anomaly detection. The anomalous behavior of the systems is determined using the values for the behavioral attributes within a specific context. An observation might be an anomaly in a given context, but an identical data instance (in terms of behavioral attributes) could be considered normal in a different context. Contextual anomalies have been most commonly explored in time-series data [9][10]. One of the earliest works in time-series anomaly detection was proposed by Fox [12]. Some of the time series anomaly detection approaches uses basic regression based models [12]. Another variant is that detects anomalies in multivariate time-series data generated by an Autoregressive Moving Average (ARMA) [12]. Any observation is tested to be anomalous by comparing it with the covariance matrix of the autoregressive process. If the observation falls outside the modeled error for the process, it is declared to be an anomaly. An extension to this technique is made by using Support Vector Regression [13]. Another example of anomaly detection in time-series data has been proposed by Basu and Meckesheimer [12]. For a given instance in a time-series the authors compare the observed value to the median of the neighborhood values.

The implemented in PlaTel-W algorithm of anomaly detection in time series benefits from multidimensional approach presented by M.Burgess [8,9] This method can be applied to detect anomalies in various types of values measured describing the service execution. The only requirement is that the values must constitute time series (e.g. how many times in a given time window a service has been executed). The more details about the implemented method and the obtained results can be found in [7].

Business logic abuses

Apart from the typical security problems related to service oriented environments, one which is crucial for further evolution of service oriented systems is business logic abuse detection. Business logic abuse is the abuse of the legitimate business logic of a website or other function that allows interaction. Business logic abuse is usually aimed to exploit in some way the system that supports certain business logic e.g. by an illicit use of a legitimate website function.

Detection of business logic abuse is difficult because the offenders are using the same functionality as the legitimate users and therefore, their actions are likely intermixed with real actions. The other problem to overcome in the context of securing business logic is that the intruder is using a legitimate flow on a website or other application, so disabling that flow would influence also the interactions of legitimate users. This is why the new and versatile methods are required to support the service oriented systems with an appropriate services that could secure them from this type of risk.

3 Security Level Evaluation

Service-Oriented Architecture (SOA) defines that services are independent, self-contained modules, which do not store state from one request to another. As services should not depend on the context or state of other services, any state-dependencies are defined using business processes, and data models. Finally, service oriented systems implements some business logic using their ability to compose applications, processes, or more composite services from other less composite services. This activity, sometimes called service composition, allows developers to compose applications and processes using services from heterogeneous environments without regard to the details and differences of those environments. In this context, business logic abuses could be detected by finding some suspicious traces in a system's or/and services' behavior.

This section presents a statistical approach to represent observations of composite services execution. The spatial relationships which are often used as a synonymous with geographical distance between objects, in this case have been redefined as the logical distance between services composing services providing selected business logic functionality. This means that the distance between services is measured in the number of executed intermediate services on the way of providing user required business logic functionality. In normal (secure) situation, a composite service execution requires the constant number of atomic services to be executed to accomplish user request. When some problems with system security appear (e.g. malicious user password guessing, hacker's search for service vulnerabilities, denial of service attacks, etc.) the change in this characteristic is expected. This assumption has been used to model, estimate and describe spatial correlations among services and to detect business logic abuses.

3.1 Experiment Description

The aim of the experiment was to prove the possibility to detect the attack aimed at business process execution in service oriented systems. The business process abuse has been illustrated by abrupt change of data flow among services constituting composite service.

The composition plan defines the relations and expected directions of dataflow among services. The composite service presented in figure 3 assumes that the composite service execution starts by running service V1 and then the results of V1 execution are taken as input by services V2 and V3, the output of V2 is taken by services V4 and V5, etc. The composite service is completed when the service V8 returns the value. All services are executed in PlaTel-R environment and their execution is monitored by corresponding module. In this example, it has been assumed that the monitoring system collects the information about the byte counts of each service output. Because there is no good reference data from the real service oriented system the experiment uses some artificially generated datasets.

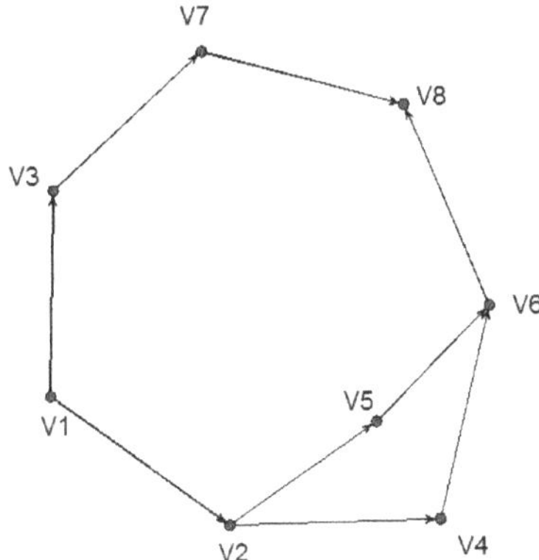

Fig. 3. Composite service execution plan

The four scenarios of composite service execution have been proposed and analyzed. The first scenario presents the situation where there is a strong correlation between the number of bytes at service input and output. It means that the number of bytes at service input determines the number of bytes at service output. The second scenario introduces some variability in relation between data quantity at input and output of services. It means that some randomly generated value reduces correlation between input and output. The third scenario presents how the results of the composite service execution analysis when there is no deterministic relation between data at input and output (it is rather unrealistic case). The last scenario shows the business logic abuse and how it influences the observable relations between services. The business logic abuse has been simulated in the fourth scenario by disrupting the correlation between input and output of the service V4. This scenario may illustrate the situation when intruder tries to execute some services out of planed order defined by the scheduled execution plan or tries to execute a service V4 using multiple and various input streams verify the service firmness.

3.2 The Spatial Analysis for Security Level Evaluation

The anomaly detection in composite service execution has been performed using some basic spatial analysis methods. The brief description of the performed steps is given below.

Moran-I test

It is the most common test for the existence of spatial autocorrelation. Moran coefficient calculates the ratio between the product of the variable of interest and its spatial lag, with the product of the variable of interest, adjusted for the spatial weights used.

$$I = \frac{n \sum_i^n \sum_j^n \omega_{ij}(x_i-\bar{x})(x_j-\bar{x})}{\sum_i^n \sum_j^n \omega_{ij} * \sum_i^n (x_i-\bar{x})^2} \tag{1}$$

Where x is the value of a variable for the ith observation, $\bar{x}$ is the mean of x and the ω_{ij} is the spatial weight of the connection between ith an jth node. Values range from -1 (dispersion) to +1 (correlation). A zero value indicates a random spatial pattern.

Local Moran test

A local Moran's test for the unit *i* can be constructed as one of the *n* components which comprise the global test:

$$I_i = \frac{(x_i-\bar{x})\sum_j^n \omega_{ij}(x_j-\bar{x})}{\frac{\sum_i^n (x_i-\bar{x})^2}{n}} \quad (2)$$

As with global statistics, it is assumed that the global mean $\bar{x}$ is an adequate representation of the variable of interest. As before, local statistics can be tested for divergence from expected values, under assumptions of normality.

Spatial correlogram

Spatial autocorrelation measures the degree to which observations are correlated to itself in space. Spatial correlogram allows to tests and visualize whether the observed value of a variable at one location is independent of values of that variable at neighboring locations. Positive spatial autocorrelation indicates that similar values appear close to each other, or cluster, in space while negative spatial autocorrelation indicates that neighboring values are dissimilar or that similar values are dispersed. Null spatial autocorrelation indicates that the spatial pattern is random. In this research the correlogram of input/output flows of the corresponding network nodes has been investigated.

3.3 Results and Interpretation

This section presents the overview of the main results of the spatial analysis applied to composite services execution in service oriented systems.

Table 1. Moran-I test results

	Scenario			
	1	2	3	4
Moran I statistic	0.14108890	0.15279006	0.5028	-0.0201311
p-value	0.03310	0.02974	-0.14499871	0.2989
Expectation	-0.1428571	-0.14285714	-0.14285714	-0.1428571
Variance	0.02389195	0.02461018	0.09145791	0.05414065

The results presented in Table 1. shows that there is a positive spatial correlation in first and second scenario. It means that the correct execution of composite service shows the spatial correlations. Contrary scenarios 3 and 4 miss this type of correlation. Especially the scenario 4 demonstrates that spatial analysis can be used to detect security breaches in composite services execution and so also business logic abuses should be detected.

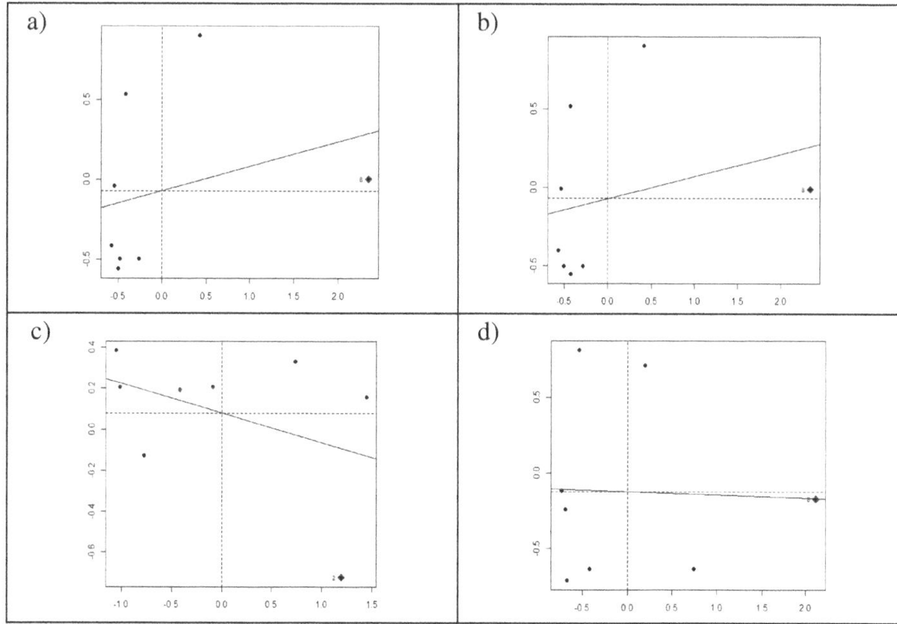

Fig. 4. Spatial correlograms

The results presented in Table 1 can be visualized as a spatial correlograms. Figure 4. presents a plot of spatial data against its spatially lagged values (axis X denotes standardized data flow volume, axis Y spatial lag). Each spatial correlogram is divided into 4 sections: Low Left (LL), Low Right (LR), High Left (HL) and High Right (HR). Points (services) located in LL and HR parts of the plot shows positive spatial correlation while points (services) form RL and HR show negative spatial correlation. The normal composite service execution is illustrated by charts fig 4.a and fig 4.b. The random services execution (no relations among services) is illustrated by chart fig 4.c. The change of the spatial correlation type is visible in chart 4.d. These results demonstrate that such form of visualization can be very helpful in business logic abuses detection.

4 Conclusions and Future Work

The paper presented an original approach to the analysis of composite service execution with the application of statistical spatial analysis methods. The proposed method is an integral element of PlaTel platform which has been designed for ICT users to support the task of service management, execution and monitoring in service oriented systems. The cooperation of all defined within PlaTel modules creates an extensive possibility to provide high quality security monitoring system. As an example application of PlaTel modules cooperation, the anomaly detection tasks has been examined. The obtained results are very promising, so the further research in this field has been planned.

The further works are intended to apply spatial analysis to real world data and to extend the analysis with some additional methods to enable reliable classification of security events. The next steps will be focused on further integration among PlaTel modules, particularly the interesting task is integration of the security analysis with composition module (PlaTel-U and PlaTel-K). Information about detected abuses should be taken into account while the next compositions of services are performed.

References

1. Magedanz, T., Blum, N., Dutkowski, S.: Evaluation of SOA concepts in telecommunications, Berlin (2007)
2. Harris, T.: SOA in Telecom (2010)
3. Strategies for enabling new serwices, http://www.tmforum.org
4. FP7 Call 1 achievements in software and services (2011)
5. Bond, G., Cheung, E., Fikouras, I., Levenhsteyn, R.: Unified telecom and web services composition: problem definition and future directions. In: IPTCOM 2009 (2009)
6. Chandola, V., Benerjee, A., Kumar, V.: Anomaly Detection, A Survey (2007)
7. Kołaczek, G.: Multiagent Security Evaluation Framework for Service Oriented Architecture Systems. In: Velásquez, J.D., Ríos, S.A., Howlett, R.J., Jain, L.C. (eds.) KES 2009, Part I. LNCS (LNAI), vol. 5711, pp. 30–37. Springer, Heidelberg (2009)
8. Burgess, M.: An Approach to Understanding Policy Based on Autonomy and Voluntary Cooperation. In: Schönwälder, J., Serrat, J. (eds.) DSOM 2005. LNCS, vol. 3775, pp. 97–108. Springer, Heidelberg (2005)
9. Burgess, M.: Two Dimensional Time-Series for Anomaly Detection and Regulation in Adaptive Systems. In: Feridun, M., Kropf, P., Babin, G. (eds.) DSOM 2002. LNCS, vol. 2506, pp. 169–180. Springer, Heidelberg (2002)
10. Gorodetski, V.I., Karsayev, O., Khabalov, A., Kotenko, I., Popyack, L.J., Skormin, V.: Agent-Based Model of Computer Network Security System: A Case Study. In: Gorodetski, V.I., Skormin, V.A., Popyack, L.J. (eds.) MMM-ACNS 2001. LNCS, vol. 2052, pp. 39–50. Springer, Heidelberg (2001)
11. Hwang, K., Liu, H., Chen, Y.: Cooperative Anomaly and Intrusion Detection for Alert Correlation in Networked Computing Systems, Technical Report, USC Internet and Grid Computing Lab, TR 2004-16 (2004)
12. Lakhina, A., Crovella, M., Diot, C.: Characterization of Network-Wide Anomalies in Traffic Flows. Technical Report BUCS-2004-020, Boston University (2004), http://citeseer.ist.psu.edu/715839.html
13. Ammeller, D., Franch, X.: Service level agreement monitor (SALMon). In: 2008 7th International Conference on Composition-Based Software Systems, pp. 224–227 (2008)

A Multi-agent Model for Image Browsing and Retrieval

Hue Cao Hong[1,2], Guillaume Chiron[3], and Alain Boucher[1]

[1] IFI, MSI team; IRD, UMI 209 UMMISCO; Vietnam National University, 42 Ta Quang Buu, Hanoi, Vietnam

[2] Hanoi Pedagogical University N^{0}2, Xuan Hoa, Phuc Yen, Vinh Phuc, Vietnam

[3] L3I, University of La Rochelle, 17042 La Rochelle cedex 1, France

Abstract. This paper presents a new and original model for image browsing and retrieval based on a reactive multi-agent system oriented toward visualization and user interaction. Each agent represents an image. This model simplifies the problem of mapping a high-dimensional feature space onto a 2D screen interface and allows intuitive user interaction. Within a unify and local model, as opposed to global traditional CBIR, we present how agents can interact through an attraction/repulsion model. These forces are computed based on the visual and textual similarities between an agent and its neighbors. This unique model allows to do several tasks, like image browsing and retrieval, single/multiple querying, performing relevance feedback with positive/nagative examples, all with heteregeneous data (image visual content and text keywords). Specific adjustments are proposed to allow this model to work with large image databases. Preliminary results on two image databases show the feasability of this model compared with traditional CBIR.

Keywords: Multi-Agent System, content-based image retrieval, attraction/repulsion forces.

1 Introduction

Content-based image retrieval (CBIR) is one of the major approaches for image retrieval and it has drawn significant attention in the past decade. CBIR uses the image visual content, extracted using various features, for searching and retrieving images. Doing so, all images are represented into a feature space, which is often high-dimensional. Most CBIR systems return the most similar images in a 1D linear format following decreasing similarities [11], which does not provide a good basis for user interaction.

Few years ago, several methods have been used in CBIR for mapping a high-dimensional feature space into a 2D space for visualization, with good results on visualization, but less on interacting with result data. Some specific similarity measures are used for visualization, like the Earth Mover's Distance (EMD) [14], for representing the distribution of images on a 2D screen. Principal Component

N.T. Nguyen et al. (Eds.): *Adv. Methods for Comput. Collective Intelligence*, SCI 457, pp. 117–126.
DOI: 10.1007/978-3-642-34300-1_11

Analysis (PCA) [7] is a basic and linear approach used for dimension reduction in image retrieval. However, with PCA only linear correlations can be detected and exploited and it does not necessarily preserve at best mutual relations between different data items [11]. Multi-dimensional scaling (MDS) [13] is a nonlinear transformation method to project a data set in a smaller space. MDS attempts to preserve the original relationships of the high-dimensional space into the low dimensional projection. MDS provides a more precise representation of relationships between images in the feature space compared to PCA. However, MDS has a very large computational cost due to its quadratic dependence. Self-organizing maps (SOM) [16] is a neural algorithm for visualizing and interpreting large high-dimensional data sets. SOM is efficient in handling large datasets [15] and it is robust even when the data set is noisy [5]. However, the structure of the SOM neural networks and the number of neurons in the Kohonen layer need to be specified a priori. Determining the number of clusters is not trivial when the characteristics of the data set are usually not known a priori [3]. In all these methods, the user can browse an image database in a more intuitive way, but interaction is still limited with these methods, especially when retrieval is based on heterogeneous features, like image features and text features, which are difficult to show together and efficiently on a 2D screen [8]. Nowadays, the problem is still unsolved.

To cope with the existing drawbacks of the existing techniques, we explore in this paper the use of a new and innovative approach for this problem, which is a multi-agent model for image browsing and retrieval. Multi-agent systems are used in CBIR mostly for distributed retrieval [10], and less for the problems of visualization and interaction. But interesting models can be found for other types of applications, on which we base our approach. For example, Renault has proposed a dynamic attraction / repulsion multi-agent model coupling [12] for the application of email classification. It aims to organize e-mails in a 2D space according to similarity, where each e-mail is represented by an agent and there is no need to specify axes or how to organize information. The model allows agents communicating with each other through virtual pheromones and to collectively auto-organize themselves in a 2D space. Without much constraints, the system can organize (cluster/classify) information and let the user intituively interacts with it. We keep the idea of using reactive agents with attraction / repulsion forces for organizing data in space and adapt it for image browsing and retrieval. For facilitating 2D display and interaction, in our model each image of the database is represented by an agent which moves in a 2D environment. Each agent interacts with other agents/images using forces. These forces are computed according to the similarities between an agent/image and its neighbors. This model allows the user to easily interact with images using an intuitive interface. The user can select (or deselect) one or many agent(s)/image(s) to perform a query and can organize these queries by moving them in the environment. The main drawback of this kind of techniques is the computation time for the multi-agent model organization and we propose some ways to cope this problem.

The rest of this paper is organized as follow. In the next section, we present in details our main contributions, which are the multi-agent model for image browsing and retrieval, the force computation model for images and the techniques to work with large image databases. Then, section 3 shows some qualitative and quantitative results of our model. Finally, section 4 gives a conclusion and some future work.

2 System Overview

2.1 Global System Model

The image database is modeled using a multi-agent system. Each agent represents an image and bases its actions on the feature vector from that image. The agents move freely in a 2D environment which has no pre-defined axes or meaning (Figure 1). They are reactive and only react to outside stimuli sent by other agents. Each agent interacts with others through forces, which can be attraction or repulsion. Forces are computed between two agents based on the the similarity of their feature vectors. Thus, agents are attracted to *similar* ones and are repulsed from *non-similar* ones. This local behavior between two agents produces at the global level an auto-organization (like clustering) of all images in the 2D space.

An image query-by-example in that model is simply an image/agent clicked by the user. An agent-query is static (does not move), but the user can move it where he wants on the screen (to organize result display). Except for that, this agent-query is behaving the same than the others, and still produces attraction / repulsion forces toward other agents. Multiple queries are possible simply by clicking multiple images. Text queries (keywords) are given by adding an agent-text, representing the keyword, in the system. The forces are computed using text similarities, when annotation are available for agent-images. Queries can be positive (normal queries) or negative (forces are inverted). This simple mechanism reproduces the *relevance feedback* behavior used in traditional CBIR systems, where the user can indicates positive or negative feedbacks to the system.

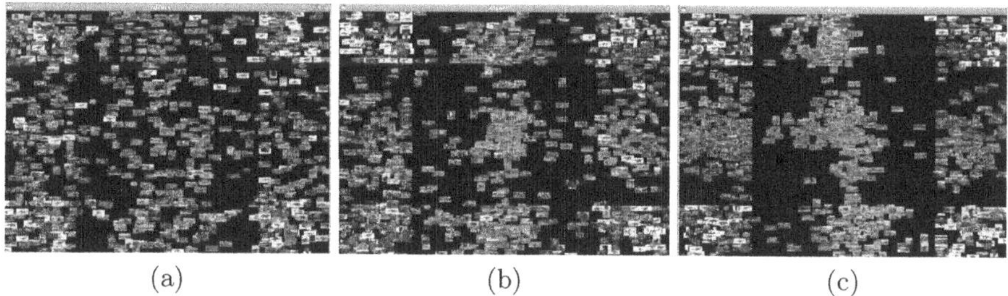

(a) (b) (c)

Fig. 1. Dynamic evolution of the system with an image query (in green in the middle). (a) The model is initialized with random placement of the images in the environment. An image is selected (in green) as a positivie query and placed in the center. (b) System state after 50 time steps. (c) After 300 time steps.

The major advantage of this model is its simplicity, allowing in a very simple and intuitive model to reproduce complex behaviors observed in traditional CBIR systems, like browsing, querying by text and/or image, interacting with user, doing relevance feedback, mapping high-dimensional and hetereogeneous features into a 2D space, etc. Moreover, where several different complex algorithms are needed to implement all these behaviors in a traditional CBIR system, in this new model, they are just consequences of the basic initial model. In the following sections we detail all these aspects.

2.2 Agent Interaction Model (Force Computation Model)

A force applied between two agents can be attractive or repulsive and is represented by a vector (magnitude and direction). If two agents are similar (image content and/or text/keyword similarities), they are attracted and if two agents are different (non-similar), they are repulsed from each other. As presented above, at each time step[1], an agent interacts with its neighbors, getting forces from them and moving reactively. The first step to compute forces is therefore to define the neighbor list for an agent, which evolves dynamically as the agent moves. Through experiments, taking neighbours that are within specific radius in space increases computation time, reduces convergence speed and result efficiency too. We prefer to implement the method of selecting N agents randomly[2] among all, which allows us to cope with these drawbacks. Thus, a *neighborhood* is more defined like a temporary (for one time step) relationship between two agents that will interacts (react through forces) and it is not based on spatial proximity. Once the neighbor list for an agent is known, then this agent can simply compute the forces it is receiving from all these neighbors and react according to them (Figure 2).

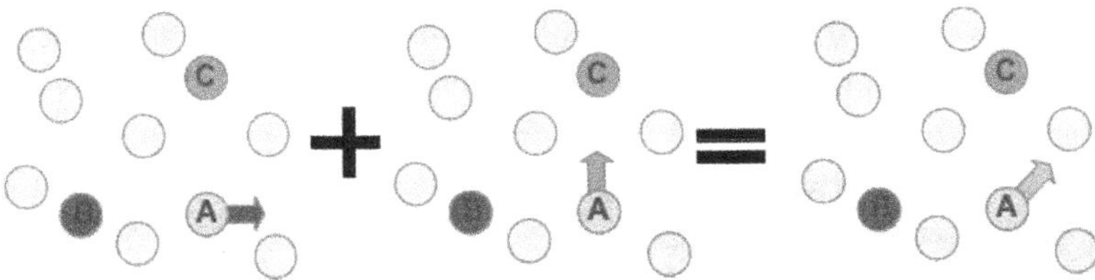

Fig. 2. Force computation. Agent A chooses (randomly) N neighbors (2 in this example) among all existing agents. The global force applied on agent A is the sum of the (repulsion) force between agents A and B and the (attraction) force between agents A and C.

[1] In our implementation, agents are executed in pseudo-parallelism managed by a scheduler, and each agent is executed for a small time unit (corresponding to one force computation loop), and so one time step in the model corresponds in reality to one iteration loop of all agents.

[2] N=20 in our experimentations.

Image-based forces. The visual similarity S between two agent-images is calculated using the Squared Chord similarity [4]: $S = \sum_{i=1}^{n}(\sqrt{v_{\text{i}}} - \sqrt{w_{\text{i}}})^2$, where $v = (v_1, v_2, ..., v_{\text{n}})$ and $w = (w_1, w_2, ..., w_{\text{n}})$ are the image feature vectors for the two images v and w. For N neighbors, an agent will calculate N similarities. All these similarities are sorted in ascending order, in order to be able to normalize them. No global normalization is done, so each agent has to locally normalize its computations based on its current information. Each similarity is normalized following:

$$S_{\text{i}} = S_{min} + i \cdot \frac{S_{max} - S_{min}}{N-1} - \frac{S_{min} + S_{max}}{2} \tag{1}$$

where S_{i} is the similarity of the ith ranked neighbor after sorting, i varying from 0 to $N-1$, S_{min} and S_{max} the minimum and maximum similarities of all neighbors. The spatial distance D_{i} between two *neighbor* agents is calculated for two points $P(x, y)$ and $Q(s, t)$ using the chessboard distance: $D(P, Q) = maximum(|x - s|, |y - t|)$. Similarly to similarities, distances are normalized using: $D_{\text{i}} = D_{\text{i}} - (D_{\text{min}} + D_{max})/2$.

Both normalizations (similarity and distance) produce negative and positive values, and the final force between an agent and a neighbor is given by Table 1. *Weak* and *strong* correspond to constant values that can be calibrated for the environment, that are used to scale the force, while *attraction* and *repulsion* correspond to the sign of the force. The force magnitude between two agents is given by:

$$F = C \cdot \frac{S_{\text{i}}}{D_{\text{i}}} \tag{2}$$

where C is a factor that can be 1 or 3 corresponding to *Weak* and *Strong* forces (Table 1).

Table 1. Magnitude and direction of a force according to the similarity and the distance between two agents

Similarity	Distance	Force
-	-	strong repulsion
+	+	strong attraction
-	+	weak repulsion
+	-	weak attraction

Text-based forces. The text similarity between an agent-image and an agent-text[3] depends on the number of keywords of each agent and the number of common keywords between the two agents: $S_{\text{i}} = (nb_{common})/(nb_{minimum})$, where nb_{common} is the number of common keywords between the two agents and $nb_{minimum}$ is the minimum of owned keywords between the two agents (i.e. if agent A has 3 keywords and agent B has 5 keywords, then $nb_{minimum} = 3$).

[3] Currently, the only agent-text is for text queries (comprising one of more keywords) given by the user, while agent-images can be queries or database images without distinction.

Text similarity can only be computed if both agents have keywords, as text query (agent-text) or as image annotations (agent-image). Like for image-based forces, the type and direction of the force between two agents follows Table 1 and the magnitude of the force is given by $F = C \cdot S_{\mathrm{i}}/D_{\mathrm{i}}$ (see above for details).

Agent global force. The final global force for an agent is simply the vectorial summation of all forces between that agent and its neighbors (Figure 2). The agent movement is induced by the final global force applied directly on itself added to an inertial force. The goal of the inertial force is to keep the results of previous computations. Various values have been experimented to determine the inertial force at time step t, the best one is to keep 80% of the final global force at time step $(t - 1)$.

2.3 Human Interaction Model

A motivation that leads the creation of this system is to provide easy and intuitive user interaction. For searching and retrieving, the user can give an image query-by-example just by clicking on the wanted image. The corresponding agent-image is then indicated as query. In the model, the query-agent is managed like all other agents except that it stops moving in the environment, letting all the other agent-images organizing themselves around it. This query can only be moved (dragged on the screen) by the user, allowing him to structure the screen space as he wants. For performing a multiple query search, the user just needs to select multiple images and organized them on the screen (Figure 3a). Through clicking on agent-queries, the user can indicate them as positive or negative. A negative query translate as an inversion of the concerned forces. Attraction forces become repulsion and repulsion forces become attraction. This behavior allows to easily perform relevance feedback, with positive and negative feedbacks (queries). Images similar to a negative query moves far from it. The user can deselect a query, i.e. let the agent returning to a normal moving behavior. Just by clicking on the agent-images the user can do different traditional tasks for browsing, querying and retrieving images, selecting positive/nagative feedbacks. The computation force model is the same for all these tasks, with the only two exceptions that a query-agent will not translate its forces into motion and that the forces can be inverted if the agent is indicated as a negative query.

Performing text queries is similar, except for the point that text-agents do not exist at beginning, so the user needs to create them. The text query allows the users to search images by text (one or several keywords), if images have a priori annotations. An interface allows the user to give one or several keywords that will be given that the newly created text-agent (Figure 3b). After that, the process is the same, this text-agent can be moved anywhere by the user, be positive or negative, and multiple text-agents can be created.

Without query, all agents move in the environment until they converge to a stable state. This stable state varies depending on the initial (random) positions of images. With one or several queries, the system will converge to the state desired by the user.

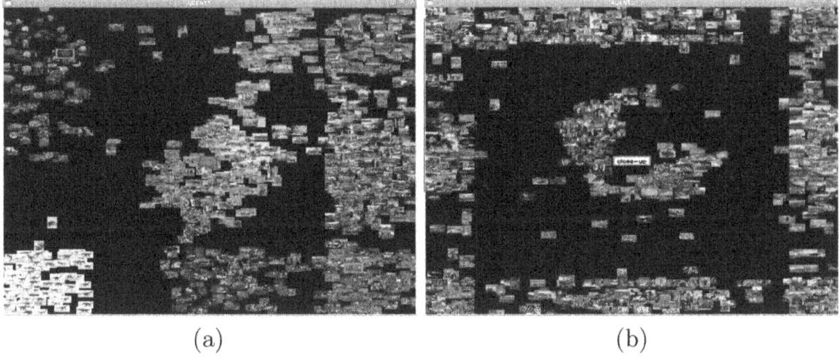

(a) (b)

Fig. 3. Different types of queries. (a) Multiples image queries, positive (2 green rectangles in opposite corners) and negative (red rectangle in the center). (b) Text query (in the center).

2.4 Working with Large Image Databases

A large image database may contain thousands or even millions of images. Computing all similarities for ranking all images can be very slow, and not very useful. In our model, visualization-centered, we are not interested to show all images and to compute all similarities, but only to show a limited number of images which are the most interesting for the user. We state the hypothesis that an user can see a maximum of 500 or 1000 images at the same time on the screen (even this number is still considered as high). If too much images are displayed on the screen, they are occluding each other and the user might loose the important ones (similar to his queries). Starting from this hypothesis, the system does not need to perform all computations, but only a limited set of them. Some adjustments are made to the model to take this into account and limit the number of computations.

The global environment is bigger than the part of the environment visible on the screen. A difference is made between the agents currently visible on the screen and the others. Precision and efficiency are needed for visible agents, but not for the others. Visible agents recompute often their forces with neighbors, while non-visible agents do this less often, from 15 to 20 times less than the others[4], the longer an agent stays in the non-visible space, the lower its priority. The neighbor random selection function only takes into account visible agents for the force computation. This allows agents to react mostly to visible images, without neglecting completely non-visible images. Forces applied to non-visible agents are bigger (the value of a *strong* force in the model described previously is bigger for them). Doing so, the positions of non-visible agents are not accurate, the

[4] In our implementation of pseudo-parallelism, this is done by creating two agent lists for the scheduler, one list of visible agents with high priority of execution, and one list of non-visible agents with a low priority. Agents can change from one list to the other depending on their current position.

important is to promote the movement between visible and non-visible sections of the environment as necessary. This dual representation reduces significantly the computation time.

3 Results

We are mainly interested in evaluating the feasibility of our multi-agent model. In order to do so, we use two image databases. The first is the Wang database [6] containing 1000 images separated in 10 classes (100 images / class). In our experiment, we use the class name has a textual annotation for each image (1 keyword/image). The second is the Corel30k database [2] containing 31,695 images divided into 320 classes (approximately 100 images / class). The images are also annotated using a total of 5587 words, with each image having from 1 to 5 keywords. In our experiment, we use the annotation keywords instead of the classes (more rich description), but we do not use the words having too few examples (<10 images/word), because these words are not generic enough to evaluate the result. For all images, we use the visual features color and texture. We compute the RGB color histogram compressed into 48 bins (16 per color) and the co-occurrence matrices with 4 features: energy, entropy, contrast and inverse difference moment. For each image, we use for our experiments a visual feature vector of 52 values and a textual vector of 1-5 keywords. For computing all results, we let the system converge completely (after 3000 time steps) to obtain stable results. The Euclidian distance is then used to rank all images with the query and produce the shown results.

Figure 4 shows the results of our system for both image databases using recall and precision curves, common evaluation tools used in information retrieval. These results are similar, close but not greater, than other more traditional CBIR systems [1,9]. These preliminary results validate the feasibility of our model as a potential new CBIR method and encourage us for pursuing to improve this model.

Our experimentations on giving different priorities for visible and non-visible agents (section 2.4) show no visible difference in terms of results, but they show a significant reduction (more than 3 times) of the computation time. Using a Dell Vostro 1310 (CPU Intel Core 2 Duo, 2 GHz, Ram 2GB) with Ubuntu 10.04 on the Corel30k database, and measuring after 3000 time steps (more than needed for full convergence of agents to a stable results), the system takes 5m35s without the priority mechanism and 1m46s with this mechanism implemented. This last time still seems very high, but one has to remind that we aim an interactive and visually intuitive system. Partial convergence[5] occurs before that time (around half) which allow the user to benefit from the results.

[5] Partial convergence in such a dynamic model means that agents are roughly at their final positions, but still oscillating and the exact distance and ranking with the query can still vary, but not too much.

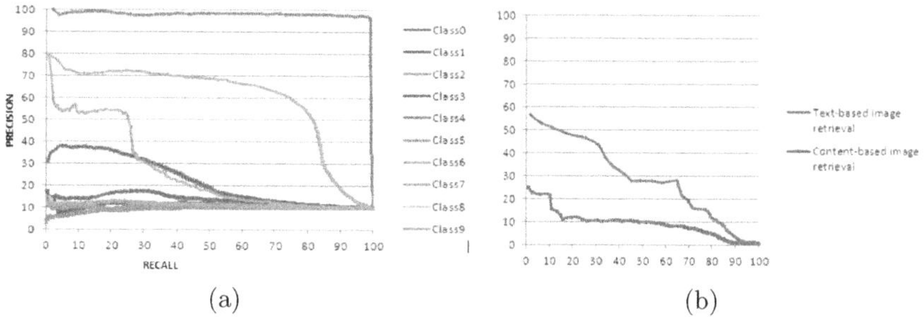

Fig. 4. Recall/precision results for our multi-agent model. (a) Results obtained for each individual class of the Wang database. (b) Global results for the Corel30k database, using only text-based or only content-based retrieval.

4 Conclusion and Future Work

In this paper we introduce a new approach for browsing and image retrieval based on a multi-agent model. It is a dynamic model where similarities between images are computed locally and over time. From a random placement of all images, convergence is obtained by the interactions between the agents and their neighbors. This model is not intended for real-time computation, but more for interactive and visually intuitive image retrieval. The main advantage of this model is its simplicity, a unique model can aggregate and unify several independent existing behaviors in CBIR, like single or multiple querying, text or image query, relevance feedback with positive and negative examples... Moreover, it simplifies the mapping of high-dimensional feature space into a 2D interface, with an auto-adaptive local model, and it allows easy human interactions, both for browsing and retrieval. The preliminary results validate the feasibility of this model, but it still needs to be improved.

More image features can be added, like in most CBIR methods, but our future work will concentrate on the force computation model, trying to simplify it even more for reducing computation time. Few experiments show us that an integer computed model can be almost as accurate as a floating model, due to local agent interactions and the incremental (converging over time) model. We aim in adding more heterogeneous features (like time stamps or image metadata) and still making it visually intuitive for the user. One potential application of this model is the supervised annotation of large image databases, where we need to minimize the number of user interaction and to maximize the annotated number of images.

References

1. Boucher, A., Dang, T.H., Le, T.L.: Classification vs recherche d'information: vers une caractérisation des bases d'images. 12èmes Rencontres de la Société Francophone de Classification (SFC), Montréal (Canada) (2005) (in French)

2. Carneiro, G., Chan, A.B., Moreno, P.J., Vasconcelos, N.: Supervised Learning of Semantic Classes for Image Annotation and Retrieval. IEEE Transactions on Pattern Analysis and Machine Intelligence 29(3), 394–410 (2007)
3. Forgáč, M.R.: Decreasing the Feature Space Dimension by Kohonen Self-Organizing Maps. In: 2nd Slovakian Hungarian Joint Symposium on Applied Machine Intelligence, Herľany, Slovakia (2004)
4. Hu, R., Ruger, S., Song, D., Liu, H., Huang, Z.: Dissimilarity mesures for content-based image retrieval. In: 2008 IEEE International Conference Multimedia and Expo (ICME), Hannover, Germany (2008)
5. Mangiameli, P., Chen, S.K., West, D.: A Comparison of SOM neural network and hierarchical clustering methods. European Journal of Operational Research 93(2), 402–417 (1996)
6. Li, J., Wang, J.Z.: Automatic linguistic indexing of pictures by a statistical modeling approach. IEEE Transactions on Pattern Analysis and Machine Intelligence 25(9), 1075–1088 (2003)
7. Moghaddam, B., Tian, Q., Lesh, N., Shen, C., Huang, T.S.: Visualization & User-Modeling for Browsing Personal Photo Libraries. International Journal of Computer Vision 56(1/2), 109–130 (2004)
8. Nguyen, N.V., Boucher, A., Ogier, J.M., Tabbone, S.: Region-Based Semi-automatic Annotation Using the Bag of Words Representation of the Keywords. In: 5th International Conference on Image and Graphics (ICIG), pp. 422–427 (2009)
9. Nguyen, N.V.: Keyword Visual Representation for Interactive Image Retrieval and Image Annotation. PhD thesis, University of La Rochelle (France) (2011) (in French)
10. Picard, D., Cord, M., Revel, A.: CBIR in distributed databases using a multi-agent system. In: IEEE International Conference on Image Processing, ICIP (2006)
11. Plant, W., Schaefer, G.: Visualising Image Database. In: IEEE International Work-Shop on Multimedia Signal Processing, pp. 1–6 (2009)
12. Renault, V.: Organisation de Sociétè d'Agents pour la Visualisation d'Informations Dynamiques. PhD thesis, University Paris 6, France (2001) (in French)
13. Rubner, Y., Guibas, L.J., Tomasi, C.: The earth movers distance, multi-dimensional scaling, and color-based image retrieval. In: APRA Image Understanding Workshop, pp. 661–668 (1997)
14. Rubner, Y., Tomasi, C., Guibas, L.J.: The Earth Mover's Distance as a Metric for Image Retrieval. International Journal of Computer Vision 40(2), 99–121 (2000)
15. Xiao, X., Dow, E.R., Eberhart, R., Miled, Z.B., Oppelt, R.J.: Gene Clustering Using Self-Organizing Maps and Particle Swarm Optimization. In: IEEE International Workshop on High Performance Computational Biology (2003)
16. Laaksonen, J., Koskela, M., Oja, E.: PicSOM – Self-organizing image retrieval with MPEG-7 content descriptors. IEEE Transactions on Neural Networks 13(4), 841–853 (2002)

An Efficient Ranging Protocol Using Multiple Packets for Asynchronous Real-Time Locating Systems

Seungryeol Go and Jong-Wha Chong

Department of Electronics and Computer Engineering
Hanyang University
Seoul, South Korea
seungryuls@nate.com, jchong@hanyang.ac.kr

Abstract. In this paper, we propose an efficient ranging protocol based on multiple packets for localization. In order to meet era of ubiquitous computing, there is a growing need for an efficient real-time location system. We assumed that multiple packets are used for ranging and proposed an efficient TWR protocol. For multiple ranging packets, the ranging performance is likely to be degraded by the frequency offset, since the total observation time increases. In the presence of a frequency offset, the ranging performance of the proposed method was analyzed and compared to those of other similar methods. Through simulations, theoretical analysis of the performance is verified.

Keywords: Ranging, Frequency Offset, SDS-TWR, RTLS.

1 Introduction

As the age of information and computing, there is a growing need for an efficient real-time location system (RTLS); for example, in place such as hospitals, shopping malls, welfare facilities, or houses, time can be saved by recognizing accurately where the nearest available products or person can be found. RTLS is gaining more interest for use in smart sensor network applications due to the increasing demand for location services in which global positioning system receivers are impractical. Accurate ranging performance between two nodes in RTLS is required in order to identify an exact location. However, due to restrictions on resource utilization, it is preferable to use multiple snapshots in place of a single snapshot for accuracy enhancement, as discussed in [1].

In this paper, we assume that there are multiple ranging packets between two nodes of a RTLS. For multiple ranging packets, the ranging performance is likely to be degraded by the frequency offset, since the total observation time increases. As a result of this offset induced error, a symmetric double-sided two-way ranging (SDS-TWR) protocol was proposed in order to remove the error. Since the SDS-TWR protocol is inefficient for multiple ranging packets, we propose a new ranging protocol for multiple ranging packets that maintains similar performance as the SDS-TWR protocol.

N.T. Nguyen et al. (Eds.): *Adv. Methods for Comput. Collective Intelligence*, SCI 457, pp. 127–134.
DOI: 10.1007/978-3-642-34300-1_12

2 Problem Formulation

This section focuses on the ranging protocol based on conventional single snapshots, and analyzes its features.

2.1 TWR Protocol

The TWR protocol [2] is the simplest of the ranging protocols; it is able to calculate the distance between two nodes using only two packets. TWR outline in Figure 1 shows the procedure of the conventional TWR protocol in a two-node scenario. As illustrated in the Figure, the propagation delay (PD) is calculated by

$$t_{p,TWR} = \frac{1}{2}\left(T_{roundA} - T_{replyB}\right) \tag{1}$$

where T_{roundA} denotes the round-trip time (RTT) at node A and T_{replyB} denotes the reply time of node B.

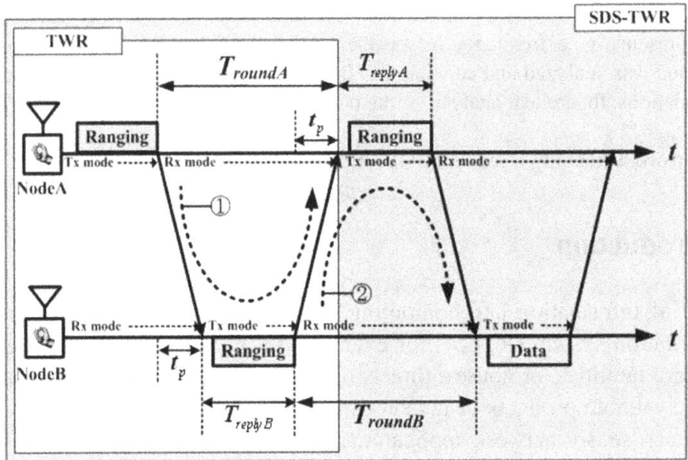

Fig. 1. Conventional TWR and SDS-TWR protocols

2.2 SDS-TWR Protocol

The SDS-TWR protocol is proposed to eliminate the offset-induced ranging error of the TWR protocol associated with T_{replyB}, as discussed in [1-3]. A minimum of four packets is required for the SDS-TWR protocol to work effectively, with the assumption that $T_{replyA}= T_{replyB}$ as determined in [3]. SDS-TWR outline in Figure 1 shows the entire process of the SDS-TWR protocol. It can be seen that two RTTs, denoted by T_{roundA} and T_{roundB}, are estimated using paths ① and ②, respectively.

$$t_{p,1} = \frac{1}{2}\left(T_{roundA} - T_{replyB}\right), \quad t_{p,2} = \frac{1}{2}\left(T_{roundB} - T_{replyA}\right) \tag{2}$$

In (2), $t_{p,1}$ and $t_{p,2}$ denote the PD through paths ① and ②, respectively. Thus, the averaged PD can be determined

$$t_{p,SDS-TWR} = \frac{1}{4}\left\{\left(T_{roundA} - T_{replyB}\right) + \left(T_{roundB} - T_{replyA}\right)\right\} \tag{3}$$

3 Proposed Protocol

The proposed TWR protocol is based on multiple ranging packets to increase packet efficiency while maintaining a ranging performance similar to that of the SDS-TWR. In Figure 2, N ranging packets are transmitted from node A to node B. Then, node B transmits a ranging packet involving ranging data to node A. The main concept behind the proposed TWR is to create N pairs of RTT paths, similar to the SDS-TWR protocol. However, unlike the SDS-TWR protocol, the RTT paths of the proposed TWR protocol are asymmetric to one another.

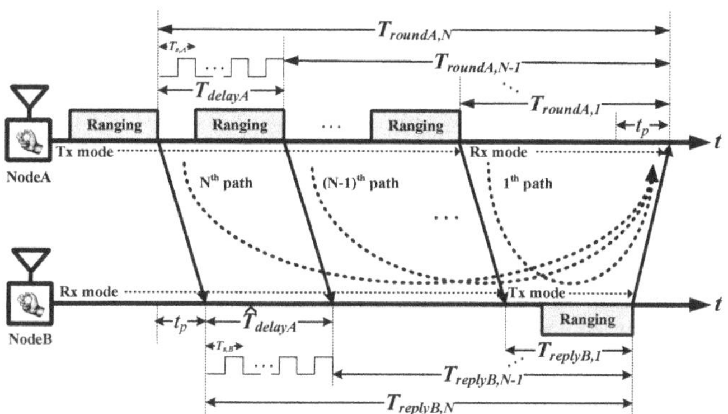

Fig. 2. Proposed TWR Protocol

In Figure 2, $T_{roundA,n}$ and $T_{replyB,n}$ represent the RTT and reply time of node B for the n-th ranging path, respectively. T_{delayA} represents the time difference between the two ranging packets, transmitted sequentially; this value varies according to the frequency offset of node A. Time-of-arrival (TOA) estimation, which is not addressed in this paper, is essentially performed for ranging [4]. Once the TOA estimation is performed on each of the received packets at node B, we can determine an estimate for T_{delayA} by the subtraction of the TOA estimation for the n-th ranging packet from that of the (n+1)-th ranging packet. The proposed TWR protocol uses $\hat{T}_{delayA}$ as the reply time of the ranging packet, i.e., $T_{replyB,1} = \hat{T}_{delayA}$, yielding the relationships for the i-th and j-th paths such that

$$
\begin{aligned}
T_{roundA,i} &= 2t_p + T_{replyB,i} = 2t_p + i\hat{T}_{delayA} \\
T_{roundA,j} &= 2t_p + T_{replyB,j} = 2t_p + j\hat{T}_{delayA}.
\end{aligned} \tag{4}
$$

As mentioned above, $\hat{T}_{delayA}$ is the value affected by the frequency offset. Thus, each of the RTTs in (4) produces a ranging error in a similar way as with the conventional TWR. However, the proposed TWR protocol makes use of the relationship between the i-th and j-th paths to remove the terms associated with $\hat{T}_{delayA}$ such that

$$
t_{i,j} = \frac{1}{2(j-i)} \times \left(j \cdot T_{roundA,i} - i \cdot T_{roundA,j} \right) \tag{5}
$$

For all ranging paths, $T_{roundA,n}$, where n=1,…,N, the averaged version of the proposed algorithm can be explained as follows:

```
Proposed TWR Algorithm
  For the received ranging packets, TOA estimation is
performed and then, T̂_delayA is obtained.
  Set T_replyB,1 = T̂_delayA and j = 1
  do : the estimation of t_roundA,1
  for i = 2 to N
    do : the estimation of t_roundA,i
    do : the estimation of t_i,j
  end for
```

4 Performance Analysis

This section analyzes the ranging performance for the TWR, SDS-TWR and the proposed TWR protocols.

4.1 TWR and SDS-TWR Protocols

For a given frequency offset at nodes A and B, denoted by ε_A and ε_B in parts per million, the parameters β_A and β_B are used to denote specific delays over a second due to frequency offset, respectively, where

$$
\beta_A = \frac{-\varepsilon_A}{1+\varepsilon_A} \text{ and } \beta_B = \frac{-\varepsilon_B}{1+\varepsilon_B} \tag{6}
$$

Thus, the ranging error induced by the frequency offset can be formulated using derivations defined in [3], such that

$$
e_{TWR} = t_p \beta_A + \frac{1}{2} \beta_A T_{replyB} \left(1+\beta_B\right). \tag{7}
$$

e_{TWR} is primarily determined by the values for T_{replyB} and β_A. However, the received signals are distorted due to additive white Gaussian noise (AWGN) leading to a TOA estimation error [4]. In consideration of the AWGN channel, the estimate for t_p can be represented by

$$\hat{t}_{p,SDS-TWR} \approx t_p + e_{TWR} + \frac{1}{2}(\delta_A + \delta_B) \tag{8}$$

where δ_A and δ_B are noise-induced TOA estimation errors at nodes A and B, respectively. Alternatively, since the SDS-TWR protocol obtains averages of PDs $t_{p,1}$ and $t_{p,2}$ as in (3), the ranging error of the SDS-TWR protocol, derived in [2], is defined as

$$e_{SDS-TWR} \approx \frac{1}{2}(\beta_A + \beta_B) \tag{9}$$

$$\hat{t}_{p,SDS-TWR} \approx t_p + e_{SDS-TWR} + \frac{1}{4}(\delta_{B,0} + 2\delta_A + \delta_{B,1}) \tag{10}$$

where $\delta_{B,0}$ and $\delta_{B,1}$ are the first and second noise-induced TOA estimation errors at node B, respectively.

4.2 Proposed TWR Protocol

In (4), $T_{roundA,i}$ and $T_{roundA,j}$ are expressed in terms of $\hat{T}_{delayA}$ and t_p. If there is the frequency offset, the time drift occurs due to $\hat{T}_{delayA}$ and t_p. However, by employing (5), the proposed TWR protocol removes the terms associated with $\hat{T}_{delayA}$ resulting in only t_p associated terms. Thus, the ranging error due to the frequency offset for the proposed TWR protocol is represented by

$$e_{prop} \approx t_p \beta_A \tag{11}$$

The ranging error of the proposed TWR protocol is similar to that of the SDS-TWR protocol in (9). For arbitrary i-th and j-th ranging packets, the estimate of t_p perturbed by the AWGN channel, based on the proposed-TWR protocol, is represented by

$$\hat{t}_{i,j} \approx t_p + e_{prop} + \frac{1}{4}(\delta_{B,0} + \delta_{B,1} + 2\delta_A) \tag{12}$$

5 Simulation Results

One thousand Monte-Carlo simulations were conducted in order to verify the effectiveness of the proposed TWR protocol. The baseband signal of the ranging packet was assumed to be chirp spread spectrum (CSS), adopted by international standards for localization such as IEEE 802.15.4a [5] and ISO/IEC 24730 [6], to model the TOA estimation process in the TWR protocol. Let us define the root mean squared error (RMSE) as $\sqrt{\frac{1}{1000}\sum_{n=1}^{1000}(\hat{t}_p - t_p)^2}$, where $\hat{T}_p$ represents the estimate of the PD for TWR, SDS-TWR and the proposed protocol, i.e., $\hat{T}_p = \hat{T}_{p,TWR}$ or $\hat{T}_p = \hat{T}_{p,SDS-TWR}$ or $\hat{T}_p = \hat{T}_{i,j}$.

5.1 Comparative Analysis

Figure 3 compares the results of analysis with the results of simulation in order to verify the protocol's performance.

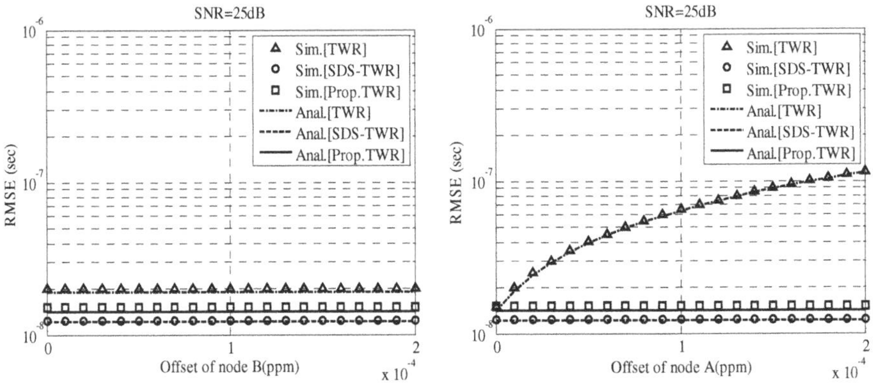

Fig. 3. Performance comparisons of relative offset: (a) fixed frequency offset of node A: 10ppm (b) fixed frequency offset of node B: 10ppm

In Figure 3-(a), ε_A is fixed and ε_B is varied. In Figure 3-(b), ε_B is fixed and ε_A is varied. When comparing the SDS-TWR and the proposed TWR protocols, the ranging error due to the frequency offset is considerably smaller compared to the TOA estimation error caused by the AWGN. The RMSEs of the SDS-TWR and the proposed TWR protocols appear to remain constant regardless of the frequency offset. For the TWR protocol, we use (7) to determine that the frequency offset of node B does not significantly affect the RMSE. Although the RMSE of the TWR protocol is fixed, irrespective of the frequency offset of B, as shown in Figure 3-(a), it changes according to the frequency offset of A, as shown in Figure 3-(b). This can be explained by (7) where the offset parameter of node A is multiplied by T_{replyB}.

5.2 Packet Efficiency of the Proposed TWR Protocol

As noted in the preceding section, TWR and SDS-TWR protocols require specific numbers of ranging packets: TWR requires two ranging packets and SDS-TWR requires four ranging packets. The proposed TWR protocol requires a minimum of three ranging packets. From the viewpoint of packet efficiency, defined as

$$P_{eff} = k / L, \tag{13}$$

where k is the ranging number and L is the total number of ranging packets used. Figure 4 compares the packet efficiency of the proposed protocol with that of the TWR and SDS-TWR protocols.

The packet efficiencies of TWR and SDS-TWR protocols are fixed at 0.5 and 0.25, respectively, regardless of L. However, the packet efficiency of the proposed TWR protocol approaches 1 as L increases. This is due to the recursive use of ranging packets, as illustrated in Figure 4.

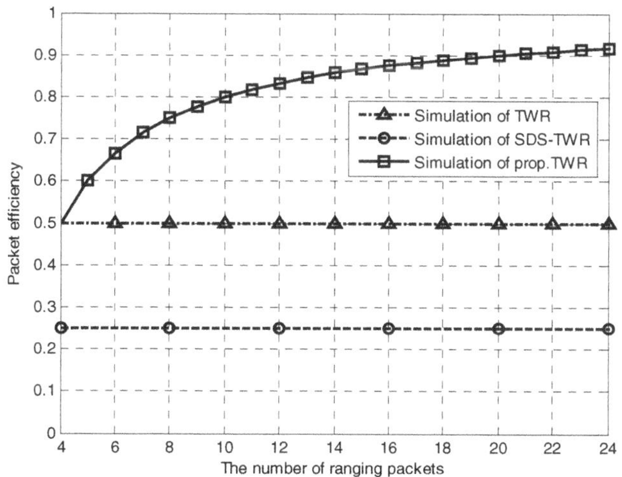

Fig. 4. Packet efficiency of the proposed TWR protocol

6 Conclusions

In this paper, we assumed that multiple packets are used for ranging and proposed an efficient TWR protocol. The proposed TWR protocol achieved the improved packet efficiency compared with conventional TWR and SDS-TWR protocols, while retaining a nearly identical tolerance to frequency offset as compared to the SDS-TWR protocol.

Acknowledgement. This work was supported by the IT R&D program of MKE/KEIT. [10035570, Development of self-powered smart sensor node platform for smart&green building

This work was supported by the Brain Korea 21 Project in 2012.

This work was sponsored by ETRI SW-SoC R&BD Center, Human Resource Development Project.

This research was supported by the MKE(The Ministry of Knowledge Economy), Korea, under the ITRC(Information Technology Research Center) support program supervised by the NIPA(National IT Industry Promotion Agency)(NIPA-2012-H0301-12-1011)

The IDEC provide research facilities for this study.

References

1. Oh, M.K., Kim, J.-Y., Lee, H.S.: Traffic-Reduced Precise Ranging Protocol for Asynchronous UWB Positioning Networks. IEEE Communications Letters 14(5) (May 2010)
2. Kim, H.: Double-sided Two-Way Ranging Algorithm to Reduce Ranging Time. IEEE Communications Letters 13(7) (July 2009)
3. Jiang, Y., Leung, V.C.M.: An asymmetric double sided two-way ranging for crystal offset. In: International Symposium on Signals, Systems and Electronics, ISSSE 2007, Montreal, pp. 525–528 (July 2007)
4. Oh, D.G., Yoon, S.H., Chong, J.W.: A Novel Time Delay Estimation Using Chirp Signals Robust to Sampling Frequency Offset for a Ranging System. IEEE Communications Letters (May 2010)
5. IEEE Standard 802.15.4a-2007, pp. 1–203 (August 2007)
6. ISO/IEC Standard 24730-5, pp. 1–72 (March 2010)

Agent-Based Model of Celtic Population Growth: NetLogo and Python

Kamila Olševičová, Richard Cimler, and Tomáš Machálek

University of Hradec Kralove, Faculty of Informatics and Management
Dept. of Information Technologies, Rokitanského 62,
Hradec Králové, 500 03, Czech Republic
{kamila.olsevicova,richard.cimler,tomas.machalek}@uhk.cz

Abstract. The agent-based model of Celtic population growth was developed using specific domain knowledge and general demographic assumptions about birth-rates and mortality. The model allows archaeologists to simulate the time series of available workforce and actual consumption of the population living in the given settlement agglomeration. Parameters of the NetLogo model were refined experimentally. The implementation in Python was created for validation and reporting. The simulated population is stable, with appropriate age distribution and growth rate. The model is used for further simulations of the settlement population dynamics and for testing hypotheses about the agricultural practices, trade and exchange etc. The final objective of our research project is to better understand the collapse of the Celtic society in Europe in the Late Iron Age.

Keywords: agent-based model, archaeology, NetLogo, population growth, Python, social simulation.

1 Introduction

Agent-based models and social simulations are being applied in archaeology for last two decades successfully, see e.g. Altaweel's realistic models of ancient Mesopotamian civilization [1] or investigation of cultural collapse of ancient Anasazi civilization [2, 6]. We intend to use the agent-based models for analyzing the conditions and circumstances of the development and collapse of the Celtic society in Europe in the Late Iron Age. Key aspects which formed the complexity of the society were settlement forms (oppida), demography, agricultural practices, producers/consumers ratio in society and the scale of work specialization, local and distant interactions (trade and exchange, monetary economy) and others [4].

The paper presents particular results related to the explanatory population growth modelling. The model was designed using domain knowledge (archaeological excavations of various oppida, regional landscape studies, demographical studies and assumptions, life-expectancy tables etc.) provided by experts. The population growth model is essential for further investigation of the carrying capacity of the settlement and available workforce.

N.T. Nguyen et al. (Eds.): *Adv. Methods for Comput. Collective Intelligence*, SCI 457, pp. 135–143.
DOI: 10.1007/978-3-642-34300-1_13

The organization of the rest of the paper is as follows. The formulation of the model and its description using the Overview-Design-Details (ODD) protocol [5] is provided in chapter 2, experiments are presented in chapter 3 and our further research directions are discussed in conclusion.

2 Formulation of the Model

Population growth is defined as the change in number of individuals over time. Natural growth is expressed as

$$\Delta P = Births - Deaths$$

It is assumed that all populations grow (or decline) exponentially (or logarithmically) unless affected by other forces. The simplest Malthusian growth model assumes the exponential growth is

$$P(t) = P_0 e^{rt}$$

where P_0 is the initial population, r is the growth rate and t is time. In case of the Celtic population living in the given settlement agglomeration, the initial number of inhabitants is said to be between 500 and 800 and the mild annual growth of population respect to high child mortality is estimated to 2%. The maximum number of inhabitants after 100-120 years is between 2000-5000. Emigration/immigration is not taken into account because massive emigration is one of possible causes of disappearance of settlement population therefore it will be investigated separately.

There are three types of peasant families: large size (approximately 20 members), medium size (approximately 10 members) and small size (4-6 members). A peasant family has got 2 adults, 1-3 children and 1 elder. Correspondingly the medium size family has got 4 adults and the large family has got 8 adults. In other words, the large family consists of approximately 7 infants (1 suckling, 3 toddlers, 3 up to 10 years), 3 older children (10-14 years), 2 young adults (15-19 years), 5 adults and 3 elderly. This information was applied in the definition of the initial age distribution of the population (see below). The same or similar age distribution of the final population is requested.

Social roles (basic family, nobility, servants, slaves etc.) as well as personal histories of individuals (marriages, children, siblings etc.) are ignored. The fertile age of women is 15-49 years and the fertility rate is 5.1. More than two children rarely survived infancy. This information was applied in the definition of the birth-rate procedure that operates with linear function (see below).

The probabilities to die are defined in abridged life-tables valid for the ancient Roman women population [7]. To obtain complete life-tables, the Elandt-Johnson estimation method [3] was applied. Experiments were performed using complete tables as well as abridged tables, and results are comparable (see below).

The model inputs are limited to:

- slider for setting the initial *number-of-years-to-be-simulated* (between 100-120),
- slider for setting the initial *number-of-inhabitants* (between 500 and 800),
- selection list of available *life-expectancy-tables*,
- selection list of available *initial-population-structure*,
- slider for setting the *birth-correction-parameter*.

There are two output variables:

- time series of *actual-consumption* of population (calories),
- time series of *actual-workforce* (number of men between 15-49 who are able to plough).

The daily calories requirements are defined (1360 for toddlers, 2000 for small children and elderly, 2500 for boys between 10 and 14 etc.) and it relates to the carrying capacity of the settlement.

According to the ODD protocol [5], the model is specified as follows:

- **Purpose:** The model is designed to explore questions about consumption and workforce of population of the Celtic settlement agglomeration.
- **Entities, state variables, scales:** Model has got one type of agents representing inhabitants. Each *inhabitant-agent* is characterized by *gender* (male, female), *age* (discrete value), *age-category* (sibling, toddler, child, older child, young adult, adult or elder). The simulations last 100-120 steps that correspond to years of the simulated time period. Auxiliary variables were added for monitoring characteristics of the whole population: *percent-of-sucklings*, *percent-of-todlers*, *percent-of-children*, *percent-of-older-children*, *percent-of-young-adults*, *percent-of-adults*, *percent-of-elders*. Summarizing variables *num-of-inhabitants*, *actual-workforce* and *actual-consumption* inform about the structure the population.
- **Process overview and scheduling:** Only one process is defined. On each time step, each *inhabitant-agent* executes *get-older* procedure. The procedure operates with selected life-table, either abridged or complete, with less or more optimistic life-expectancy (4 tables were available). The *inhabitant-agents* representing women between 15 and 49 years execute also the *birth-rate* procedure that operates with the *birth-correction-parameter*:

```
ask inhabitants with
[age>14 and age<50 and gender="female"]
    [ if random 100<((-9/17)*age+560/17)+birth-correction
      [  hatch 1 ]
    ]
```

Parameters of this linear function were refined experimentally with respect to domain knowledge. The function expresses the decreasing chance of pregnancy in relation to the age of woman *inhabitant-agent*.

- **Design concepts:** The basic principle addressed by the model is that the consumption and workforce of the whole population depend on properties of individuals and on specific rules defined by domain experts. Stochasticity (randomness) is used to represent the sources of variability such as particular age distribution in the initial population, life expectancy of individual *inhabitant-agents* etc. Other agent-based models' design concepts such as sensing, observation or emergence are not taken into account.
- **Initialization:** The initial population consist of seven age-groups (*sucklings, toddlers, children, older children, young adults, adults* and *elders*), the proportional distribution of each age-group was defined in NetLogo as follows. Parameters of the random function were identified experimentally:

```
let local-random random 35
if local-random<4 [report random 2]     ; sucklings
if local-random<6 [report 2+random 2]   ; todlers
if local-random<11 [report 4+random 6]  ; children
if local-random<15 [report 10+random 5] ; older-children
if local-random<18 [report 15+random 5] ; young-adults
if local-random<31 [report 20+random 30]; adults
if local-random<35 [report 49+random 20]; elder
```

- **Input data:** There are no input data during the run of simulation. All parameters are initiated at the beginning.
- **Submodels:** Different formulas can be defined inside *birth-rate* and *get-older* procedures. Together with parameters (*birth-correction*, *life-table*) it defines the submodels.

3 Experimental Results

See fig. 1 for graph and monitors from a single run of the simulation in NetLogo [8]. The exponential growth of population (all age groups respectively) is noticeable. Although the NetLogo interface is user friendly when defining the simulation inputs, for presentation of simulation outputs it is better to export data and process them in spreadsheet (we used MS Excel).

Our primary objective was to find a stable reliable model of the Celtic population growth, with matching initial and final age distributions. To achieve this, many different setups of the model were tested to refine the parameters in formulas. See fig. 2 and 3 for initial and final distributions for two sizes of the initial population (500 and 800 inhabitant-agents).

The effect of the *birth-correction-parameter* can be seen on fig 4 (the annual growth), fig. 5 (number of children per woman) and fig 6. (number of inhabitants). These characteristics differ in their sensitivity to changes of the parameter. The final population shown on fig. 6 is very sensitive to increasing value of the *birth-correction*. The results differ more than three times (1693 to 5476) in case of *birth-correction* values 6 and 14. The annual growth of the population and the number of children per woman are not so sensitive to the parameter.

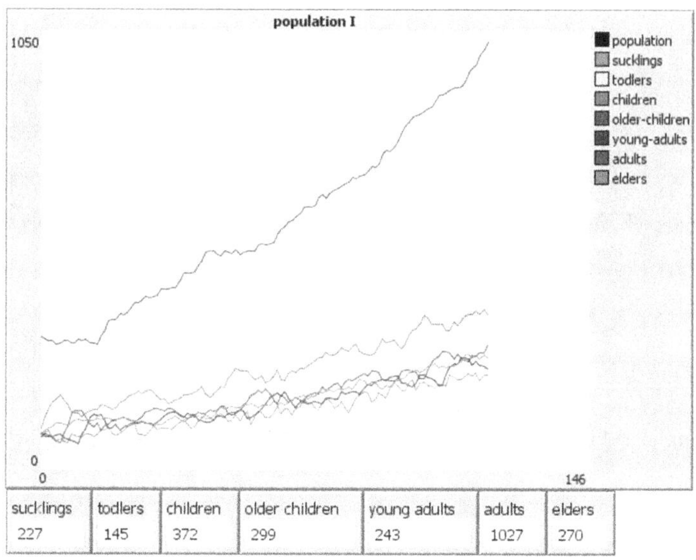

Fig. 1. Population growth during 120 years (NetLogo)

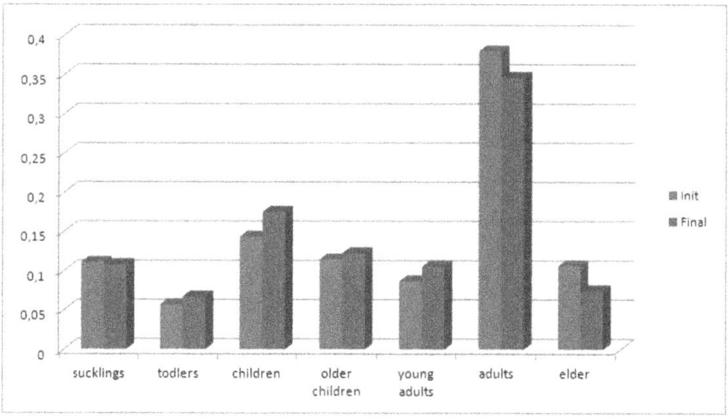

Fig. 2. Initial and final age distribution (initial *number-of-inhabitants* = 500) (NetLogo)

During the development of model, we have been also using Python language and environment along with packages SciPy (Scientific Python) and NumPy (Numeric Python) as a prototyping, validation and reporting tool. Due to Python's object oriented and modular nature, it was easy to decompose the model into smaller components which were developed and tested independently. Python has been also used to calculate extended versions of abridged life tables. Data were collected from 1000 repetitions of simulation.

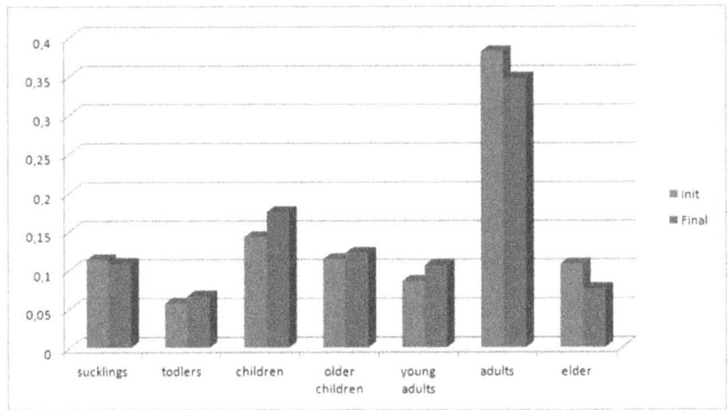

Fig. 3. Initial and final age distribution (initial *number-of-inhabitants* = 800) (NetLogo)

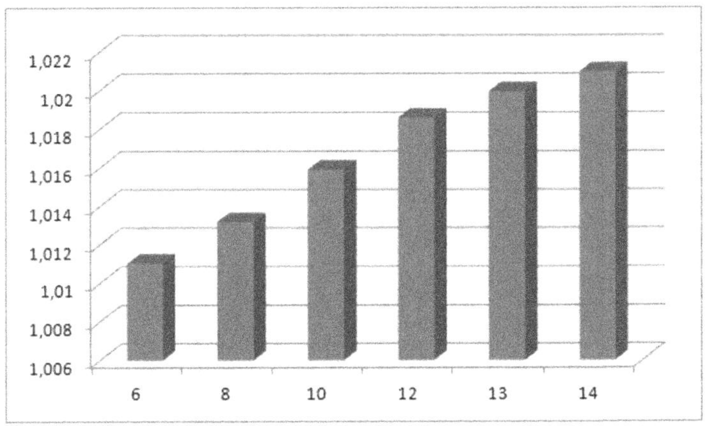

Fig. 4. Annual growth of the population related to *birth-correction* parameter (NetLogo)

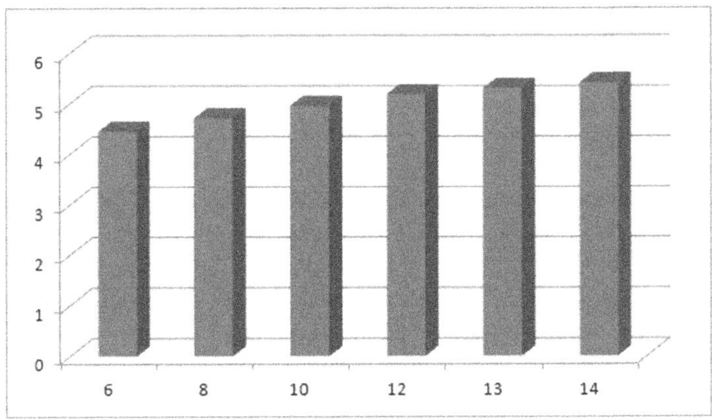

Fig. 5. Number of children per woman related to *birth-correction* parameter (NetLogo)

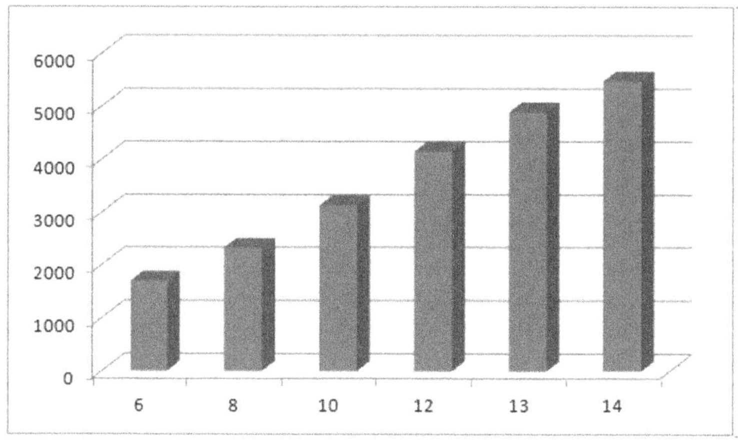

Fig. 6. Number of inhabitants related to *birth-correction* parameter (NetLogo)

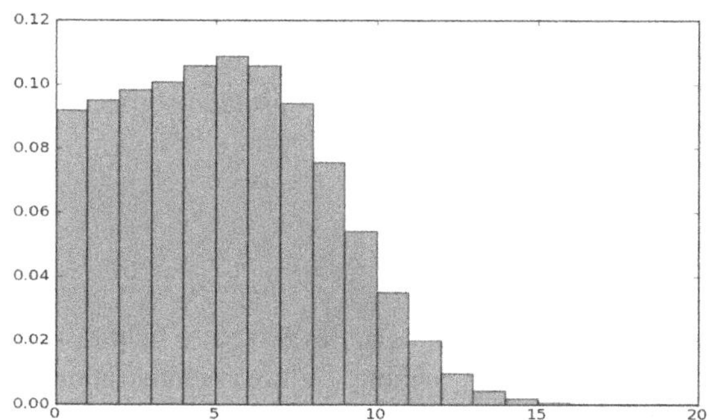

Fig. 7. Woman's probability of having specific number of children (Python)

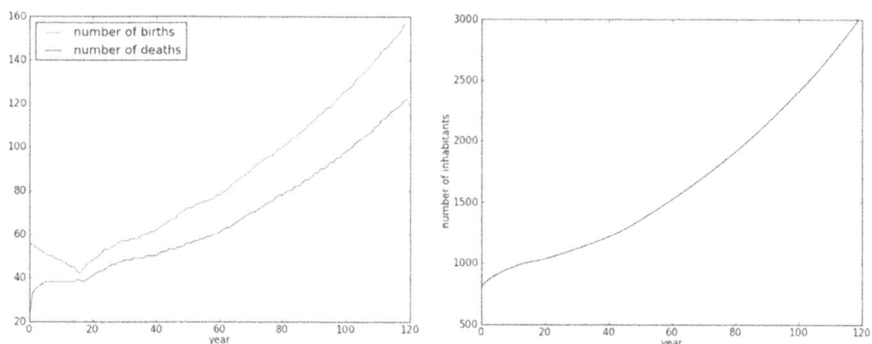

Fig. 8. Numbers of deaths and births per year and total number of inhabitants (Python)

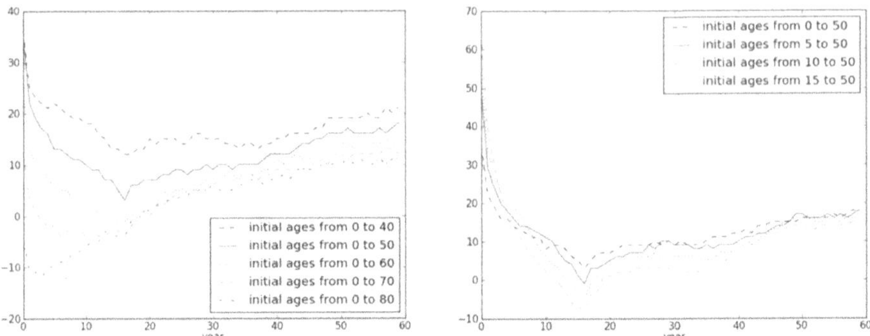

Fig. 9. Absolute annual population growths for miscellaneous initial configurations (Python)

See fig. 7 for Python model result for *birth-correction parameter* set to 6. Average number of children per fertile woman is 4.7 and median is 5, which is comparable with the NetLogo model.

See fig. 8 for Python model result for numbers of deaths and births per year (on the left) and total number of inhabitants (on the right), both for *birth-correction* = 6. The absolute annual growth of the population for miscellaneous configurations (all with uniform distribution of ages) is presented in fig. 9.

4 Conclusion

The agent-based models and simulations are expected to help to better understand the development and collapses of the ancient Celtic agglomerations in Bohemia. Our NetLogo model of the Celtic population growth is essential for further simulations of optional agricultural practices and economic mechanism. The model provides realistic time series of consumption, workforce and age distributions of population.

Acknowledgement. The support of the Czech Science Foundation under Grant P405/12/0926 *Social modelling as a tool for understanding Celtic society and cultural changes at the end of the Iron Age* grant is kindly acknowledged.

References

1. Altaweel, M.: Investigating agricultural sustainability and strategies in northern Mesopotamia: results produced using socio-ecological modeling approach. Journal of Archaeological Science 35, 821–835 (2008)
2. Axtell, R.L., et al.: Population Growth and Collapse in a Multi-Agent Model of the Kayenta Anasazi in Long House Valley. Proceedings of the National Academy of Sciences 99, 7275–7279 (2002)
3. Baili, P., Micheli, A., Montanari, A., Capocaccia, R.: Comparison of Four Methods for Estimating Complete Life Tables from Abridged Life Tables Using Mortality Data Supplied to EUROCARE-3. Mathematical Population Studies 12, 183–198 (2005)

4. Danielisová, A.: The oppidum of České Lhotice and its hinterland. Ph.D. dissertation (2008) (unpublished)
5. Grimm, V., et al.: The ODD protocol: A review and first update. Ecological Modelling 221, 2760–2768 (2010)
6. Janssen, M.: Understanding Artificial Anasazi. Journal of Artificial Societies and Social Simulation 12(4), 13 (2009)
7. Saller, R.: Patriarchy, Property and Death in the Roman Family, Cambridge (1994)
8. Wilensky, U.: NetLogo. Center for Connected Learning and Computer-Based Modeling, Northwestern University, Evanston, IL (1999), `http://ccl.northwestern.edu/netlogo/`

Part IV

Data Mining Methods and Applications

Ant Colony Inspired Clustering Based on the Distribution Function of the Similarity of Attributes

Arkadiusz Lewicki[1], Krzysztof Pancerz[1], and Ryszard Tadeusiewicz[2]

[1] University of Information Technology and Management in Rzeszów, Poland
{alewicki,kpancerz}@wsiz.rzeszow.pl
[2] AGH University of Science and Technology, Kraków, Poland
rtad@agh.edu.pl

Abstract. The paper presents results of research on the clustering problem on the basis of swarm intelligence using a new algorithm based on the normalized cumulative distribution function of attributes. In this approach, we assume that the analysis of likelihood of the occurrence of particular types of attributes and their values allows us to measure the similarity of the objects within a given category and the dissimilarity of the objects between categories. Therefore, on the basis of the complex data set of attributes of any type, we can successfully raise a lot of interesting information about these attributes without necessity of considering their real meaning. Our research shows that the algorithm inspired by the mechanisms observed in nature may return better results due to the modification of the neighborhood based on the similarity coefficient.

Keywords: ant colony clustering analysis, ant colony optimization, swarm intelligence, self-organization, unsupervised clustering, data mining, distribution function.

1 Introduction

Clustering became a very important field of research in data mining. This type of problems concerns identification of natural groups, where objects similar to each other are placed in one group while objects varying significantly are placed in different groups. It is interesting for researchers in the fields of statistics, machine learning, pattern recognition, knowledge acquisition and databases [6], [9], [18]. This issue includes the aspects of data processing, determining the similarity and dissimilarity of objects as well as methods for searching optimal solutions. In mathematical terms, clustering methods are based on searching for the partition minimizing a given criterion function. The various methods differ on reaching this requirement [3], [4], [14], [15]. Having defined the function of proximity specified for each pair of objects, we can use the algorithm to create groups.

N.T. Nguyen et al. (Eds.): *Adv. Methods for Comput. Collective Intelligence*, SCI 457, pp. 147–156.
DOI: 10.1007/978-3-642-34300-1_14

In terms of the overall mechanism, clustering methods can be divided into several categories such as flat clustering methods, hierarchical clustering methods, density-based clustering methods, graph-based clustering methods, nature inspired clustering methods or others which do not correspond to any of listed categories. Most of clustering algorithms requires determining a number of clusters (classes, categories) to which objects will be allocated. One of the more promising solutions, which does not require a predefined number of clusters, is the use of nature inspired clustering methods. This paper shows that the ant colony clustering algorithms are very efficient in this regard. We propose here to extend the idea originally published in [17].

2 The Approach

The main advantage of the base algorithm created previously is data prioritization. This feature eliminates the need to predetermine a number of clusters. However, sometimes we have a situation that the resulting number of clusters is less than prospective one for data characterized by a large number of small classes. Another disadvantage of this algorithm is a lack of stability of obtained solutions, especially in case of infrequent clusters, when there is a high probability of destroying them even if data have been identified correctly. Therefore, this algorithm was the basis for us to modify the strategy for searching the decision space. The basic idea was to obtain a clearer separation among individual groups of data, because if the distance between groups of objects is insufficient or there are connections among them, then the hierarchical algorithm interprets them as a single cluster. We can improve this situation by modifying the neighborhood function (similarity function) and the correlation of parameters. The analysis of probabilities of occurrences of individual attribute values, conditional probabilities and other statistical dependencies can deliver us a lot of interesting information about the attributes without covering the real meaning of each attribute.

The similarity of two objects x and y belonging to a finite set X can be estimated as the similarity between the distribution functions $g(y)$:

$$g(y) = P(y|x) = \frac{f_{xy}}{\sum\limits_{y' \in X} f_{xy'}} \tag{1}$$

where f_{xy} is a number of occurrences of x and y.

In order to achieve the smallest variance within a group of similar objects and the most intergroup variance, we can use similarity of distribution of attributes. In this case, we can calculate:

– The Jaccard coefficient:

$$d(o_i, o_j) = 1 - \frac{\sum\limits_{k=1}^{n} o_{ik} o_{jk}}{\sum\limits_{k=1}^{n} o_{ik}^2 + \sum\limits_{k=1}^{n} o_{jk}^2 - \sum\limits_{k=1}^{n} o_{ik} o_{jk}}. \tag{2}$$

In view of the fact that, in the modified algorithm, the degree of dissimilarity of objects is taken into account, the formula must be converted to the form:

$$d(o_i, o_j) = 1 - \frac{1}{2} \frac{\sum_{k=1}^{n} o_{ik} o_{jk}}{\sum_{k=1}^{n} o_{ik}^2 + \sum_{k=1}^{n} o_{jk}^2 - \sum_{k=1}^{n} o_{ik} o_{jk}}. \tag{3}$$

– The Dice coefficient:

$$d(o_i, o_j) = 2 \frac{\sum_{k=1}^{n} o_{ik} o_{jk}}{\sum_{k=1}^{n} o_{ik}^2 \sum_{k=1}^{n} o_{jk}^2}. \tag{4}$$

In both coefficients:

- o_i and o_j are objects represented by the corresponding n element vectors of features,
- o_{ik} and o_{jk} denote the k-th element (feature) of the vectors o_i and o_j, respectively, for $k = 1, \ldots, n$.

Referring to the basic ant clustering algorithm published previously [11], [12], [17], we can determine the neighborhood function as:

$$f(o_i) = \begin{cases} \delta \sum_{o_j \in L} \left(1 - \frac{d(o_i, o_j)}{\alpha}\right) & \text{if } f(o_i) > 0 \text{ and } \forall_{o_j \in L} \left(1 - \frac{d(o_i, o_j)}{\alpha}\right) > 0, \\ 0 & \text{otherwise,} \end{cases} \tag{5}$$

where δ is the factor defining the size of the tested neighborhood, α is a parameter scaling the dissimilarities within the neighborhood function, L represents a neighborhood area of searching.

Our new strategy has been tested on different quantity data sets. In this paper, data sets are named DS1, DS2 and DS3 and they contain 780, 200 and 65 objects, respectively. Their vectors of features are described by values of financial data. These data have been obtained by courtesy of Laboratory of Adapting Economic Innovations in Information Technology Facilities at the University of Information Technology and Management in Rzeszów, Poland.

In order to determine the influence of control parameters in the implemented algorithm, we have examined three sets of the most promising parameters verified for similar types of problems [1], [2], [3], [5], [7], [10], [13], [16], [17], [21], [22]. They were marked with P1, P2 and P3, respectively, and their values are presented in Table 1.

To tune these parameters, there has been applied a method of auto-tuning their values determined individually for each thread on the basis of a number f of failed operations of dropping objects during last steps.

In the clustering process, we expect a state enabling us to get the new partition internally homogeneous, but externally heterogeneous. If the partition is characterized by high quality solutions, then it means that the number of clusters

Table 1. The adopted values of the initialization parameters of the implemented algorithm

Parameter	Parameter values P1	Parameter values P2	Parameter values P3
Number of ants	150	200	500
Memory size	20	30	50
Step size	50	100	120
α	0.45	0.75	0.9
β	2	2	2

is fixed properly. Quality assessment methods of grouping data can be related to the verification of the proposed solution on the basis of internal criteria, external criteria, and end-user criteria. The first type of evaluation is associated with a specially selected measure, which examines properties of the solution. The most popular measure in this case is the Dunn index [8]. Evaluation results obtained with the use of an external criterion answer the question: "how well does the proposed solution give a solution created by man?". Here, we use the Rand index [19] and F-measure [20]. The above-mentioned factors for the creation of the qualitative characteristics of solutions have been also adopted by us for the proposed solution. The first factor (Formula 6) uses a minimum distance d_{min} between two objects from different groups as well as the maximum distance d_{max} between two objects within a given group. It verifies compact and well-separated clusters.

$$D_N = \frac{d_{min}}{d_{max}}. \tag{6}$$

The Rand index, measuring accuracy of two partitions X and Y by comparing each pair of test objects, can be presented as:

$$R = \frac{a+b}{a+b+c+d}, \tag{7}$$

where

- a is the number of pairs of objects belonging to the same set in X and the same set in Y,
- b is the number of pairs of objects belonging to different sets in X and different sets in Y,
- c is the number of pairs of objects belonging to the same set in X and different sets in Y,
- d is the number of pairs of objects belonging to different sets in X and the same set in Y.

The F-measure is based on two components:

- the precision coefficient p:

$$p = \frac{tp}{tp+fp} \tag{8}$$

where tp means a number of correct results (true positive) in a classification process and fp means unexpected results (false positive) in a classification process,

- the recall coefficient r:

$$r = \frac{tp}{tp + fn} \tag{9}$$

where fn means a number of missing results (false negative) in a classification process.

This means that, in a considered clustering problem, we can extract four types of decisions (on the basis of classification):

- a decision is true positive, if a tested pair of objects is in a pattern group and in a generated group,
- a decision is true negative, if a tested pair of objects is located together neither in a pattern group nor in a generated group,
- a decision is false negative, when we divide a pair of objects from the pattern collection to different groups,
- a decision is false positive, when we put a pair of objects in one group, which are not in any pattern group.

The F-measure, which is the harmonic mean of precision and recall, is calculated using the formula:

$$F(r,p) = \frac{2pr}{p + r}. \tag{10}$$

A special case of the general F-measure is a situation where we use a non-negative real value β:

$$F_\beta(r,p) = \frac{(\beta^2 + 1)pr}{\beta^2 p + r}. \tag{11}$$

The proposed algorithm based on the presented concept can be written as pseudo-instructions (see Algorithm 1).

The presented idea should guarantee that the operations of raising and dropping objects under consideration will be deterministic for a very little value of the density function in the first case and a very high value in the second case. It is connected with the process accelerating the formation of clusters, but only in areas which consist of a large number of objects with similar attribute vectors.

The proposed approach has been compared to other solutions known in this area, i.e., the k-means algorithm as well as the ATTA algorithm created by Handl and Knowles [10], [11], [12], [16].

3 Results

Test results have been obtained by 50 runs. Multiple empirical verification of the quality of the results on a sample iris data set showed that the best set of parameters implemented in the algorithm is the set P2 (presented in Table 1). The values are in accordance with those suggested by the authors of the ATTA

Algorithm 1. Ant Colony Clustering Algorithm based on the normalized distribution function

```
for each object o_i do
|  Place o_i randomly on the grid;
end
for each agent (ant) a_i do
|  Select randomly the object o_i;
|  Pick up the object o_i by a_i;
|  Put a_i in a random place on the grid;
end
for t = 1 to tmax do
|  Select randomly the agent a_i;
|  Move the agent a_i;
|  Calculate the probability for object picking;
|  o=object carried by the agent a_i;
|  Calculate the probability for object dropping;
|  dropped=try to drop the object o;
|  if dropped = false then
|  |  raised = false;
|  |  while raised = false do
|  |  |  o_i=select randomly one of the free objects;
|  |  |  raised=try to pick up the object o_i;
|  |  end
|  end
end
Print object positions;
```

algorithm. The results of assessment of the proposed algorithm with the Jaccard coefficient and with the Dice coefficient as well as the ATTA algorithm and the k-means algorithm using three indexes for examined data sets are presented in Table 2.

The results show that the new approach with the Jaccard coefficient works better for smaller data sets. For this type of collection, the algorithm is the best possible one. For larger collections, the k-means algorithm works better. This situation concerned a case where data sets consisted of 200 and 780 objects, respectively. For the largest sets, the algorithm with the Dice coefficient achieved better results than the ATTA algorithm and the algorithm with the Jaccard coefficient.

Studies of proposed modifications in relation to the standard version of the ant clustering algorithm showed the impact of the proposed solution on the linear correlation of the main parameters of the algorithm. The obtained results are included in Table 3. Analysis of these values indicates that the adopted approach allowed reduction of used factors in the proposed algorithm in relation to the ATTA algorithm. In this case, the range was from 0.10 to 0.13. The ATTA

Table 2. Evaluation of the quality of the new approach in comparison to the ATTA algorithm and the k-means algorithm

Data set	Index	New approach with Jaccard coefficient	New approach with Dice coefficient	ATTA algorithm	k-means algorithm
DS1	Dunn index	0.954	0.965	0.886	**0.986**
DS1	Rand	0.903	0.963	0.857	**0.978**
DS1	F-measure	0.846	0.871	0.865	**0.892**
DS2	Dunn index	0.708	0.598	0.694	**0.962**
DS2	Rand	0.552	0.463	0.504	**0.564**
DS2	F-measure	0.574	0.481	0.473	**0.678**
DS3	Dunn index	**1.391**	0.917	1.227	0.770
DS3	Rand	**0.912**	0.743	0.836	0.862
DS3	F-measure	**0.782**	0.757	0.764	0.747

Table 3. The impact of the proposed solution on the linear correlation of the main parameters of the algorithm

Algorithm	Data set	Correlation coefficient
New approach with Jaccard coefficient	DS1	0.132473
New approach with Jaccard coefficient	DS2	0.114033
New approach with Jaccard coefficient	DS3	0.124522
New approach with Dice coefficient	DS1	0.132473
New approach with Dice coefficient	DS2	0.106767
New approach with Dice coefficient	DS3	0.104678
ATTA	DS1	0.624273
ATTA	DS2	0.565561
ATTA	DS3	0.498221

algorithm had the range from 0.49 to 0.62. Therefore, we can conclude that the algorithm with the Jaccard coefficient and the algorithm with the Dice coefficient are suitable for the use in case where the additional knowledge about groups of objects contained in the data set is not necessary.

The best results for most collections have been achieved by the k-means algorithm. However, in case of the proposed approach, a number of groups was unknown. Therefore, this property is an important advantage of the proposed approach.

4 Conclusions

An attempt to find the proper partition of any set of objects described by vectors of features (attributes) is one of the most difficult and very complex tasks. In fact, data clustering is not often associated only with a lack of information on a number of output classes, but also with the necessity of interpretation and standardization of input data. Existing and implemented solutions, both deterministic and non-deterministic, are effective for small spaces. Therefore, there has been proposed in the paper a new approach to the clustering problem, which is based on the verified mechanism of swarm intelligence, i.e., collective activity of agents (ants). They carry out a searching process of available solutions in the field with the specified heuristic rules for determining a function of similarity of adjacent objects. It has a significant impact on the probability of picking up or dropping the object. Previously implemented and verified heuristic algorithms take into account the applicable measure of a distance between the examined objects in a space of solutions and the starting point is the most common, i.e., designing a matrix using the Minkowski metric tensor for this purpose. Meanwhile, as our experience demonstrates, in many cases, the better approach is the modified hierarchical ant colony clustering algorithm taking into account a function of similarity of attributes using the distribution function based on the Jaccard coefficient or the Dice coefficient. The obtained results confirmed its greater usefulness in case of vectors being the standardized values because of the dominance of the attributes associated with large values over those associated with smaller values. In this case, the proposed modification of the heuristic approach delivers us a faster algorithm, which requires less iterations to obtain the same result and the solution that rarely stops at the local optimums. Such a solution will be appropriate in case of non-structured data representing a significant problem with mapping images.

The proposed approach can be successfully implemented in case of data sets with quantitative features. However, in practice, there may be a situation where the test data will include both quantitative and qualitative features. In such a situation, there should be considered separate grouping of objects for each criterion using a proper measure of similarity. Another issue is consideration of data containing partial and incomplete descriptions. Therefore, such cases will be the next step of our research which will allow us to gain the knowledge on the possibilities of using the current direction and its modifications depending on the different types of analyzed data.

Acknowledgments. This paper has been partially supported by the grant No. N N519 654540 from the National Science Centre in Poland.

References

1. Abbass, H., Hoai, N., McKay, R.: AntTAG: A new method to compose computer programs using colonies of ants. In: Proceedings of the IEEE Congress on Evolutionary Computation, Honolulu (2002)
2. Azzag, H., Monmarche, N., Slimane, M., Venturini, G.: AntTree: a new model for clustering with artificial ants. In: Proceedings of the 2003 Congress on Evolutionary Computation, Beijing, China, pp. 2642–2647 (2003)

3. Berkhin, P.: Survey of clustering data mining techniques. Tech. rep. Accrue Software, Inc., San Jose, California (2002)
4. Bin, W., Zhongzi, S.: A clustering algorithm based on swarm intelligence. In: Proceedings of 2001 International Conferences on Info-tech and Info-net, Beijing, China, pp. 58–66 (2001)
5. Boryczka, U.: Ant clustering algorithm. In: Proceedings of the Conference on Intelligent Information Systems, Zakopane, Poland, pp. 377–386 (2008)
6. Deneubourg, J., Goss, S., Franks, N., Sendova-Franks, A., Detrain, C., Chrétien, L.: The dynamics of collective sorting: Robot-like ants and ant-like robots. In: Proceedings of the First International Conference on Simulation of Adaptive Behaviour: From Animals to Animats 1, pp. 356–365. MIT Press, Cambridge (1991)
7. Dorigo, M., Di Caro, G., Gambardella, L.M.: Ant algorithms for discrete optimization. Artificial Life 5(2), 137–172 (1999)
8. Dunn, J.: A fuzzy relative of the ISODATA process and its use in detecting compact well-separated clusters. Journal of Cybernetics 3(3), 32–57 (1973)
9. Han, Y., Shi, P.: An improved ant colony algorithm for fuzzy clustering in image segmentation. Neurocomputing 70(4-6), 665–671 (2007)
10. Handl, J., Knowles, J., Dorigo, M.: Ant-based clustering and topographic mapping. Artificial Life 12(1), 35–62 (2006)
11. Handl, J., Knowles, J., Dorigo, M.: Ant-based clustering: a comparative study of its relative performance with respect to k-means, average link and 1d-som. Tech. rep., IRIDIA (2003)
12. Handl, J., Knowles, J., Dorigo, M.: Strategies for the Increased Robustness of Ant-based Clustering. In: Di Marzo Serugendo, G., Karageorgos, A., Rana, O.F., Zambonelli, F. (eds.) ESOA 2003. LNCS (LNAI), vol. 2977, pp. 90–104. Springer, Heidelberg (2004)
13. Lewicki, A.: Generalized non-extensive thermodynamics to the ant colony system. In: Świątek, J., Borzemski, L., Grzech, A., Wilimowska, Z. (eds.) Information Systems Architecture and Technology: System Analysis Approach to the Design, Control and Decision Support, Wroclaw (2010)
14. Lewicki, A.: Non-euclidean metric in multi-objective ant colony optimization algorithms. In: Świątek, J., Borzemski, L., Grzech, A., Wilimowska, Z. (eds.) Information Systems Architecture and Technology: System Analysis Approach to the Design, Control and Decision Support, Wroclaw (2010)
15. Lewicki, A., Tadeusiewicz, R.: The recruitment and selection of staff problem with an ant colony system. In: Proceedings of the 3rd International Conference on Human System Interaction, Rzeszów, Poland, pp. 770–774 (2010)
16. Lewicki, A., Tadeusiewicz, R.: An Autocatalytic Emergence Swarm Algorithm in the Decision-Making Task of Managing the Process of Creation of Intellectual Capital. In: Hippe, Z.S., Kulikowski, J.L., Mroczek, T. (eds.) Human – Computer Systems Interaction, Part I. AISC, vol. 98, pp. 271–285. Springer, Heidelberg (2012)
17. Lewicki, A., Pancerz, K., Tadeusiewicz, R.: The Use of Strategies of Normalized Correlation in the Ant-Based Clustering Algorithm. In: Panigrahi, B.K., Suganthan, P.N., Das, S., Satapathy, S.C. (eds.) SEMCCO 2011, Part I. LNCS, vol. 7076, pp. 637–644. Springer, Heidelberg (2011)
18. Ouadfel, S., Batouche, M.: An efficient ant algorithm for swarm-based image clustering. Journal of Computer Science 3(3), 162–167 (2007)
19. Rand, W.: Objective criteria for the evaluation of clustering methods. Journal of the American Statistical Association 66(336), 846–850 (1971)

20. van Rijsbergen, C.J.: Information Retrieval. Butterworth, London (1979)
21. Scholes, S., Wilson, M., Sendova-Franks, A.B., Melhuish, C.: Comparisons in evolution and engineering: The collective intelligence of sorting. Adaptive Behavior - Animals, Animats, Software Agents, Robots, Adaptive Systems 12(3-4), 147–159 (2004)
22. Vizine, A., de Castro, L., Hruschka, E., Gudwin, R.: Towards improving clustering ants: An adaptive ant clustering algorithm. Informatica 29(2), 143–154 (2005)

Maintenance of IT-Tree for Transactions Deletion

Thien-Phuong Le[1], Bay Vo[2], Tzung-Pei Hong[3], Bac Le[4], and Jason J. Jung[5]

[1] Faculty of Technology, Pacific Ocean University, NhaTrang City, Vietnam
phuonglt@pou.edu.vn
[2] Information Technology College, Ho Chi Minh, Vietnam
vdbay@itc.edu.vn
[3] Department of CSIE, National University of Kaohsiung, Kaohsing City, Taiwan, R.O.C.
tphong@nuk.edu.tw
[4] Department of Computer Science, University of Science, Ho Chi Minh, Vietnam
lhbac@fit.hcmus.edu.vn
[5] Department of Computer Engineering, Yeungnam University, Republic of Korea
j2jung@intelligent.pe.kr

Abstract. Zaki et al. designed a mining algorithm based on the IT-tree structure, which traverses an IT-tree in depth-first order, generates itemsets by using the concept of equivalence classes, and rapidly computes the support of itemsets using tidset intersections. However, the transactions need to be processed batch-wise. In real-world applications, transactions are commonly changed. In this paper, we propose an algorithm for the management of the deleted transactions based on the IT-tree structure and pre-large concepts. Experimental results show that the proposed algorithm has a good performance.

Keywords: Data mining, frequent itemset, incremental mining, pre-large itemsets, equivalence class, IT-tree.

1 Introduction

Many algorithms for mining association rules from transaction databases have been proposed. Most of them are based on the Apriori algorithm [1], which generates-and-tests candidate itemsets at each level. However, the databases may need to be iteratively scanned, which leads to a high computational cost. In addition to Apriori-based algorithms, some new approaches have been proposed. Han et al. [3] proposed the frequent-pattern tree (FP-tree) structure for efficiently mining the frequent itemsets (FIs) without candidate generation. Zaki et al. [13], [14] proposed the itemset-tidset tree (IT-tree) structure and designed an algorithm for the fast mining of all FIs. Their approach uses the depth-first search technique for the generation of itemsets and the tidset for the fast computation of the supports of itemsets. They showed this algorithm outperforms the original Apriori algorithm.

All Apriori, IT-tree-based and FP-tree-based algorithms use batch mining. They must thus process all the transactions batch-wise. In real-world applications, transactions are commonly inserted and deleted. Therefore, designing an efficient

N.T. Nguyen et al. (Eds.): *Adv. Methods for Comput. Collective Intelligence*, SCI 457, pp. 157–166.
DOI: 10.1007/978-3-642-34300-1_15

algorithm for the maintenance of association rules as a database grows is critically important. The first incremental mining algorithm was the Fast-UPdate algorithm (called FUP) [2]. Like Apriori-based algorithms, FUP generates candidates and repeatedly scans the database, although it avoids a lot of unnecessary checking. Hong et al. [4] proposed the pre-large itemsets to further reduce the need for rescanning the original database. Besides, Le et al. also [9] proposed the Pre-FUT algorithm (FUP algorithm using a Trie data structure and pre-large itemsets) to efficiently handle newly inserted transactions based on pre-large concept. This algorithm does not require the original database to be rescanned until a number of new transactions has been inserted. The maintenance cost is thus reduced with the prelarge-itemset concept.

Some different algorithms were proposed as well [6], [11]. These methods are mainly based on the FP-tree structure. Hong et al. modified the FP-tree structure and designed a fast updated frequent-pattern tree (FUFP-tree) [5] for handling newly inserted transactions. Besides, Lin et al. [7] proposed the structure of pre-large trees to update and re-build the FUFP-tree based on the concept of pre-large itemsets when a number of old transactions have been deleted. In addition to the FP-tree, the IT-tree structure is also an efficient data structure for mining FIs. Recently, Le et al. [10] also proposed the Pre-FUIT algorithm (Fast-Update algorithm based on the IT-tree structure and the pre-large concept) for the efficient handling of the insertion of transactions.

In this paper, we propose an algorithm for handling of the deletion of transactions based on the concept of pre-large itemsets and the IT-tree structure. With the IT-tree structure, the support values of the candidate itemsets can be computed rapidly using tidset intersections.

2 Review of Related Works

In this section, studies related to the IT-tree structure and pre-large-itemset algorithm are briefly reviewed.

2.1 IT-Tree Structure

Zaki et al. [13], [14] proposed the IT-tree structure and designed the ECLAT (Equivalence CLAss Transformation) algorithm for mining FIs. The algorithm, which is based on equivalence classes, scans the database only once. The IT-tree structure is a prefix tree; each node in an IT-tree is represented by an itemset-tidset pair $X \times t(X)$, where X is an item or itemset and $t(X)$ is the list of transactions with X. All the children of a given node with X belong to its equivalence class as they share the same prefix X. The main difference between the Apriori and ECLAT algorithms lies in how they traverse the prefix tree and how they determine the support of an itemset. The former traverses the prefix tree in breadth-first order. It first checks itemsets of size 1, and then it checks itemsets of size 2, and so on. It determines the support of an itemset by scanning the whole database. The ECLAT algorithm, instead traverses the prefix tree in depth-first order. That is, it extends an itemset prefix until it reaches the boundary between the frequent and infrequent itemsets. Then, it backtracks to search on the next prefix (in the fixed order of items). It determines the support of a new itemset by computing the intersection between tidsets of two itemsets. More details can be found in [13], [14].

2.2 Pre-large-itemset Algorithm

The pre-large concept was proposed by Hong et al. [4]. It uses two thresholds, namely the upper threshold and the lower threshold, to set the pre-large itemsets. Hong et al. also proposed the pre-large-itemset algorithm. It is based on a safety number f of inserted transactions to reduce the need for rescanning the original databases for the efficient maintenance of the frequent itemsets.

Considering an original database and some transactions that are to be deleted by the two support thresholds. Lin et al. [8] proposed the following formula for computing f in the case of deleted transactions.

$$f = \left\lfloor \frac{(S_u - S_l)\,d}{S_u} \right\rfloor \quad (1)$$

where S_u is the upper threshold, S_l is the lower threshold, and d is the number of original transactions. When the number of deleted transactions is smaller than f, the algorithm did not need to rescan the original database.

3 Proposed Algorithm

The notation used in the proposed maintenance algorithm are given below.

- D the original database
- T the set of deleted transactions
- U the entire updated database, i.e., $D - T$
- d the number of transactions in D
- t the number of transactions in T
- S_l the lower support threshold for pre-large itemsets
- S_u the upper support threshold for large itemsets, $S_u > S_l$
- X an item
- Tr an IT-tree stores the set of pre-large, and large itemsets from D
- T?(X) tidset of item X in T
- I_T the set of items in T
- $sup^{Tr}(X)$ the support of X in Tr
- $[P]$ a equivalence class of node P in Tr
- $[\emptyset]$ the equivalence class of the root in Tr, which is $\emptyset$

3.1 Proposed Algorithm

The main idea of the proposed algorithm is to traverse an IT-tree in depth-first order to update the supports and tidsets of the itemsets stored in the IT-tree. The algorithm uses the downward-closure property [1] to prune unpromising itemsets while traversing the IT-tree. The details of the algorithm are given below. The rescanned itemsets in the IT-tree are the itemsets that are neither pre-large nor large in the original database but small in the deleted transactions. When the number of deleted transactions exceeds the safety number *f*, the algorithm rescans the original database to handle the rescanned itemsets.

INPUT: A lower support threshold S_l, an upper threshold S_u, an IT-tree *Tr* that stores large and pre-large itemsets derived from the original database *D* consisting of (*d-c*) transactions (*c* is the number of deleted transactions from the last rescan to now), and a set of *t* deleted transactions.

OUTPUT: A new IT-tree *Tr* that stores large and pre-large itemsets, and rescanned itemsets from *U*.

BEGIN

$f = \left\lfloor \frac{(S_u - S_l)d}{S_u} \right\rfloor$

if $(t + c > f)$\{
    rescan the original database to determine whether the rescanned itemsets *Tr* in are large or pre-large \}
for each item $X \in I_T$\{ //I_T is the 1-itemsets in *T*
    if (*X* does not exist in *Tr*)\{
        if $\left(\frac{|T^-(X)|}{t} < S_l\right)$\{
            add *X* to *Tr* and mark *X* as a rescanned itemset
            $sup^{Tr}(X) = |T^-(X)|$ \}//end if
\}//end for
for each item $X \in Tr$\{
    if $(X \in I_T)$\{
        if $\left(\frac{sup^{Tr}(X) - |T^-(X)|}{d - t - c} \geq S_l\right)$
            $sup^{Tr}(X) = sup^{Tr}(X) - |T^-(X)|$
        else remove *X* from *Tr*
    \}**else if** $\left(\frac{|sup^{Tr}(X)|}{d - t - c} \geq S_l\right)$
            $T^-(X) = \emptyset$
        else remove *X* from *Tr*
\}//end for
ENUMERATE_FIs (*Tr*, [∅])
if $(t + c > f)$ \{
    $d = d - t - c$
    $c = 0$ \}
else $c = c + t$

END

```
PROCEDURE ENUMERATE_FIs(Tr, [P])
BEGIN PROCEDURE
24. for each itemset X_i ∈ [P] and X_i is not a rescanned itemset{
25.     for each itemset X_j ∈ [P]  with j>i  and X_j is not a rescanned itemset {
26.         Z = X_i ∪ X_j
27.         T⁻(Z) = T⁻(X_i) ∩ T⁻(X_j)
28.         if (Z exists in Tr)
29.         if ( (sup^Tr(Z) - |T⁻(Z)|) / (d - t - c) ≥ S_l ) {
30.             sup^Tr(Z) =  sup^Tr(Z) - |T⁻(Z)|
31.             [P_i] = [P_i] ∪ {Z×T⁻(Z) }
32.         else remove Z from Tr
33.     else if ( |T⁻(Z)| / t < S_l ) {
34.             Add Z to Tr and mark Z as a rescanned itemset
35.             sup^Tr(Z) = |T⁻(Z)| }
36.     }// end for
37. ENUMERATE_FIs(Tr, [P_i])
38. }//end for
END PROCEDURE
```

Lines 4 to 9 process the 1-itemsets which exist in I_T but are not retained in *Tr* and lines 10 to 18 process the 1-itemsets which are pre-large or large in the original database and are retained in I_T. After line 18, the total count and the tidset of each 1-itemset in *Tr* have been updated and the rescanned itemsets have been found.

The procedure **ENUMERATE_FIs(*Tr*, [*P*])** will update the support of the itemsets stored in IT-tree. This task is recursive. The final results are the new IT-tree that stores all the large and pre-large itemsets.

3.2 An Example

In this section, an example is given to illustrate the proposed incremental data mining algorithm. Assume that the initial data set includes the 10 transactions shown in Table 1. For S_l = 30% and S_u = 50%, the sets of large and pre-large itemsets for the given data are shown in Table 2. An IT-tree *Tr* that stores the set of pre-large and large itemsets in the original database is shown in Fig. 1.

Table 1. An example of original database

Original database	
TID	***Items***
1	ACE
2	ABDE
3	BCDE
4	ACE
5	ACE
6	ABC
7	BDE
8	ABCE
9	ABCD
10	CEF

Table 2. Large and pre-large itemsets that are derived from the original database

Large itemsets					
1 item	*Count*	*2 items*	*Count*	*3 items*	*Count*
A	7	AC	6		
B	6	AE	5		
C	8	CE	6		
E	8				
Pre-large itemsets					
1 item	*Count*	*2 items*	*Count*	*3 items*	*Count*
D	4	AB	4	ABC	3
		BC	4	ACE	4
		BD	4	BDE	3
		BE	4		
		DE	3		

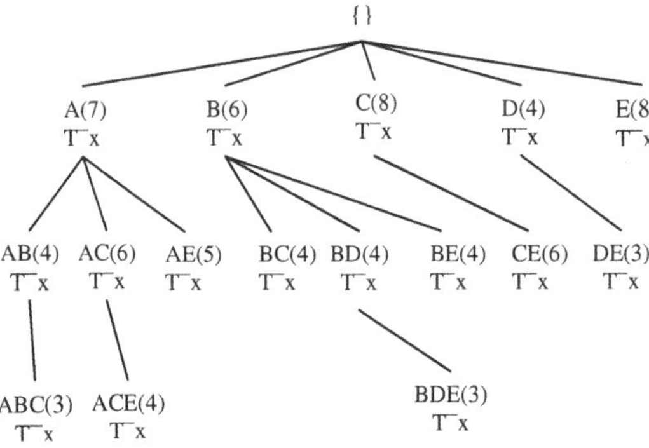

Fig. 1. IT-tree *Tr* that stores the set of pre-large and large itemsets in the original database

Assume that the last two transactions (with TIDs 9 and 10, respectively) are deleted from the original database. Table 3 shows the vertical layout of the two deleted transactions.

Table 3. Vertical data format of two deleted transactions

Items	*TIDs*
A	9
B	9
C	9 10
D	9
E	10
F	10

For the above data, the proposed maintenance algorithm proceeds as follows. The variable *c* is initially set at 0.

From Line 1 of the algorithm

$$f = \left\lfloor \frac{(S_u - S_l)d}{S_u} \right\rfloor = \left\lfloor \frac{(0.5 - 0.3)10}{0.5} \right\rfloor = 4$$

In Lines 2 to 4, since $t + c = 2 + 0 = 2 \leq f$, rescanning the original database is unnecessary, so nothing is done. Lines 4 to 18 then update the support and tidset of each 1-itemset in *Tr* based on the tidsets of items in the two deleted transactions. In this example, item $\{F{:}1\}$ is a rescanned itemset as $\{F\}$ is small in the original database and in the two deleted transactions. The results are shown in Fig. 2.

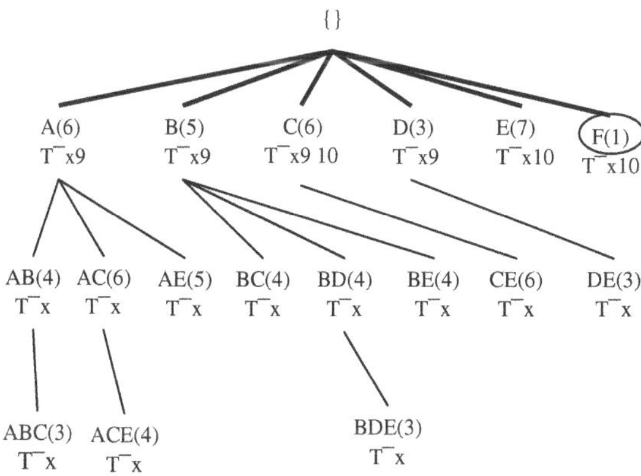

Fig. 2. IT-tree *Tr* after the 1-itemsets have been processed

After Line 18, the algorithm calls the procedure **ENUMERATE_FIs(*Tr*, [∅])** with $[\varnothing] = \{A, B, C, D, E\}$. The result after $[A]$ has been processed is shown in Fig. 3. The node $A(6)$ and T^{-}x9 in Fig. 3 shows that the final support count of itemset $\{A\}$ is 6 and the tidset of $\{A\}$ in the two deleted transactions is $\{9\}$.

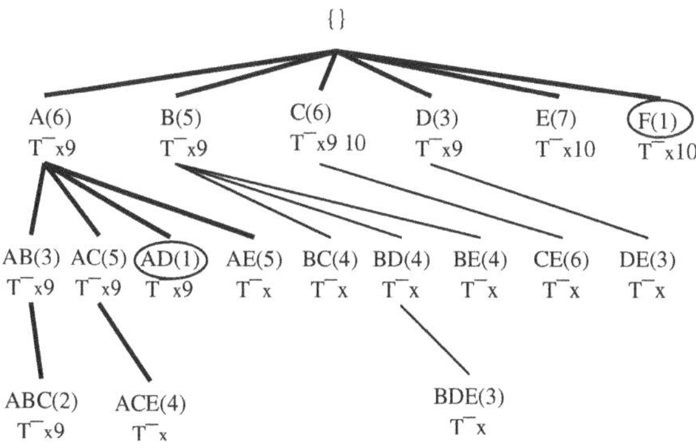

Fig. 3. IT-tree *Tr* after $[A]$ has been processed

The final IT-tree *Tr* after [∅] has been processed is shown in Fig. 4.

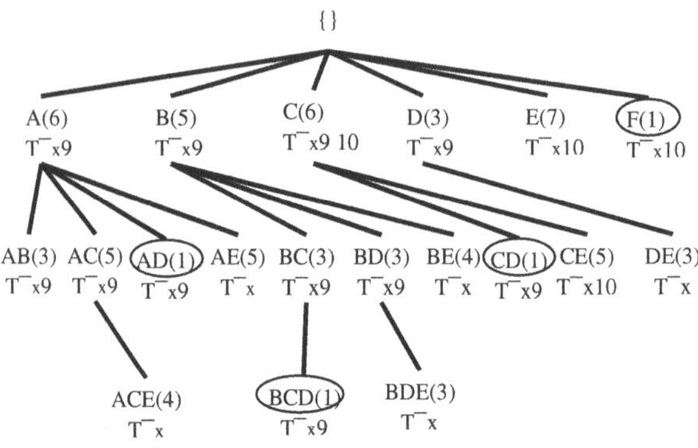

Fig. 4. Final IT-tree *Tr*

After the procedure **ENUMERATE_FIs(*Tr*, [∅])** has ended, lines 20 to 23 are executed. Since $t + c = 2 + 0 = 2 \leq f$, $c = c + t = 2 + 0 = 2$. The result of the algorithm is the IT-tree *Tr*, which stores the pre-large and the large itemsets for all updated databases, as well as the rescanned itemsets. The IT-tree *Tr* can then be used for processing the next deleted transactions.

Note that the final value of c is 2 in this example and $f - c = 2$. This implies that two more transactions can be deleted without rescanning the original database.

4 Experimental Results

Experiments were conducted to show the performance of the proposed algorithm. All the algorithms were implemented on a PC with a Core 2 Duo (2 x 2 GHz) CPU and 2 GB of RAM running Windows 7. All the algorithms were coded in C++. The well-known kosarak (with 990002 transactions) is used in the experiments.

The performance of the proposed algorithm was compared with that of the pre-large-tree-based algorithm [8] when a number of transactions were deleted from the database. A total of 990002 transactions were also used to build an initial IT-tree and pre-large tree. Sets of 2000 transactions were then also sequentially used as deleted transactions for the experiment. The upper and the lower support thresholds were set to 0.5% and 1%, respectively. Fig. 5 shows the execution times of the two algorithms for processing each set of the 2000 deleted transactions. For the experiment with the incremental threshold values, the lower support threshold was set at 0.5% to 2.5% (in 1% increments) and the upper support threshold was set 1% to 3% (in 1% increments). Sets of 2000 transactions were used as the deleted transactions for the experiments. Fig. 6 shows the comparison of the execution times for various threshold values. It can be observed that the proposed maintenance algorithm runs faster than the pre-large-tree-proposed algorithm.

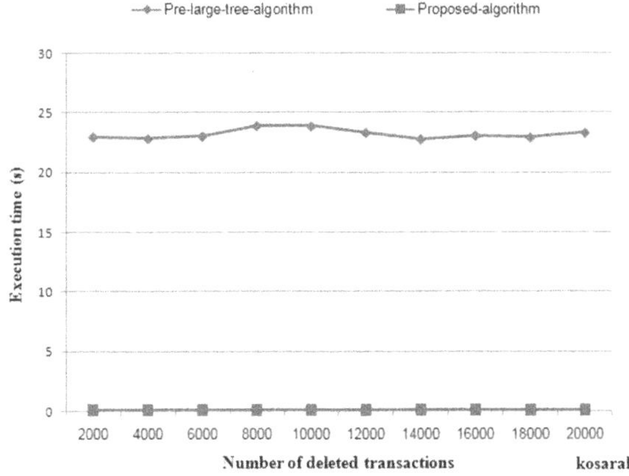

Fig. 5. Comparison of the execution times for sequentially deleted transactions

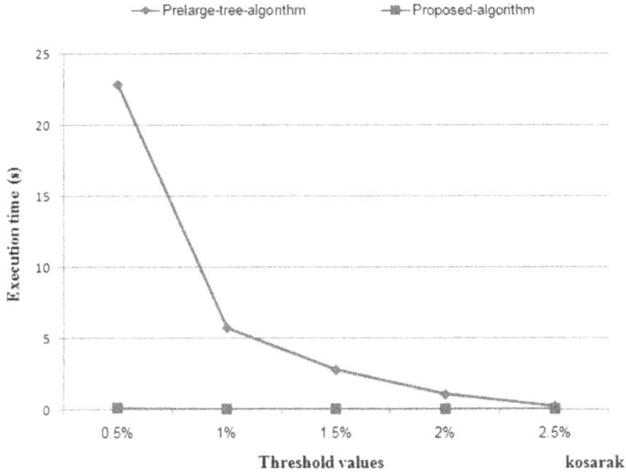

Fig. 6. Comparison of execution times for various threshold values

5 Conclusion

This paper proposed an approach based on pre-large concepts for the maintenance of an IT-tree during transactions deletion. The proposed algorithm uses the IT-tree structure to facilitate tree traversal and the updating of itemset supports. The supports of the candidate itemsets can be rapidly computed using tidset intersections. Pre-large itemsets are used for to reduce the number of database scans. User-specified upper and lower support thresholds are used to avoid the small items directly becoming large in the updated database when transactions are deleted. With these strategies, the execution times of the proposed approach is lower than that of a pre-large-tree-based algorithm [8].

Dynamic Bit-Vector (DBV) [12] is an efficient data structure for mining FIs. DBV can be used to compress a database in one scan and shorten the length of the tidset, speeding up the DBV intersections process. In the future, the authors will apply DBV for the maintenance of the IT-tree for transactions deletion.

References

[1] Agrawal, R., Srikant, R.: Fast algorithm for mining association rules. In: The International Conference on Very Large Data Bases, pp. 487–499 (1994)

[2] Cheung, D.W., Han, J., Ng, V.T., Wong, C.Y.: Maintenance of discovered association rules in large databases: An incremental updating approach. In: The Twelfth IEEE International Conference on Data Engineering, pp. 106–114 (1996)

[3] Han, J., Pei, J., Yin, Y.: Mining frequent patterns without candidate generation. In: The 2000 ACM SIGMOD International Conference on Management of Data, pp. 1–12 (2000)

[4] Hong, T.P., Wang, C.Y., Tao, Y.H.: A new incremental data mining algorithm using pre-large itemsets. Intelligent Data Analysis 5(2), 111–129 (2001)

[5] Hong, T.P., Lin, C.W., Wu, Y.L.: Incrementally fast updated frequent pattern trees. Expert Systems with Applications 34(4), 2424–2435 (2008)

[6] Koh, J.-L., Shieh, S.-F.: An Efficient Approach for Maintaining Association Rules Based on Adjusting FP-Tree Structures1. In: Lee, Y., Li, J., Whang, K.-Y., Lee, D. (eds.) DASFAA 2004. LNCS, vol. 2973, pp. 417–424. Springer, Heidelberg (2004)

[7] Lin, C.W., Hong, T.P., Lu, W.H.: The Pre-FUFP algorithm for incremental mining. Expert Systems with Applications 36(5), 9498–9505 (2009)

[8] Lin, C.W., Hong, T.P., Lu, W.H.: Maintenance of the prelarge trees for record deletion. In: The Twelfth WSEAS International Conference on Applied Mathematics, pp. 105–110 (2007)

[9] Le, T.P., Vo, B., Hong, T.P., Le, B.: Incremental mining frequent itemsets based on the trie structure and the prelarge itemsets. In: The 2011 IEEE International Conference on Granular Computing, pp. 369–373 (2011)

[10] Le, T.P., Hong, T.P., Vo, B., Le, B.: An efficient incremental mining approach based on IT-tree. In: The 2012 IEEE International Conference on Computing & Communication Technologies, Research, Innovation, and Vision for the Future, pp. 57–61 (2012)

[11] Li, X., Deng, Z.-H., Tang, S.: A Fast Algorithm for Maintenance of Association Rules in Incremental Databases. In: Li, X., Zaïane, O.R., Li, Z. (eds.) ADMA 2006. LNCS (LNAI), vol. 4093, pp. 56–63. Springer, Heidelberg (2006)

[12] Vo, B., Hong, T.P., Le, B.: DBV-Miner: A Dynamic Bit-Vector approach for fast mining frequent closed itemsets. Expert Systems with Applications 39(8), 7196–7206 (2012)

[13] Zaki, M.J., Parthasarathy, S., Ogihara, M., Li, W.: New algorithms for fast discovery of association rules. In: The Third International Conference on Knowledge Discovery and Data Mining, pp. 283–286 (1997)

[14] Zaki, M.J., Hsiao, C.J.: Efficient algorithms for mining closed itemsets and their lattice structure. IEEE Transactions on Knowledge and Data Engineering 17(4), 462–478 (2005)

Dimensionality Reduction by Turning Points for Stream Time Series Prediction

Van Vo[1,2], Luo Jiawei[1,*], and Bay Vo[3]

[1] School of Information Science and Engineering, Hunan University, China
luojiawei@hnu.edu.cn
[2] Faculty of Information Technology, Ho Chi Minh University of Industry, Vietnam
vothithanhvan@hui.edu.vn
[3] Information Technology College, Ho Chi Minh, Vietnam
vdbay@itc.edu.vn

Abstract. In recent years, there has been an explosion of problems concerned with mining time series databases and dimensionality reduction is one of the important tasks in time series data mining analysis. These approaches are very useful to pre-process the large dataset and then use it to analyze and mine. In this paper, we propose a method based on turning points to reduce the dimensions of stream time series data, this task helps the prediction process faster. The turning points which are extracted from the maximum or minimum points of the time series stream are proved more efficient and effective in preprocessing data for stream time series prediction. To implement the proposed framework, we use stock time series obtained from Yahoo Finance, the prediction approach based on Sequential Minimal Optimization and the experimental results validate the effectiveness of our approach.

Keywords: Stream Mining, Time Series Dimensionality Reduction, Turning Points, Stream Time Series, Time Series Prediction.

1 Introduction

One of the important tasks of data mining is the process of finding and extracting potential information and knowledge hidden in the large dataset. Lately, time series related studies [1] have received serious attention by many scientists for wide scientific and other applications. It is not similar with the traditional achieved data, time series stream have their own special characteristics such as large in data size, high dimensionality and necessary to update continuously.

In the context of time series data mining, the fundamental problem is how to represent the time series data [2]. Another approach is in transforming the time series to another domain for dimensionality reduction [3-5], [13] followed by an indexing mechanism.

[*] Corresponding author.

N.T. Nguyen et al. (Eds.): *Adv. Methods for Comput. Collective Intelligence*, SCI 457, pp. 167–176.
DOI: 10.1007/978-3-642-34300-1_16

The recent researches [6] analyze the movement of a stock based on a selected number of points from the time series. These points include Important Points (IPs), Perceptually Important Points (PIPs) and Turning Points (TPs). Important points [7] are local minimum and maximum points in a time series. Perceptually important points are used in the identification of frequently appearing technical (analysis) patterns. PIPs usually contain a few noticeable points and they are used for technical pattern matching in stock applications [8].

Prediction systems use historical and other available data to predict a future event. Time series prediction is an interesting and challenging task in the field of data mining. Generally, time series prediction has two main tendencies, statistical and computational intelligence. Statistical approaches include moving average, autoregressive moving average, autoregressive integrated moving average, linear regression and multiple regression models [1]. Jin-Fang Yang et al. [9] provide the approach of support vector regression (SVR) based on the Sequential Minimal Optimization (SMO) algorithm to build the model and predict single time series by mining computation.

We are interested in dimensionality reduction in order to reduce the number of data before applying the prediction techniques that are suitable for the stream data environment, which is in support of low prediction, training cost and high accuracy of the future values. In this paper, we implement the TPs technique as preprocessing data apply to the SMO prediction algorithm. The experiment has two parts, first we remove the less important points and then we evaluate the accuracy of future values depending on the number of histories used for prediction. To attest the effectiveness of our approach we use the stock price data set obtained from Yahoo Finance.

This paper is organized as follows. We define the problem and propose the framework in Section 2. Section 3 presents our research-on time series dimensionality reduction by the turning points approach and time series prediction by SMO. Section 4 provides evaluation of our experiments. Finally, we conclude our work in Section 5.

2 Problem Statement and Framework

A time series is a set of consecutive observations, each one being recorded at a specific time. A discrete time series is one where the set of times at which observations are made is a discrete set. Continuous time series are obtained by recording observations continuously over some time interval.

With our approach, three main processes in the framework are shown in Fig. 1 and the notations in Table 1.

Definition 1: Let's assume that we have n time series $T_1, T_2, \ldots,$ *and* T_n in the stream data environment, each T_i containing m ordered values at the current timestamp $(m-1)$, that is, $T_i = \{t_{i0}, \ldots, t_{i(m-1)}\}$ where t_{ij} is the value at timestamp j in T_i.

Definition 2: Suppose that n stream time series only receive data after F timestamps, after applying perceptually important points and turning points techniques we have p time series. In other words, for each time series T_p', the future values $t_{ip}, t_{i(p+1)}, \ldots,$ *and* $t_{i(p+F-1)}$ fitting to timestamps $p, (p+1), \ldots,$ *and* $(p+F-1)$, respectively, arrive in a batch manner at the same timestamp $(p+F)$. At the period from timestamp p to $(p+F-1)$, the system does not know about F future values in each time series.

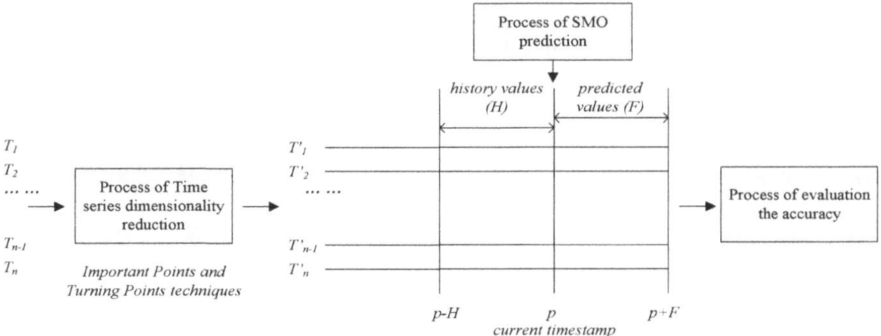

Fig. 1. Our proposed framework

Table 1. Symbols and their descriptions

Symbol	***Description***
T_i	the ith time series ($T_i, T_i, \ldots T_n$ with i=1..n)
T_i'	the ith time series after dimensionality reduction techniques
n,	the number of time series
m	the number of data points in the stream ($T_{i0}, T_{i1}, \ldots T_{i(m-1)}$)
p	the number of data points after dimensionality reduction
H	the number of historical data used to predict
F	the number of future predicted values

Definition 3: The objective of the prediction system is to efficiently predict the $n \times F$ values for the n time series streams with the prediction error as low as possible and the accuracy as high as possible. The prediction error is defined as the difference between the actual value and predicted value for the corresponding period.

$$E_t = A_t - F_t \tag{1}$$

where E, A and F are the prediction error, actual value and predicted value at period t.

3 Our Approach

3.1 Turning Points

A time series consists of a sequence of local maximal/minimal points and several of them mirror the information of data trend reversals. These local maximum and minimum points are called critical points; in another way we can say that a time series is comprised of a sequence of critical points. These critical points are often called *turning points* because they show the change in the trend of the time series stream.

In our approach, we only consider the critical points of each time series streams in certain period. The turning points in time series are defined as the points that separate two adjacent trends and have the shortest distance from the release time of announcements. Only some of the critical points are preserved (those critical points which are considered as interference factors are moved).

All the local maximal/minimal points in an original time series stream are extracted to form the initial critical point series. After constructing the initial critical point series T_i', a critical point selection criteria is applied to filter out the critical points corresponding to noise. The T_i and T_i' are called original and arrived at after pre-processing time series stream.

We consider that the first and the last data point in the original time series T_i are preserved as the first and last points in T_i'. The direction for selection is also based on the fluctuation threshold λ_v and the time duration threshold λ_t. The λ_v is defined as the Equation (2) below:

$$\lambda_v = \frac{1}{k} \times \sum |t_k| \tag{2}$$

where t_k is the kth value of time series stream

In a stream environment, for a given time series $T_{ij} = \{t_{i1}, t_{i2}, , t_{im}\}$, a turning point (peak or trough) in the time series T_{ij} is defined as any time period j of the ith stream such that the change (decrease or increase) in the observations of the time series after consideration both specific thresholds of fluctuation and time duration, it is not sure that the points located on the upward or downward trend of each time series stream.

Our strategies for eliminating the points that are not important are shown in detail below. The time duration threshold λ_t in our approach is five successive points.

```
for each time series stream
 if ((x_i to x_i+4 are successive points) and y_i > y_i+1)
   if ((y_i+1 < y_i+3) and (y_i, y_i+2, y_i+4 < λ_v))
       choose points x_i, x_i+3 and x_i+4 (eliminate x_i+1, x_i+2)
   else if ((y_i+1 > y_i+3) and (y_i, y_i+2, y_i+4 < λ_v))
     choose points x_i, x_i+1 and x_i+4 (eliminate x_i+2, x_i+3)
 end if
end for
```

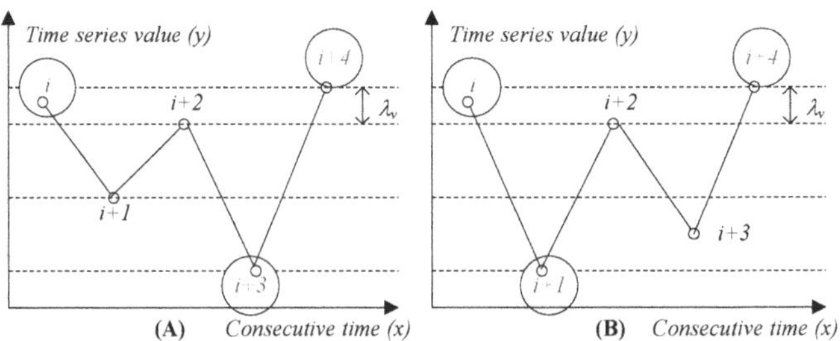

Fig. 2. First strategy for eliminating points

```
for each time series stream
 if ((x_i to x_{i+4} are successive points) and y_i < y_{i+1})
   if ((y_{i+1} < y_{i+3}) and (y_i, y_{i+2}, y_{i+4} < λ_v))
     choose points x_i, x_{i+3} and x_{i+4} (eliminate x_{i+1}, x_{i+2})
   else if ((y_{i+1} > y_{i+3}) and (y_i, y_{i+2}, y_{i+4} < λ_v))
     choose points x_i, x_{i+1} and x_{i+4} (eliminate x_{i+2}, x_{i+3})
 end if
end for
```

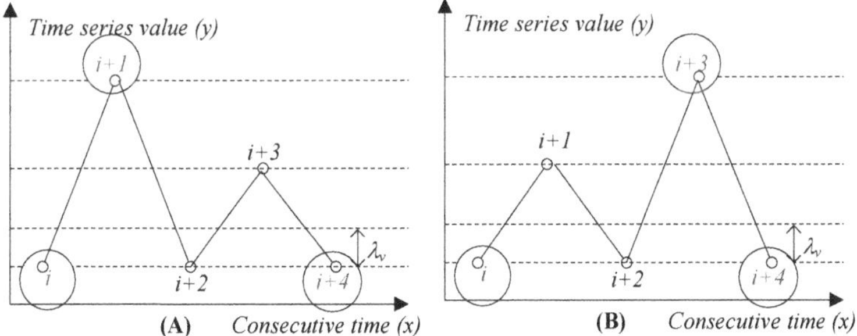

Fig. 3. Second strategy for eliminating points

```
for each time series stream
 if ((x_i to x_{i+4} are successive points) & y_i<y_{i+1}<y_{i+2}<y_{i+3})
   if ((y_{i+3}>y_{i+4}) and (|y_i-y_{i+1}|< λ_v and |y_{i+1}-y_{i+2}|<λ_v))
     eliminate x_{i+1}, x_{i+2}, choose x_i, x_{i+3} and x_{i+4}
   else if ((y_{i+1} > y_{i+3}) and (y_i, y_{i+2}, y_{i+4} < λ_v))
     eliminate x_{i+2}, x_{i+3}, choose x_i, x_{i+1} and x_{i+4}
 end if
end for
```

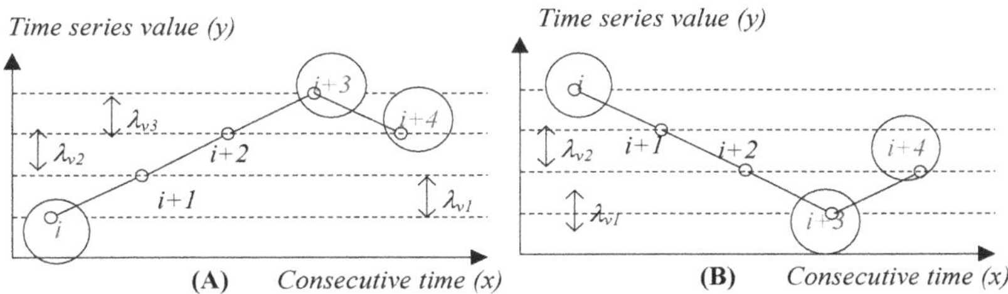

Fig. 4. Third strategy for eliminating points

In order to ensure the change in the observations of the turning points in term of both the time and value, we set a step range for eliminating the points that are not important. To avoid a false turning point, that locates on the upward or downward trend of the time series, in being identified as a true one, our strategy is to ensure that the type of the previously identified turning point is opposite *(downward or upward)* to that of the current point. It means that a peak must follow a valid trough closely and no additional peak is located between them.

3.2 Applying SMO Algorithm for Prediction

Working with the large data set environment, the prediction using the Support Vector Machine algorithm [9-10] makes the operation speed slower. Especially, in the stream time series environment with many time series streams $T_1, T_2, \ldots, T_n$ with $T_i = \{t_{i0}, t_{i1}, t_{i2}, \ldots t_{i(m-1)}\}$, if each one of the time series puts its own kernel matrix into the main memory, the primary memory will overflow.

We chose the SMO prediction algorithm [11] because it is based on the SVM. In additional, the SMO algorithm just calls the kernel matrix iteration therefore the performance is improved significantly.

SMO is an iterative algorithm for solving the optimization problem of the SVM. SMO algorithm implements the dividing problem into a series of smallest possible sub-problems, which are then solved analytically. Because of the linear equality constraint involving the Lagrange multipliers, the smallest possible problem involves two such multipliers. Then, for any two multipliers ξ and ξ^*, the constraints are reduced to:

$$\begin{aligned} 0 \le \xi, \xi^* \le C \\ y_1\xi + y_1\xi^* = k \end{aligned} \tag{3}$$

where C is an SVM hyper parameter and this reduced problem can be solved analytically.

Our algorithm proceeds as below for each time series:

```
1. Find a Lagrange multiplier ξ that violates the Karush-
   Kuhn-Tucker [11] conditions for the optimization problem.
2. Pick a second multiplier ξ*, optimize the pair (ξ,ξ*).
3. Repeat steps 1 and 2 until convergence.
```

When all the Lagrange multipliers ξ, ξ^* satisfy the KKT conditions (within a user-defined tolerance), it means the problem has been solved. Although the SMO algorithm is guaranteed to converge, heuristics are used to choose the pair of multipliers to accelerate the rate of convergence.

We select the first Lagrange multiplier by using the external loop of the SMO algorithm to enable the Lagrange multiplier to optimize. Choosing the second Lagrange multiplier is according to maximizing the step length of the learning during joint optimization. $|E_1 - E_2|$ is used to approximate the step size in SMO [9-10].

3.3 Prediction Accuracy

We use statistics to evaluate the simulation effect and predictive validity of the prediction model. There are several types of evaluations such as mean absolute error (MAE), mean absolute percentage error (MAPE), root mean squared error (RMSE), etc.

In order to ensure the accuracy of prediction results, the model must be evaluated accurately and its performance generalized before it can be used to predict. The prediction error [12] is defined as the difference between the actual value and the predicted value for the corresponding period as in the Equation (1). In this paper, the accuracy of an n stream time series T_1, T_2... *and* T_n will use the average of each time series. The average is computed with the Equation (3):

$$\bar{E} = \frac{1}{N}\sum_{i=1}^{N} E_i \tag{4}$$

4 Experimental Evaluation

4.1 Experimental Environment and Dataset

The experimental dataset used was the financial stock time series data. The stock data set contains daily stock prices obtained from Yahoo Finance. The experiment was performed on a 2G AMD PC with 2GB of main memory and running the Windows XP operating system. We tested our approaches with four different stock companies (HBC, ORCL, MSFT and UN) and 125 stock ticker's daily closing price during half of the year.

4.2 Experimental Results and Analysis

In this section, we talk about the experiments and the results in this study to asses our proposed method. Our discussion is constructed with the following two aspects. First, we reduce the dimensions of the stream time series in the preprocessing step. Second, we consider the relationship between the number of history price stock values (after filtering the turning points) and the predicted number of future values.

In Fig. 5(a), we show the original closing stock data of four companies. As shown, the data is does not have a large variability within the nearest 6 months. The characteristic of the daily stock data does not exist on the weekends and some special days. As observed in the figure, the variances of ORCL, MSFT and UN streams are not very high; while HBC stream has a higher fluctuation. In Fig. 5(b), we present the results data after implementing the turning point proposed approach. After dimensionality reduction, the numbers of data point are difference, 97 points for HBC, 91 points for ORCL, 101 points for MSFT and 113 for UN.

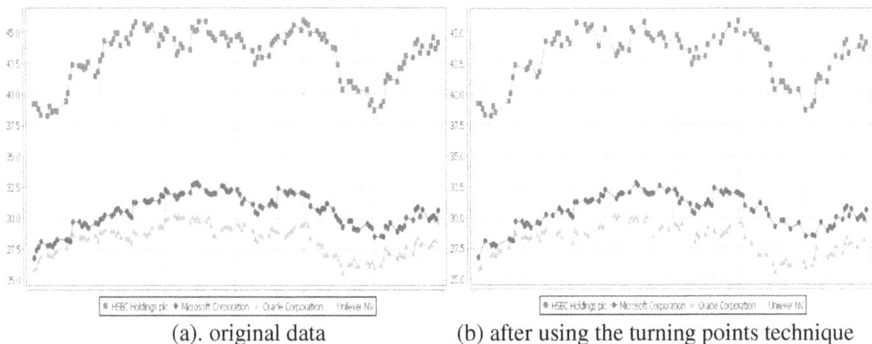

(a). original data (b) after using the turning points technique

Fig. 5. Stock streams of HBC, MSFT, ORCL and UN (obtained from Yahoo Finance)

In Fig. 6, we show both history and future prediction values of stock stream in the case of using the turning points technique. The results of training data evaluated by RMSE of both original data and data after dimensionality reduction technique are shown in Table 2. For prediction of next 20 days, we need 20 steps ahead for the training process (we only give five first steps ahead).

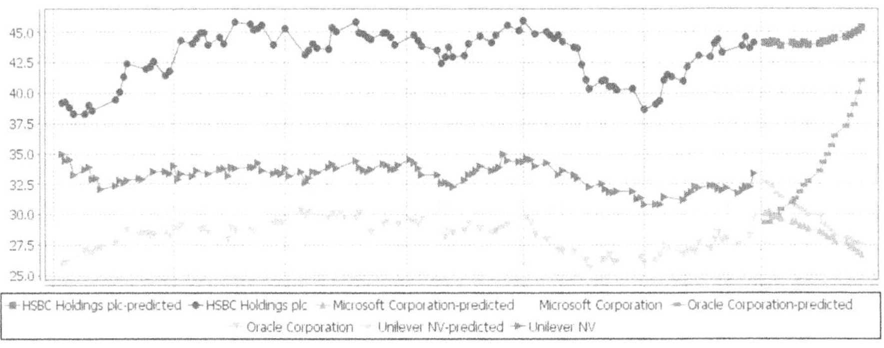

Fig. 6. Future prediction of stock price streams after using turning points techniques

Table 2. Evaluation on training data of stock price streams

Step(s)-ahead	*1*	*2*	*3*	*4*	*5*
HBC	*0.7237*	*0.9091*	*1.0324*	*1.1458*	*1.2428*
ORCL	*0.5113*	*0.6266*	*0.7198*	*0.823*	*0.897*
MSFT	*0.4415*	*0.5233*	*0.6087*	*0.6688*	*0.7094*
UN	*0.3652*	*0.436*	*0.5112*	*0.5611*	*0.5908*
$\overline{E}$	*0.510425*	*0.62375*	*0.718025*	*0.799675*	*0.86*

A comparison of our approach with other dimensionality reduction techniques is shown in Fig. 7 (Fig. 7(a) shows the original time series data, Fig. 7(b) and Fig. 7(c) display the result of PAA [13] dimensionality reduction and our approach of the original data). PAA implements the dimensionality reduction of a time series from N in the original space to N' in the reduced space by segmenting the time series into equal-sized frames and representing each segment by the mean of the data points that lie within that frame. The PAA method is simple and straightforward but the result does not retain the shape of the change in the time-series stock data. In spite of this, the PAA method has been successfully used as a competitive method. However, our approach for dimensionality reduction still endures one deficiency, that is, it complicated to implement and its computational complexity is higher than that of the PAA technique.

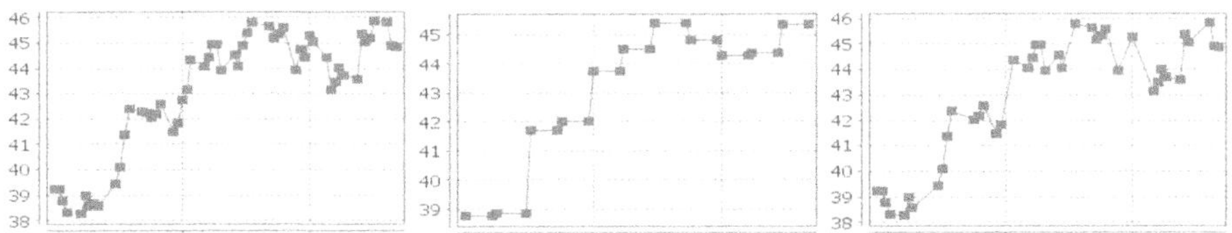

(a) original data (b) dimensionality reduction with PAA (c) our approach.

Fig. 7. The comparison of our approach and the PAA technique

5 Conclusions and Future Work

We have studied the problem of time series prediction with dimensionality reduction of large historical data to a smaller data set and then using the data mining technique to compute future values.

In this paper, we have shown and theoretically explained the use of the turning points approach to reduce dimensions and used the SMO based on the SVM prediction, and provided the evaluation indicators of accuracy and generalization. After the dimensionality reduction method based on the turning points process, the time series generated by our approach still maintains the shape of the original trends. The approach is very meaningful in the environment where the data set is large. Therefore, we applied this work on the stock price data set obtained from Yahoo Finance.

In the future, we propose to use the model for calculating the knowledge of collectives [14] as a process in the analysis of time series because the knowledge of a collective is more proper than the knowledge of its members.

Acknowledgement. The work is supported by the National Natural Science Foundation of China (Grant no. 60873184) and the Science and Technology Planning Project of Hunan Province (Grant no. 2011FJ3048). We would like to thank The University of Waikato for providing machine learning open source [15].

References

1. Fu, T.: A review on time series data mining. Engineering Applications of Artificial Intelligence 24(1), 164–181 (2011)
2. Keogh, E., Chakrabarti, K., Pazzani, M., Mehrotra, S.: Dimensionality reduction for fast similarity search in large time series databases. Knowledge and Information Systems 3(3), 263–286 (2001)
3. Lendasse, A., Lee, J., Bodt, E.D., Wertz, V., Verleysen, M.: Dimension reduction of technical indicators for the prediction of financial time series - Application to the BEL20 Market Index. European Journal of Economic and Social Systems 15(2), 31–48 (2001)
4. Wang, Q., Megalooikonomou, V.: A dimensionality reduction technique for efficient time series similarity analysis. Information Systems 33(1), 115–132 (2008)
5. Zhang, Z., Li, J., Wang, H., Wang, S.: Study of principal component analysis on multi-dimension stock data. Chinese Journal of Scientific Instrument 26(8), 2489–2491 (2005)
6. Fu, T.-C., Chung, F.-L., Luk, R., Ng, C.-M.: Representing financial time series based on data point importance. Engineering Applications of Artificial Intelligence 21(2), 277–300 (2008)
7. Bao, D.: A generalized model for financial time series representation and prediction. Applied Intelligence 29(1), 1–11 (2008)
8. Bao, D., Yang, Z.: Intelligent stock trading system by turning point confirming and probabilistic reasoning. Expert Systems with Applications 34(1), 620–627 (2008)
9. Yang, J.F., Zhai, Y.J., Xu, D.P., Han, P.: SMO algorithm applied in time series model building and forecast. In: Proc. of the 6th International Conference on Machine Learning and Cybernetics, Hong Kong, vol. 4, pp. 2395–2400 (2007)
10. Osuna, E., Freund, R.: An improved training algorithm for support vector machines. In: Proceedings of the 1997 IEEE Neural Networks for Signal Processing VII, New York, pp. 276–285. IEEE (1997)
11. Platt, J.C.: Sequential Minimal Optimization: A fast algorithm for training support vector machines. Advances in kernel methods, pp. 185–208. MIT Press, Cambridge (1999)
12. Chen, Z., Yang, Y.: Assessing Forecast Accuracy Measures (2004), `http://www.stat.iastate.edu/preprint/articles/2004-10.pdf`
13. Keogh, E., Chakrabarti, K., Pazzani, M., Mehrotra, S.: Locally Adaptive Dimensionality Reduction for Indexing Large Time Series Databases. In: Proceedings of ACM SIGMOD Conference on Management of Data, pp. 151–162 (2001)
14. Nguyen, N.T.: Inconsistency of Knowledge and Collective Intelligence. Cybernetics and Systems 39(6), 542–562 (2008)
15. Hall, M., Frank, E., Holmes, G., Pfahringer, B., Reutemann, P., Ian, H.: The WEKA data mining software: an update. SIGKDD Explorations 11(1), 10–18 (2009)

Clustering as an Example of Optimizing Arbitrarily Chosen Objective Functions

Marcin Budka

Bournemouth University, BH12 5BB Poole, UK
mbudka@bournemouth.ac.uk, mbudka@gmail.com
http://www.budka.co.uk/

Abstract. This paper is a reflection upon a common practice of solving various types of learning problems by optimizing arbitrarily chosen criteria in the hope that they are well correlated with the criterion actually used for assessment of the results. This issue has been investigated using clustering as an example, hence a unified view of clustering as an optimization problem is first proposed, stemming from the belief that typical design choices in clustering, like the number of clusters or similarity measure can be, and often are suboptimal, also from the point of view of clustering quality measures later used for algorithm comparison and ranking. In order to illustrate our point we propose a generalized clustering framework and provide a proof-of-concept using standard benchmark datasets and two popular clustering methods for comparison.

Keywords: clustering, cluster analysis, optimization, genetic algorithms, particle swarm optimization, general-purpose optimization techniques.

1 Introduction

Clustering, also known as cluster analysis, is a well-known unsupervised learning task in data mining. The application areas of clustering are virtually endless, from customer segmentation in Customer Relationship Management systems, to image processing, to information retrieval or mining unstructured data [15]. Although the research on clustering dates back to the mid-20th century and there are now dozens of various clustering algorithms one can find in the literature [15], the problems we are facing today are essentially still the same, only on a larger scale due to the recent explosion of the amounts of generated data. Most typical of these problems concern the number of clusters (groups) in the data, the choice of an appropriate similarity measure or the choice of the clustering algorithm itself. We argue that these concerns stem from more fundamental issues of what constitutes a cluster or how to assess the quality of the obtained clustering result.

Clustering is also a good example of a common practice of optimizing arbitrarily chosen criteria in the hope that they are well correlated with the criterion actually used for assessment of the result. We perceive it as an interesting contradiction if for example one is using the mean absolute percentage error to select the best model, but mean squared error to fit the candidate models, while

N.T. Nguyen et al. (Eds.): *Adv. Methods for Comput. Collective Intelligence*, SCI 457, pp. 177–186.
DOI: 10.1007/978-3-642-34300-1_17

neither of them adequately reflects the problem being addressed, in effect leading to spurious and puzzling outcomes [18]. This of course stems from the fact that the latter is much easier to optimize, but as we demonstrate in this study, direct optimization of the assessment criterion is often still a viable approach, especially that the computational power is nowadays cheap and abundant.

In the following sections we address the above problems, pointing out the gaps in the current body of research and proposing an alternative view on clustering, which unifies all the issues mentioned above under a comprehensive, generalized framework. This work should be perceived as a proof of concept or position paper and hence the experimental evaluation of the proposed approach is by no means exhaustive in terms of algorithms and datasets used.

2 The Notion of a Cluster

Clustering is the process of discovering structure in data by identifying homogenous (or natural) groups of patterns [8]. The whole procedure is hence based on a similarity or distance function. The general agreement is that the patterns within a cluster should be more similar to each other than to patterns belonging to different clusters [15].

Definition of a cluster as a homogenous or natural group of patterns is however still not precise enough, in a sense that it does not specify in detail the features that a clustering algorithm should have. Hence many popular notions of clusters exist e.g. well separated groups with low distances among cluster members (e.g. K-means clustering algorithm [19]), dense areas of the input space (e.g. Expectation Maximization clustering algorithm [6]) or groups of patterns having particular information-theoretic properties (e.g. Cauchy-Schwarz divergence clustering algorithm [16]). The central issue is that in most cases these notions of clusters result from the characteristics of the clustering algorithm rather than being consciously designed to fit a particular application or problem, which often stems from the lack or ignorance of domain knowledge.

The choice of a similarity measure, which is inherent in the definition of what constitutes a cluster, is a good example of the above. Suppose that in a database of driving licence candidates there are the following two binary attributes: 'blind in left eye' and 'blind in right eye'[1]. Without knowing the context, one might be tempted to use the Hamming distance [13] (i.e. the number of positions at which the corresponding input vectors are different) as a measure of (dis)similarity. This would however lead to a situation in which a person blind in left eye is more similar to a person blind in both eyes than to a person blind in right eye, producing spurious results[2].

In our view the notion of a cluster and other related choices (e.g. similarity measure) should therefore be application-driven. This is especially important in

[1] This excellent example has been taken from [8].

[2] In some countries this could also possibly lead to the developer of the system being sued for discrimination if people blind in the left eye and people blind in the right eye were not treated equally.

unsupervised learning as the obtained results are often used to define a supervised learning problem (e.g. customer segmentation for developing a classification system for new customers), where the usual GIGO[3] principle applies. Also it is difficult to think about a 'natural grouping of patterns' without a context of a particular application – in the driving licence candidate example, grouping people blind in left eye and blind in right eye together is certainly more 'natural' than assigning them to the 'blind in both eyes' cluster.

3 The Number of Clusters

Every standard clustering algorithm requires the user to choose the number of clusters to be identified in the data [15]. This in a sense contradicts the whole idea of discovering structure in data, as the choice of the number of clusters imposes a limit on what can actually be discovered, especially if this parameter needs to be specified in advance. Although in hierarchical clustering [20] the user at least has a chance to explore the results at different levels of granularity, this is only feasible when dealing with relatively small datasets.

Of course there are approaches which try to provide guidance in this respect (see [7] for example) but they are usually very specific heuristics, tailored towards particular clustering algorithms and certainly suboptimal.

Recognizing the need for a more systematic and general approach we hence propose to cast clustering as an optimization problem within a generalized framework, allowing to free the users of the burden of 'guesstimating' the initial parameters of clustering algorithms and effectively artificially and considerably limiting the space of possible solutions.

4 Clustering as an Optimization Problem

The central issue in solving the clustering problem (and in fact any other computational problem) is to be able to quantify how good or bad a given result is. This in turn allows to compare various results and choose the one(s), which meet the modeler's expectations, in the context of a particular application.

Clustering can thus be viewed as an optimization problem, where the set of parameters that can be manipulated is not limited to the cluster memberships of all input patterns, but rather also includes the number of clusters, the similarity measure as well as its parameters. The whole difficulty now comes down to designing an appropriate objective function (criterion), reflecting the knowledge of the application area and expectations of the designer. We have deliberately used the word 'design' to emphasize that the objective function shouldn't be chosen arbitrarily from the vast set of existing functions, but should rather be carefully engineered to have the required properties and capture relevant factors from the point of view of a given learning problem.

[3] Garbage In, Garbage Out.

Most popular general-purpose clustering methods have necessarily been designed the other way round, with the clustering algorithm as a starting point and the objective function as a result or by-product of the algorithm design choices. In some cases, it is even not clear what objective a given algorithm actually optimizes, although the authors usually argue and demonstrate (using synthetic data, carefully crafted for this occasion) the superiority of their approach.

As a result the choice of a clustering algorithm to solve a particular problem is often motivated by factors like the familiarity of a user with an algorithm, its popularity or speed. Although in many cases it is difficult to undermine this last motivation (i.e. speed), on a closer look it may turn out that an approach only marginally slower can produce much better results, because it better fits the problem at hand.

Through this kind of uninformed choices people often either optimize 'something' regardless of its fit to their problem, or optimize 'something else' hoping that it is somehow correlated with their objective. The latter issue is actually typical for other learning problems, with a number of recent data mining competitions, where the criteria of success are often quite different from the objectives that the methods used by the contestants try to optimize, being one example.

Another good and in fact somewhat paradoxical example is the routine use of various clustering quality measures (like the Davies-Bouldin index [5] or the Dunn index [10]) to compare and rank different clustering algorithms. Each of these algorithms optimize a different criterion, very often only loosely related to the clustering quality measure used. The paradox here is that the very quality measures used would themselves make good objective functions (as long as their calculation does not require to know the 'ground truth', which is for example the case with a popular Jaccard similarity coefficient [14]), especially that they have been designed to capture various, in one way or another relevant, properties of the clustering result. Going a step further, performing simultaneous, multi-objective optimization of a set of quality measures, which are then used to compare the algorithms, definitely seems like a better initial choice than starting with an arbitrarily selected, popular clustering algorithm. We return to this issue in Section 6, where the two cluster quality measures mentioned above are used as optimization criteria in the experimental validation of the presented ideas.

5 Generalized Clustering Framework

It is well known that there are no 'one-fits-all' solutions as stated by the 'no free lunch' theorem [8]. Hence the power to solve any problem efficiently lies in the possibility of using a variety of diverse tools or simply a framework. In this section we thus propose a generalized clustering framework, allowing to construct and solve clustering problems using a variety of clustering and optimization algorithms, suited to the level of knowledge about the application domain.

The framework, which has been depicted in Figure 1, consists of two libraries: (1) the Optimization Method Library (OpML) and (2) the Objective Function Library (ObFL). All objective functions in the ObFL are associated with at least

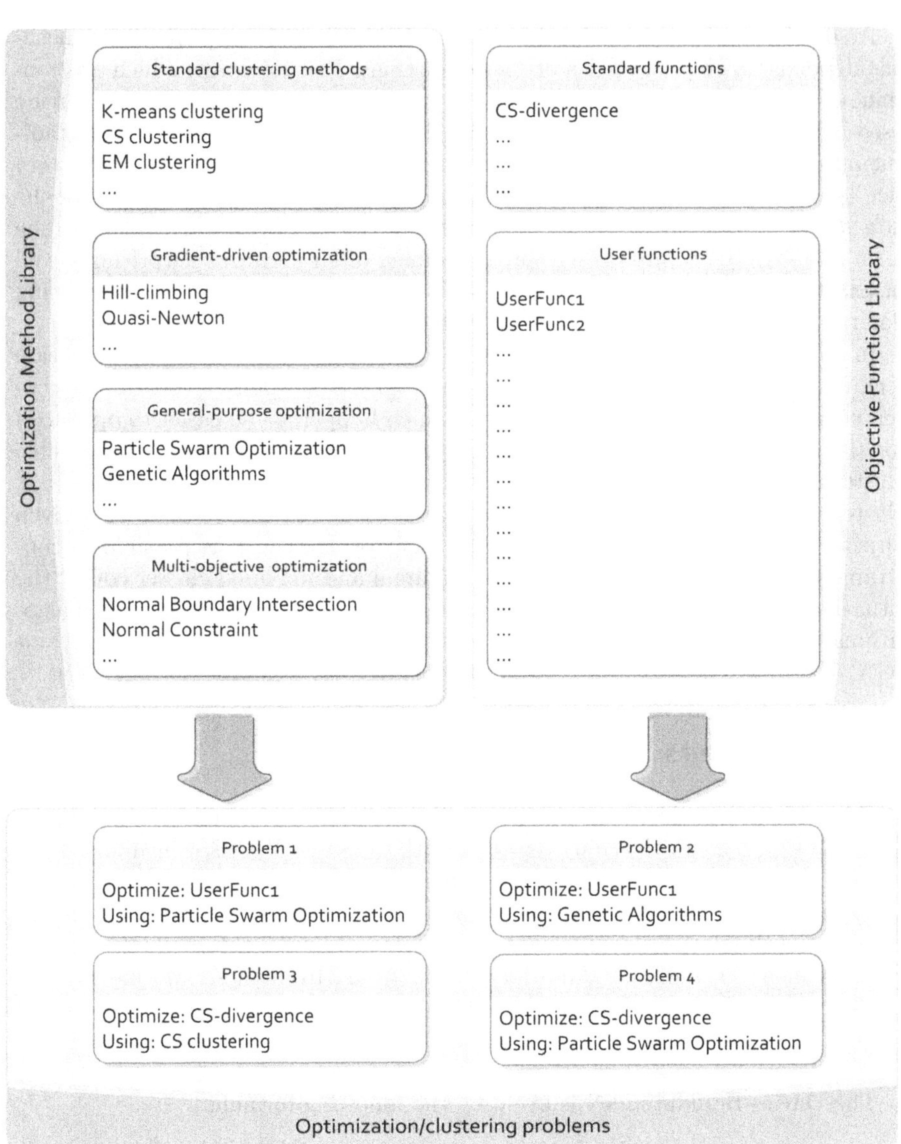

Fig. 1. Generalized clustering framework

one optimization method in the OpML – although this has not been shown in Figure 1 for clarity, not all optimization algorithms are suitable for all objective functions. The ObFL can also include user-defined objective functions, which can be designed freely, for example also as combinations of other functions.

Association of the user-defined functions with appropriate optimization methods depends on the properties of these functions. For objectives which are continuous and differentiable, any gradient-driven optimization technique can be used. If the gradient formula cannot be derived analytically, numerical gradient approximation might be the solution – the optimization literature is very rich (see [11] for example). If possible, an optimization scheme dedicated for the non-standard objective function can also be created. This usually takes on the form of a greedy optimization approach, which often performs surprisingly well in practice (see [3] and [4] for example), with the additional benefit of being deterministic and hence providing reproducibility of the results.

In more complex cases one should revert to the general-purpose optimization techniques, always having in mind that the goal is to optimize a consciously chosen criterion (and not any other criteria which happen to be easier to optimize) for the results to be meaningful. Due to the fact that the space of all possible cluster assignments, combined with all possible values of other parameters (e.g. cluster count) is immense, exhaustive search for an optimum value of a given objective function is computationally prohibitive. Even if it is possible to constrain this space, e.g. by specifying the minimal and maximal cluster count, the situation does not usually improve much. This however is an area in which algorithms based on the exploration-exploitation paradigm, like Genetic Algorithms (GA) [12] or Particle Swarm Optimization (PSO) [17], can prove their merit.

6 Experiments

We have performed the proof-of-concept experiments on two well-known datasets from the UCI machine learning repository [1]: IRIS and WINE, which have been chosen for illustrative purposes. In the experiments we compare the performance of a number of standard clustering methods with the approach proposed in this paper using the two clustering quality measures mentioned before i.e. the Davies-Bouldin index [5] and the Dunn index [10]. Apart from these two indexes, we made the algorithms also optimize the cluster count, which for K-means and EM-clustering meant an exhaustive search in the range between 2 and 9[4] for both datasets.

The Davies-Bouldin index is given by the following formula:

$$DB = \frac{1}{c}\sum_{i=1}^{c} \max_{i \neq j} \left(\frac{\sigma_i + \sigma_j}{\delta(C_i, C_j)} \right) \tag{1}$$

where c is the number of clusters, $\delta(C_i, C_j)$ is the distance between the ith and j^{th} cluster (in our case the Euclidean distance between their centroids C_i and

[4] This number was chosen for convenience of binary representation in GA (3 bits).

C_j), and σ_i represents the average distance of all elements in the ith cluster to its centroid (C_i). The Davies-Bouldin index favours clusterings with low intra-cluster and high inter-cluster distances and the lower its value, the better.

The Dunn index is given by:

$$DI_m = \min_{1 \leq i \leq c} \left\{ \min_{1 \leq j \leq c, j \neq i} \left\{ \frac{\delta(C_i, C_j)}{\max_{1 \leq k \leq c} \Delta_k} \right\} \right\} \tag{2}$$

where Δ_k represents the intra-cluster distance (in our case maximal pairwise Euclidean distance between the objects in the cluster). The Dunn index also favours clusterings with low intra-cluster and high inter-cluster distances, although the compactness of the clusters is assessed in a different way. The Dunn index should be maximized.

The experimental results have been reported in Table 1. Although as mentioned before the quality measures used should favour the K-means algorithm, as the idea behind it is to find compact, well separated clusters, it is not the case. As it can be seen, direct optimization of the clustering quality measures, which is the approach postulated in this paper, is vastly beneficial in terms of both DB and DI_m, for both datasets.

The high values of the Dunn index and low values of the Davies-Bouldin index are out of reach of the standard clustering methods, with the Particle Swarm Optimization (PSO) algorithm being an absolute leader here. Intuitively, for

Table 1. Dunn and Davies-Bouldin index optimization

Clustering method	IRIS				WINE			
	DI_m		DB		DI_m		DB	
	c	**index**	c	**index**	c	**index**	c	**index**
'Ground truth'	3	0.3815	3	0.8635	3	0.3757	3	1.1985
K-means clustering[a]	2	0.7120	2	0.4595	2	0.5202	3	1.0589
EM-clustering[b]	7	0.5296	3	0.7563	6	0.4407	3	1.1411
Direct optimization with GA[c]	2	0.8142	2	0.4390	2	0.5412	3	**1.0381**
Direct optimization with PSO[d]	2	**0.8875**	2	**0.4385**	2	**0.5689**	3	**1.0381**

[a] PRTools [9] implementation has been used with the maximum number of iterations (i.e. random selections of the initial cluster centres) set to 1000; the experiment has been repeated 100 times and the best result has been reported; default values of the remaining parameters have been used.

[b] PRTools implementation has been used; the experiment has been repeated 100 times and the best result has been reported; default values of the remaining parameters have been used.

[c] Built-in MATLAB implementation has been used; the population size has been set to 1000; default values of the remaining parameters have been used.

[d] Particle Swarm Optimization Toolbox [2] for MATLAB has been used; the population size has been set to 1000; default values of the remaining parameters have been used.

both GA and PSO this should come at the price of increased computational requirements as the number of times the optimization criterion alone had to be calculated is orders of magnitude higher than in the case of standard clustering methods. As it turns out, the computational time needed to calculate the criteria was negligible and in fact the running time of GA was comparable to that of EM-clustering, mostly due to the exhaustive search strategy for the optimal cluster count that had to be employed in the latter case. For the PSO, the clustering has been performed in about $70 - 80\%$ of the time required by GA. Although we recognize that some running time difference might be a result of differences in implementations, standard library versions have been used to alleviate this effect as much as possible.

An interesting thing to note are the values of DB and DI_m for the 'ground truth' i.e. the situation when the clusters correspond with the classes given with both datasets. Regardless of the clustering algorithm used, in all cases the obtained clusterings are scored better than the original classes in the data. This means that the original classes do not admit to the notions of compactness and separation, as defined by the two criteria used. In other words if one is after discovering the original classes in the data, neither DB nor DI_m is an appropriate choice in this case, as apparently neither of them is able to reflect the true class structure. This confirms the importance of choosing a proper, application-specific clustering criterion as postulated in this study.

7 Conclusions

In this paper we have proposed a unified view of clustering as an optimization problem. We have investigated the importance of conscious selection of an objective function, which is appropriate in the context of a particular application and should be used not only for assessing the clustering quality but also as an optimization criterion. We have also presented a generalized clustering framework and provided a proof-of-concept, which validated the above ideas.

During this study a number of challenges has been identified. The most notable of these challenges are:

1. Identification of exact objective functions optimized by popular clustering algorithms as well as properties of the clustering results produced. As mentioned before, it is not clear what objective functions are optimized by many popular clustering algorithms or if it is even possible to express those in a closed form. This kind of knowledge would however allow to better understand the characteristics of problems for which a given clustering algorithm is most suitable.
2. Development of advanced heuristics to constrain the search space and overcome high computational requirements of general-purpose optimization methods. The size of the search space when viewing clustering as an optimization problem is immense. On the other hand it is also immense in many other problems yet there are techniques which successfully deal with this issue. In case of clustering, good performance of PSO proves that optimization

is a viable approach and we believe that there still is a lot of room for improvement.
3. Development of new objective functions and clustering quality measures fitting different types of applications, together with optimization approaches dedicated for these objective functions. Although there is no 'one-fits-all' solution, there certainly exist groups of problems which have similar properties. Identification of these groups and development of tailored clustering objective functions is an interesting challenge.

We hope that this study, which is a first step towards designing of a new breed of clustering approaches will stimulate and inspire further research in this exciting area of data mining.

Acknowledgments. The research leading to these results has received funding from the EU 7th Framework Programme (FP7/2007-2013) under grant agreement no. 251617.

References

1. Asuncion, A., Newman, D.: UCI Machine Learning Repository (2007)
2. Birge, B.: PSOt – a particle swarm optimization toolbox for use with Matlab. In: Proceedings of the 2003 IEEE Swarm Intelligence Symposium SIS03 Cat No03EX706, pp. 182–186 (2003)
3. Budka, M., Gabrys, B.: Correntropy-based density-preserving data sampling as an alternative to standard cross-validation. In: The 2010 International Joint Conference on Neural Networks (IJCNN), pp. 1–8 (July 2010)
4. Budka, M., Gabrys, B.: Ridge regression ensemble for toxicity prediction. Procedia Computer Science 1(1), 193–201 (2010)
5. Davies, D., Bouldin, D.: A cluster separation measure. IEEE Transactions on Pattern Analysis and Machine Intelligence (2), 224–227 (1979)
6. Dempster, A., Laird, N., Rubin, D.: Maximum Likelihood from Incomplete Data via the EM Algorithm. Journal of the Royal Statistical Society. Series B (Methodological) 39(1), 1–38 (1977)
7. Dubes, R.: How many clusters are best?-an experiment. Pattern Recognition 20(6), 645–663 (1987)
8. Duda, R., Hart, P., Stork, D.: Pattern Classification, 2nd edn. John Wiley & Sons, New York (2001)
9. Duin, R., Juszczak, P., Paclik, P., Pekalska, E., de Ridder, D., Tax, D., Verzakov, S.: PR–Tools 4.1. A MATLAB Toolbox for Pattern Recognition (2007), `http://prtools.org`
10. Dunn, J.: Well-separated clusters and optimal fuzzy partitions. Journal of Cybernetics 4(1), 95–104 (1974)
11. Fletcher, R.: Practical methods of optimization, 2nd edn. Wiley (2000)
12. Fraser, A.: Simulation of genetic systems by automatic digital computers vi. epistasis. Australian Journal of Biological Sciences 13(2), 150–162 (1960)
13. Hamming, R.: Error detecting and error correcting codes. Bell System Technical Journal 29(2), 147–160 (1950)

14. Jaccard, P.: Etude comparative de la distribution florale dans une portion des Alpes et du Jura (1901)
15. Jain, A., Murty, M., Flynn, P.: Data clustering: a review. ACM Computing Surveys (CSUR) 31(3), 264–323 (1999)
16. Jenssen, R., Erdogmus, D., Hild, K.E., Príncipe, J.C., Eltoft, T.: Optimizing the Cauchy-Schwarz PDF Distance for Information Theoretic, Non-parametric Clustering. In: Rangarajan, A., Vemuri, B.C., Yuille, A.L. (eds.) EMMCVPR 2005. LNCS, vol. 3757, pp. 34–45. Springer, Heidelberg (2005)
17. Kennedy, J., Eberhart, R.: Particle swarm optimization. In: Proceedings of IEEE International Conference on Neural Networks, vol. 4, pp. 1942–1948. IEEE (1995)
18. Kohavi, R., Deng, A., Frasca, B., Longbotham, R., Walker, T., Xu, Y.: Trustworthy online controlled experiments: Five puzzling outcomes explained. In: KDD 2012, Beijing China, August 12-16 (2012)
19. MacQueen, J., et al.: Some methods for classification and analysis of multivariate observations. In: Proceedings of the Fifth Berkeley Symposium on Mathematical Statistics and Probability, California, USA, vol. 1, p. 14 (1967)
20. Sibson, R.: Slink: an optimally efficient algorithm for the single-link cluster method. The Computer Journal 16(1), 30–34 (1973)

IPMA: Indirect Patterns Mining Algorithm

Tutut Herawan[1], A. Noraziah[1], Zailani Abdullah[2],
Mustafa Mat Deris[3], and Jemal H. Abawajy[4]

[1] Faculty of Computer System and Software Engineering
Universiti Malaysia Pahang
[2] Department of Computer Science
Universiti Malaysia Terengganu
[3] Faculty of Science Computer and Information Technology
Universiti Tun Hussein Onn Malaysia
[4] Scholl of Information Technology
Deakin University
{tutut,noraziah}@ump.edu.my, zailania@umt.edu.my,
mmustafa@uthm.edu.my, jemal.abawajy@deakin.edu.au

Abstract. Indirect pattern is considered as valuable and hidden information in transactional database. It represents the property of high dependencies between two items that are rarely occurred together but indirectly appeared via another items. Indirect pattern mining is very important because it can reveal a new knowledge in certain domain applications. Therefore, we propose an Indirect Pattern Mining Algorithm (IPMA) in an attempt to mine the indirect patterns from data repository. IPMA embeds with a measure called Critical Relative Support (CRS) measure rather than the common interesting measures. The result shows that IPMA is successful in generating the indirect patterns with the various threshold values.

Keywords: Indirect patterns, Mining, Algorithm, Critical relative support.

1 Introduction

Data mining is the process of extracting some new nontrivial information from large data repository. In other definitions, data mining is about making analysis convenient, scaling analysis algorithms to large databases and providing data owners with easy to use tools in helping the user to navigate, visualize, summarize and model the data [1]. In summary, the ultimate goal of data mining is more towards knowledge discovery. One of the important models and extensively studies in data mining is association rule mining.

Since it was first introduced by Agrawal *et al.* [2] in 1993, Association Rule Mining (ARM) has been studied extensively by many researchers [3-21]. It is also known as "basket analysis". The general aim of ARM is at discovering interesting relationship among a set of items that frequently occurred together in transactional database [22]. However, under this concept, infrequent items are automatically considered as

N.T. Nguyen et al. (Eds.): *Adv. Methods for Comput. Collective Intelligence*, SCI 457, pp. 187–196.
DOI: 10.1007/978-3-642-34300-1_18

not important and pruned out. In certain domain applications, least items may also provide useful insight about the data such as competitive product analysis [23], text mining [24], web recommendation [25], biomedical analysis [26], etc.

Indirect association [27] refers to a pair of items that are rarely occurred together but their existences are highly depending on the presence of mediator itemsets. It was first proposed by Tan *et al.* [23] for interpreting the value of infrequent patterns and effectively pruning out the uninteresting infrequent patterns. Recently, the problem of indirect association mining has become more and more important because of its various domain applications [27-31]. Generally, the studies on indirect association mining can be divided into two categories, either focusing on proposing more efficient mining algorithms [24,27,31] or extending the definition of indirect association for different domain applications [5,27,28]

The process of discovering indirect association is a nontrivial and usually relies on the existing interesting measures. Moreover, most of the measures are not designed for handling the infrequent patterns. Therefore, in this paper we propose an Indirect Patterns Mining Algorithm (IPMA) by utilizing the strength of Least Pattern Tree (LP-Tree) data structure [11]. In addition, Critical Relative Support (CRS) measure [32] is also embedded in the model to mine the indirect association rules among the infrequent patterns.

The rest of the paper is organized as follows. Section 2 describes the related work. Section 3 explains the proposed method. This is followed by performance analysis through two experiment tests in section 4. Finally, conclusion and future direction are reported in section 5.

2 Related Work

Indirect association is closely related to negative association, they are both dealing with itemsets that do not have sufficiently high support. The negative associations' rule was first pointed out by Brin *et al.* [33]. The focused on mining negative associations is better that on finding the itemsets that have a very low probability of occurring together. Indirect associations provide an effective way to detect interesting negative associations by discovering only frequent itempairs that are highly expected to be frequent.

Until this recent, the important of indirect association between items has been discussed in many literatures. Tan *et al.* [23] proposed INDIRECT algorithm to extract indirect association between itempairs using the famous Apriori technique. There are two main steps involved. First, extract all frequent items using standard frequent pattern mining algorithm. Second, find the valid indirect associations from the candidate indirect association from candidate itemsets.

Wan *et al.* [24] introduced HI-Mine algorithm to mine a complete set of indirect associations. HI-Mine generates indirect itempair set (IIS) and mediator support set (MSS), by recursively building the HI-struct from database. The performance of this algorithm is significantly better than the previously developed algorithm either for synthetic or real datasets. IS measure [34] is used as a dependence measure.

Chen *et al.* [27] proposed an indirect association algorithm that was similar to HI-mine, namely MG-Growth. The aim of MG-Growth is to discover indirect association patterns and its extended version is to extract temporal indirect association patterns. The differences between both algorithms are, MG-Growth used the directed graph and bitmap to construct the indirect itempair set. The corresponding mediator graphs are then generated for deriving a complete set of indirect associations. In this algorithm, temporal support and temporal dependence are used in this algorithm.

Kazienko [25] presented IDARM* algorithm to extracts complete indirect associations rules. In this algorithm, both direct and indirect rules are joined together to form a useful of indirect rules. Two types of indirect associations are proposed named partial indirect association and complete ones. The main idea of IDARM* is to capture the transitive page from user-session as part of web recommendation system. A simple measure called Confidence [2] is employed as dependence measure.

Lin *et al.* [35] proposed GIAMS as an algorithm to mine indirect associations over data streams rather than static database environment. GIAMS contains two concurrent processes called PA-Monitoring and IA-Generation. The first process is to set off when the users specify the required window parameters. The second process is activated once the users issues queries about current indirect associations. In term of dependence measure, IS measure [34] is again adopted in the algorithm.

3 The Proposed Method

3.1 Association Rule

Throughout this section the set $I = \{i_1, i_2, \cdots, i_{|A|}\}$, for $|A| > 0$ refers to the set of literals called set of items and the set $D = \{t_1, t_2, \cdots, t_{|U|}\}$, for $|U| > 0$ refers to the data set of transactions, where each transaction $t \in D$ is a list of distinct items $t = \{i_1, i_2, \cdots, i_{|M|}\}$, $1 \le |M| \le |A|$ and each transaction can be identified by a distinct identifier TID.

Definition 1. *A set* $X \subseteq I$ *is called an itemset. An itemset with k-items is called a k-itemset.*

Definition 2. *The support of an itemset* $X \subseteq I$, *denoted* $\text{supp}(X)$ *is defined as a number of transactions contain X.*

Definition 3. *Let* $X, Y \subseteq I$ *be itemset. An association rule between sets X and Y is an implication of the form* $X \Rightarrow Y$, *where* $X \cap Y = \phi$. *The sets X and Y are called antecedent and consequent, respectively.*

Definition 4. *The support for an association rule* $X \Rightarrow Y$, *denoted* $\text{supp}(X \Rightarrow Y)$, *is defined as a number of transactions in D contain* $X \cup Y$.

Definition 5. *The confidence for an association rule* $X \Rightarrow Y$, *denoted* $\text{conf}(X \Rightarrow Y)$ *is defined as a ratio of the numbers of transactions in D contain* $X \cup Y$ *to the number of transactions in D contain X. Thus*

$$\text{conf}(X \Rightarrow Y) = \frac{\text{supp}(X \Rightarrow Y)}{\text{supp}(X)}$$

Definition 6. (Least Items). *An itemset X is called least item if* $\text{supp}(X) < \alpha$, *where* α *is the minimum support (minsupp)*

The set of least item will be denoted as Least Items and

$$\text{Least Items} = \{X \subset I \mid \text{supp}(X) < \alpha\}$$

Definition 7. (Frequent Items). *An itemset X is called frequent item if* $\text{supp}(X) \geq \alpha$, *where* α *is the minimum support.*

The set of frequent item will be denoted as Frequent Items and

$$\text{Frequent Items} = \{X \subset I \mid \text{supp}(X) \geq \alpha\}$$

3.2 Indirect Association Rule

Definition 8. *An item pair* $\{X\ Y\}$ *is indirectly associated via a mediator M, if the following conditions are fulfilled:*

a. $\text{supp}(\{X,Y\}) < t_s$ (itempair support condition)
b. There exists a non-empty set *M* such that:
   a) $\text{supp}(\{X\} \cup M) \geq t_m$ and $\text{supp}(\{Y\} \cup M) \geq t_m$ (mediator support condition)
   b) $\text{dep}(\{X\}, M) \geq t_d$ and $\text{dep}(\{Y\}, M) \geq t_d$, where $\text{dep}(A, M)$ is a measure of dependence between itemset *A* and *M* (mediator dependence measure)

The user-defined thresholds above are known as itempair support threshold (t_s), mediator support threshold (t_m) and mediator dependence threshold (t_d), respectively. The itempair support threshold is equivalent to *minsupp* (α). Normally, the mediator support condition is set to equal or more than the itempair support condition $(t_m \geq t_s)$

The first condition is to ensure that (X, Y) is rarely occurred together and also known as least or infrequent items. In the second condition, the first-sub-condition is to capture the mediator *M* and for the second-sub-condition is to make sure that *X* and *Y* are highly dependence to form a set of mediator.

Definition 9. (Critical Relative Support). *A Critical Relative Support (CRS) is a formulation of maximizing relative frequency between itemset and their Jaccard similarity coefficient.*

The value of Critical Relative Support denoted as CRS and

$$\mathrm{CRS}(A,B)=\max\left(\left(\frac{\mathrm{supp}(\mathrm{A})}{\mathrm{supp}(\mathrm{B})}\right),\left(\frac{\mathrm{supp}(\mathrm{B})}{\mathrm{supp}(\mathrm{A})}\right)\right)\times\left(\frac{\mathrm{supp}(A\Rightarrow B)}{\mathrm{supp}(\mathrm{A})+\mathrm{supp}(\mathrm{B})-\mathrm{supp}(A\Rightarrow B)}\right)$$

CRS value is between 0 and 1, and is determined by multiplying the highest value either supports of antecedent divide by consequence or in another way around with their Jaccard similarity coefficient. It is a measurement to show the level of CRS between combination of the both Least Items and Frequent Items either as antecedent or consequence, respectively. Here, Critical Relative Support (CRS) is employed as a dependence measure for 2(a) in order to mine the desired Indirect Association Rule.

3.3 Algorithm Development

Determine Minimum Support. Let I is a non-empty set such that $I=\{i_1,i_2,\cdots,i_n\}$, and D is a database of transactions where each T is a set of items such that $T\subset I$. An itemset is a set of item. A k-itemset is an itemset that contains k items. From Definition 6, an itemset is said to be least (infrequent) if it has a support count less than α.

Construct LP-Tree. A Least Pattern Tree (LP-Tree) is a compressed representation of the least itemset. It is constructed by scanning the dataset of single transaction at a time and then mapping onto a new or existing path in the LP-Tree. Items that satisfy the α (Definitions 6 and 7) are only captured and used in constructing the LP-Tree.

Mining LP-Tree. Once the LP-Tree is fully constructed, the mining process will begin using bottom-up strategy. Hybrid 'Divide and conquer' method is employed to decompose the tasks of mining desired pattern. LP-Tree utilizes the strength of hash-based method during constructing itemset in support descending order.

Construct Indirect Patterns. The pattern is classified as indirect association pattern if it fulfilled the two conditions. The first condition is elaborated in Definition 8 where there are three sub-conditions. One of them is mediator dependence measure. CRS from Definition 9 is employed as mediator dependence measure between itemset in discovering the indirect patterns. Figure 1 shows a complete procedure to construct the Indirect Patterns Mining Algorithm (IPMA).

```
IPMA
Input: Dataset
Output: Indirect Patterns
Read dataset, D
Set Minimum Support, minsupp
for (I_i ∈ D) do
    Determine support count, ItemSupp
end for loop
```

Fig. 1. A Complete Procedure of Indirect Patterns Mining Algorithm

Sort *ItemSupp* in descending order, *ItemSuppDesc*
for $(ISD_i \in ItemSuppDesc)$ **do**
  if $\text{supp}(\text{ISD}_\text{i}) \geq \text{minsupp}$ **then**
    $FItems \leftarrow ISD_i$
  end if
end for loop
for $(ISD \in ItemSuppDesc)$ **do**
  if $\text{supp}(\text{ISD}_\text{i}) < \text{minsupp}$ **then**
    Generate Least items, $(LItems \leftarrow ISD)_i$
  end if
end for loop
Construct Frequent and Least Items, $(FLItems \leftarrow FItems \bigcup LItems)$
for $(t_i \in T)$ **do**
  if $(LItems \bigcap i_i \in T > 0)$ **then**
    if $(t_i \in FLItems)$ **then**
      Construct items in transaction in descending order, $(TItemsDesc \leftarrow t_i)$
    end if
  end if
end for loop
for $(ID_i \in ItemsDesc)$ **do**
  Construct LP-Tree
end for loop
for $(pp_i \in LP\_Tree)$ **do**
  Construct Conditional Items, $(CondItems \leftarrow pp_i)$
end for loop
for $(CI_i \in CondItems)$ **do**
  Construct Conditional LP-Tree, $(CondLPT \leftarrow CI_i)$
end for loop
for $(CLPT_i \in Conditional\ \text{LP\_Tree})$ **do**
  Construct Patterns, $(P \leftarrow CLPT_i)$
end for loop
for $(p_i \in P)$ **do**
  Construct Candidate Itemspair, $(cIP_j \leftarrow P_i < t_s)$
  Construct Candidate Mediator, $(cM_j \leftarrow P_i > t_m)$
  if $(\text{supp}(cIP_j \cup cM_j) > t_m \ \&\ CRS(cIP_j \cup cM_j) \geq t_d)$ **then**
    Generate Indirect Patterns
  end if
end for

Fig. 1. *(continued)*

4 Experiment Test

In this section, the performance analysis is made by comparing the total number of patterns being extracted based on the predefined thresholds using our proposed algorithm, IPMA. Only 3 items are involved in forming a complete pattern; two items as an antecedent and one item as a consequence. The mediator is appeared as a part of antecedent. We conducted our experiment using two datasets. The experiment has been performed on Intel® Core™ 2 Quad CPU at 2.33GHz speed with 4GB main memory, running on Microsoft Windows Vista. All algorithms have been developed using C# as a programming language.

The first dataset is 2008/2009 intake students in computer science program (UPU-Intake dataset). The data was obtained from Division of Academic, Universiti Malaysia Terengganu in a text file and Microsoft excel format. There were 160 students involved and their identities were removed due to the confidentiality agreement. In the original set of data, it consists of 35 attributes and the detail information were explained in 10 tables which provided in Microsoft excel format.

Figure 2 shows the performance analysis againts UPU-Intake dataset. Minimum Support (*minsupp or* α) and Mediator Support Threshold (t_m) are set to 30% and 40%, respectively. Varieties of minimum CRS (min-CRS) were employed in the experiment. During the performance analysis, 46 patterns were produced. The trend is, the total number of indirect patterns were kept reducing when the values of min-CRS were kept increasing. However, there is no change in term of total indirect patterns when the min-CRS values were suddenly reached at 0.3 until 0.5.

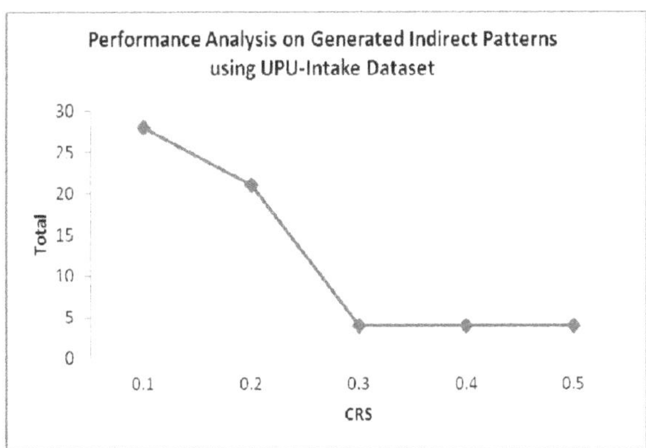

Fig. 2. Performance Analysis of Generated Indirect Patterns against Different CRS values against UPU-Intake Dataset

The second dataset is Car Evaluation from UCI Machine Learning Repository [36]. This dataset contains 1728 instances and 6 attributes. It was derived from a simple hierarchical decision model originally developed for the demonstration of DEX. The model evaluates cars according to six input attributes named buying, maint, doors, persons, lug_boot and safety.

Figure 3 shows the performance analysis of benchmarked Car Evaluation dataset. The datasets was downloaded from UCI Machine Learning Repository. The Minimum Support (*minsupp or* α) and Mediator Support Threshold (t_m) are fixed similar as used by the first experiment. Various minimum CRS (min-CRS) were also utilized during the experiment. From the analysis, 125 patterns were generated. The total number of indirect patterns were kept reducing when the values of min-CRS were kept increasing. However, there is no different in term of total indirect patterns being generated for the values of min-CRS at 0.2 and 0.3, respectively. Indeed, indirect patterns were not generated when the min-CRS value was reached at 0.6.

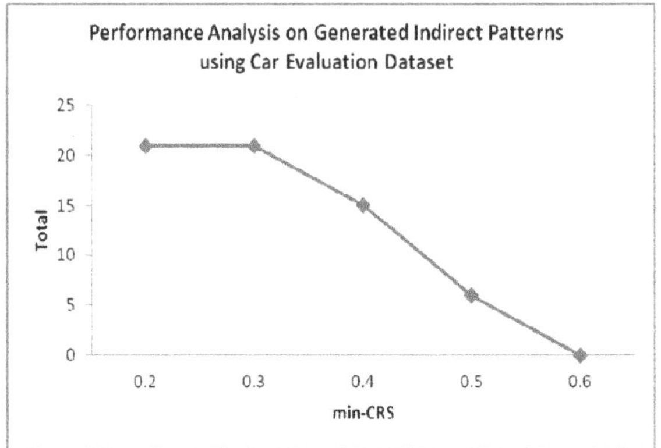

Fig. 3. Performance Analysis of Generated Indirect Patterns against Different CRS values against Car Evaluation Dataset

5 Conclusion

Mining indirect pattern is a very important study especially when dealing with rare events because it may contribute into discovering of new knowledge. Indirect pattern represents the property of high dependencies between two items that rare occurs together in the same transaction but actually is occurring indirectly via another itemset. Therefore, we proposed an algorithm called Indirect Pattern Mining Algorithm (IPMA) to extract the hidden indirect patterns from the data repository. IPMA embeds with a measure called Critical Relative Support (CRS) measure rather than the common interesting measures. We conducted the experiment based on a real dataset and one benchmarked dataset. The result shows that IPMA can successfully produce the different number of indirect patterns against the different threshold values. In the near future, we plan to apply IPMA into several benchmarked datasets and real datasets.

Acknowledgement. This work is supported by the research grant from Research Management Center of Universiti Malaysia Pahang.

References

1. Fayyad, U., Piatetsky-Shapiro, G., Smyth, P.: From Data Mining to Knowledge Discovery in Databases, pp. 37–54 (1996)
2. Agrawal, R., Srikant, R.: Fast Algorithms for Mining Association Rules in Large Databases. In: Proceedings of the 20th International Conference on Very Large Data Bases, pp. 487–499 (1994)
3. Mannila, H., Toivonen, H., Verkamo, A.I.: Discovery of Frequent Episodes in Event Sequences. Data Mining and Knowledge Discovery 1, 259–289 (1997)
4. Park, J.S., Chen, M.S., Yu, P.S.: An Effective Hash-based Algorithm for Mining Association Rules. In: Proceedings of the ACM-SIGMOD Int. Conf. Management of Data (SIGMOD 1995), pp. 175–186. ACM Press (1995)
5. Savasere, A., Omiecinski, E., Navathe, S.: An efficient algorithm for mining association rules in large databases. In: Proceedings of the 21st International Confenference on Very Large Data Bases (VLDB 1995), pp. 432–443. ACM Press (1995)
6. Fayyad, U., Patesesky-Shapiro, G., Smyth, P., Uthurusamy, R.: Advances in Knowledge Discovery and Data Mining. MIT Press, MA (1996)
7. Bayardo, R.J.: Efficiently Mining Long Patterns from Databases. In: Proceedings of the ACM-SIGMOD International Conference on Management of Data (SIGMOD 1998), pp. 85–93. ACM Press (1998)
8. Zaki, M.J., Hsiao, C.J.: CHARM: An efficient algorithm for closed itemset mining. In: Proceedings of the 2002 SIAM Int. Conf. Data Mining, pp. 457–473. SIAM (2002)
9. Agarwal, R., Aggarwal, C., Prasad, V.V.V.: A tree projection algorithm for generation of frequent itemsets. Journal of Parallel and Distributed Computing 61, 350–371 (2001)
10. Liu, B., Hsu, W., Ma, Y.: Mining Association Rules with Multiple Minimum Support. In: Proceedings of the 5th ACM SIGKDD International Conference on Knowledge Discovery and Data Mining, pp. 337–341. ACM Press (1999)
11. Abdullah, Z., Herawan, T., Deris, M.M.: Scalable Model for Mining Critical Least Association Rules. In: Zhu, R., Zhang, Y., Liu, B., Liu, C. (eds.) ICICA 2010. LNCS, vol. 6377, pp. 509–516. Springer, Heidelberg (2010)
12. Abdullah, Z., Herawan, T., Deris, M.M.: Mining Significant Least Association Rules Using Fast SLP-Growth Algorithm. In: Kim, T.-h., Adeli, H. (eds.) AST/UCMA/ISA/ACN 2010. LNCS, vol. 6059, pp. 324–336. Springer, Heidelberg (2010)
13. Abdullah, Z., Herawan, T., Noraziah, A., Deris, M.M.: Extracting Highly Positive Association Rules from Students' Enrollment Data. Procedia Social and Behavioral Sciences 28, 107–111 (2011)
14. Abdullah, Z., Herawan, T., Noraziah, A., Deris, M.M.: Mining Significant Association Rules from Educational Data using Critical Relative Support Approach. Procedia Social and Behavioral Sciences 28, 97–101 (2011)
15. Abdullah, Z., Herawan, T., Deris, M.M.: An Alternative Measure for Mining Weighted Least Association Rule and Its Framework. In: Zain, J.M., Wan Mohd, W.M.b., El-Qawasmeh, E. (eds.) ICSECS 2011, Part II. CCIS, vol. 180, pp. 480–494. Springer, Heidelberg (2011)
16. Abdullah, Z., Herawan, T., Deris, M.M.: Visualizing the Construction of Incremental Disorder Trie Itemset Data Structure (DOSTrieIT) for Frequent Pattern Tree (FP-Tree). In: Badioze Zaman, H., Robinson, P., Petrou, M., Olivier, P., Shih, T.K., Velastin, S., Nyström, I. (eds.) IVIC 2011, Part I. LNCS, vol. 7066, pp. 183–195. Springer, Heidelberg (2011)
17. Herawan, T., Yanto, I.T.R., Deris, M.M.: Soft Set Approach for Maximal Association Rules Mining. In: Ślęzak, D., Kim, T.-h., Zhang, Y., Ma, J., Chung, K.-i. (eds.) DTA 2009. CCIS, vol. 64, pp. 163–170. Springer, Heidelberg (2009)

18. Herawan, T., Yanto, I.T.R., Deris, M.M.: SMARViz: Soft Maximal Association Rules Visualization. In: Badioze Zaman, H., Robinson, P., Petrou, M., Olivier, P., Schröder, H., Shih, T.K. (eds.) IVIC 2009. LNCS, vol. 5857, pp. 664–674. Springer, Heidelberg (2009)
19. Herawan, T., Deris, M.M.: A soft set approach for association rules mining. Knowledge Based Systems 24(1), 186–195 (2011)
20. Herawan, T., Vitasari, P., Abdullah, Z.: Mining Interesting Association Rules of Student Suffering Mathematics Anxiety. In: Zain, J.M., Wan Mohd, W.M.b., El-Qawasmeh, E. (eds.) ICSECS 2011, Part II. CCIS, vol. 180, pp. 495–508. Springer, Heidelberg (2011)
21. Abdullah, Z., Herawan, T., Deris, M.M.: Efficient and Scalable Model for Mining Critical Least Association Rules. In a special issue from AST/UCMA/ISA/ACN 2010. Journal of the Chinese Institute of Engineer 35(4), 547–554 (2012)
22. Leung, C.W., Chan, S.C., Chung, F.: An Empirical Study of a Cross-level Association Rule Mining Approach to Cold-start Recommendations. Knowledge-Based Systems 21(7), 515–529 (2008)
23. Tan, P.N., Kumar, V., Srivastava, J.: Indirect Association: Mining Higher Order Dependences in Data. In: Proceedings of the 4th European Conference on Principles and Practice of Knowledge Discovery in Databases, pp. 632–637. Springer, Heidelberg (2000)
24. Wan, Q., An, A.: An Efficient Approach to Mining Indirect Associations. Journal Intelligent Information Systems 27(2), 135–158 (2006)
25. Kazienko, P.: Mining Indirect Association Rules for Web Recommendation. International Journal of Applied Mathematics and Computer Science 19(1), 165–186 (2009)
26. Tsuruoka, Y., Miwa, M., Hamamoto, K., Tsujii, J., Ananiadou, S.: Discovering and Visualizing Indirect Associations between Biomedical Concepts. Bioinformatics 27(13), 111–119 (2011)
27. Chen, L., Bhowmick, S.S., Li, J.: Mining Temporal Indirect Associations. In: Ng, W.-K., Kitsuregawa, M., Li, J., Chang, K. (eds.) PAKDD 2006. LNCS (LNAI), vol. 3918, pp. 425–434. Springer, Heidelberg (2006)
28. Cornelis, C., Yan, P., Zhang, X., Chen, G.: Mining Positive and Negative Association from Large Databases. In: Proceedings of the 2006 IEEE International Conference on Cybernatics and Intelligent Systems, pp. 1–6. IEEE (2006)
29. Kazienko, P., Kuzminska, K.: The Influence of Indirect Association Rules on Recommendation Ranking Lists. In: Proceeding of the 5th International Conference on Intelligent Systems Design and Applications, pp. 482–487 (2005)
30. Tseng, V.S., Liu, Y.C., Shin, J.W.: Mining Gene Expression Data with Indirect Association Rules. In: Proceeding of the 2007 National Computer Symposium (2007)
31. Wu, X., Zhang, C., Zhang, S.: Efficient Mining of Positive and Negative Association Rules. ACM Transaction on Information Systems 22(3), 381–405 (2004)
32. Abdullah, Z., Herawan, T., Noraziah, A., Deris, M.M.: Mining Significant Association Rules from Educational Data using Critical Relative Support Approach. Procedia Social and Behavioral Sciences 28, 97–191 (2011)
33. Brin, S., Motwani, R., Ullman, J., Tsur, S.: Dynamic itemset counting and implication rules for market basket data. In: Proceedings of the International ACM SIGMOD Conference, pp. 255–264. ACM Press (1997)
34. Tan, P., Kumar, V., Srivastava, J.: Selecting the Right Interestingness Measure for Association Patterns. In: Proceedings of the 8th International Conference on Knowledge Discovery and Data Mining, pp. 32–41 (2002)
35. Lin, W.-Y., Wei, Y.-E., Chen, C.-H.: A Generic Approach for Mining Indirect Association Rules in Data Streams. In: Mehrotra, K.G., Mohan, C.K., Oh, J.C., Varshney, P.K., Ali, M. (eds.) IEA/AIE 2011, Part I. LNCS, vol. 6703, pp. 95–104. Springer, Heidelberg (2011)
36. UCI Machine Learning Repository: Car Evaluation Data Set, `http://archive.ics.uci.edu/ml/datasets/`

Local Region Partitioning for Disguised Face Recognition Using Non-negative Sparse Coding

Khoa Dang Dang and Thai Hoang Le

University of Science, Vietnam National University, Ho Chi Minh City
{ddkhoa,lhthai}@fit.hcmus.edu.vn

Abstract. In this paper, three initializing methods for the Non-negative Sparse Coding are proposed for the disguised face recognition task in two scenarios: sunglasses or scarves. They aim to overcome previous sparse coding methods' difficulty, which is the requirement for a comprehensive training set. This means spending much more effort for collecting images and matching, which is not practical in many real world applications. To build a training set from a limited database containing one neutral facial images per person, a number of training images are derived from one image in the database using one of the three following partitioning methods: (1) grid-based partitioning, (2) horizontal partitioning and (3) geometric partitioning. Experiment results will show that these initialization methods facilitate Non-negative sparse coding algorithm to converge much faster compared to previous methods. Furthermore, trained features are more localized and more distinct. This leads to faster recognition time with comparable recognition results.

Keywords: Occluded face recognition, non-negative sparse coding, local region features.

1 Introduction

For many years, occluded face recognition has been a challenge to researchers in the field. The main difficulty is due to obscured face regions that cause the matching between training and testing samples to fail.

Recently, Sparse representation (SR) or Sparse coding (SC) has proven its ability in dealing with noised and incomplete data. A research [1] showed that sparse coding performs in a very similar way with human vision system. An image is represented by basic features, which are independent from each other. Moreover, authors also pointed out shortcomings of the Principle Component Analysis (PCA) that could be resolved by SC. State of the art approaches usually utilize comprehensive databases covering most various facial conditions (various expressions, illumination, ...) of each person. Given a test image, these methods perform matching on all training images, the more facial conditions available in the training set, the more accurate it is. This is not practical in some real-time applications when only one or two images per person presented. Besides, the matching phase could be slow due to a large amount of training images needed to be compared.

N.T. Nguyen et al. (Eds.): *Adv. Methods for Comput. Collective Intelligence*, SCI 457, pp. 197–206.
DOI: 10.1007/978-3-642-34300-1_19

To resolve the mentioned problems, it's required to find more generalized features from a limited database. In order to combine the strength of the sparse coding and the ability of local features in dealing with disguised faces, Shastri and Levine [2] have proposed the Non-negative sparse coding (NNSC) to learn local facial features. The "non-negative"term means coefficient sets and basic vectors are always positive. Based on a research in [3], human vision system combines signals together, not eliminate each other. Authors did not mention about how to set up a training set and how long to train basic vectors in order to archive such results as reported in the paper. So these experiments are re-performed based on the original method. However, the result was not as expected (the highest performance as when re-implemented following the paper was 21.5%). We argue this is because the random initialization step in the algorithm makes the training converge very slowly, and final basic vectors are not localized enough for recognition.

In this paper, three methods are proposed for constructing training sets by partitioning facial images into multiple initial basic images (represented as basic vectors), each contains a part of one's face. In this way, basic vectors converge more quickly and result in more localized features representing prominent facial regions. Moreover, only one image per person is needed to set up a training set. Three initialization methods are studied in this paper: (1) grid-based partitioning means dividing an image into a 3×3 grid, each cell corresponds to a region of one's face and used to construct an initial basic vector; (2) horizontal partitioning means an image is split into three horizontal regions, this is based on the analysis that occlusions (sunglasses or scarf) always cover horizontal parts of a face, eliminating that part may help to increase the chance to recognize a disguised face; (3) geometric partitioning means a facial image is split based on geometric characteristics of human face, [4] suggested dividing a face image into six geometric regions, and [5] integrated this technique successfully with PCA and LDA. Their method is included in this paper with a minor modification. Three proposed methods will be evaluated to study how the initialization step affects the recognition performance.

The rest of this paper is structured as following: Section 2 is a review of selective methods. A brief introduction of NNSC will be made in section 3. Next, three proposed initialization methods are presented. Experiments are evaluated in section 5 and an overall conclusion is drawn in the final section of this paper.

2 Related Works

Sparse coding method usually goes with local features to increase recognition performance against disguise (sunglasses or scarves). Wright et. al [6] successfully applied sparse coding on the AR database [7] with 97.5% and 93.5% for faces wearing sunglasses and scarves, respectively. Their method splits an image into a n×m grid and then performs sparse coding on each set of cells, results are combined to make a final decision. Moreover, their method also predicts occluded parts of an image, which could help to improve the recognition rate.

Derived from that, Nguyen [8] used an image pyramid including 4 sub images resized from an original image with different scales 1:1, 1:2, 1:4 and 1:8 and split images into 16, 8, 4 and 2 parts corresponding to the scales. Their highest results is 98.5% on both wearing-sunglasses and wearing-scarves data sets. Meanwhile, Yang et. al. [9] argued that the practical face recognition problem is not always follow Gaussian or Laplacian distribution, and they proposed to use the maximum likelihood estimation to be more robust to outliers. In another approach, Elhamifar and Vidal [10] modeled the face recognition problems as structured sparse representation involving in finding a representation for a test sample based on the minimum number of blocks from the dictionary.

3 An Introduction on Sparse Coding

3.1 Sparse Coding

Let $\boldsymbol{W} = [\boldsymbol{w}_1, \boldsymbol{w}_2, ...\boldsymbol{w}_n] \in \mathbb{R}^{d\times n}$ is a matrix with n columns corresponding to n basic images, each represented by a basic vector. $\boldsymbol{w}_i \in \mathbb{R}^{d\times 1}$. The sparse coding finds a sparse coefficient set: $\boldsymbol{h} \in \mathbb{R}^{1\times n}$ representing an image $\boldsymbol{y} \in \mathbb{R}^{d\times 1}$ so that $\boldsymbol{h}$ is sparest, it means:

$$\boldsymbol{h} = \min_{\boldsymbol{h}'} \|\boldsymbol{h}'\|_0 \text{ s.t. } \boldsymbol{y} = \boldsymbol{W}\boldsymbol{h} \ . \quad (1)$$

with$\|\bullet\|_0$ is the number of non-zero elements in a vector. [11] proved that solving this equation is an NP-hard problem. Authors also showed if vector $\boldsymbol{h}$ is sparse enough (only a few non-zero elements), (1) can be approximated the following equation:

$$\boldsymbol{h} = \min_{\boldsymbol{h}'} \|\boldsymbol{h}'\|_1 \text{ s.t. } \boldsymbol{y} = \boldsymbol{W}\boldsymbol{h} \ . \quad (2)$$

In real problems, 2 is hardly satisfied. So the problem now turns to find a vector $\boldsymbol{h}$ minimizing the following objective function:

$$J(\boldsymbol{W}, \boldsymbol{h}) = \|\boldsymbol{y} - \boldsymbol{W}\boldsymbol{h}\|_2^2 + \lambda \|\boldsymbol{h}\|_1 \ . \quad (3)$$

the scalar value λ is the trade-off between the sparsity of vector $\boldsymbol{h}$ and matching errors.

3.2 Non-negative Sparse Coding

Derived from the sparse coding, the NNSC applies a non-negative condition on matrix $\boldsymbol{W}$ and vector $\boldsymbol{h}$, so (2) becomes:

$$\min_{\boldsymbol{h}} \|\boldsymbol{y} - \boldsymbol{W}\boldsymbol{h}\|_2^2 + \lambda \|\boldsymbol{h}\|_1 \ . \quad (4)$$

so that $\forall i, j, k : \boldsymbol{W}_{ji} \geq 0, \boldsymbol{h}_k \geq 0,$ and $\|\boldsymbol{w}_i\|_2 = 1$, j and i are row and column indexes of matrix $\boldsymbol{W}$, $\boldsymbol{h}_k$ is the k-th coefficient of vector h and $\boldsymbol{w}_i$ is the i-th column vector of matrix $\boldsymbol{W}$.

This non-negative condition was proved in Hoyer's research[12]. Human visual system tends to combine received signals rather than 'cancel each other out'. He also proposed a method for learning non-negative sparse representation.

The learning phase of this method is to find a matrix $\boldsymbol{W} \in \mathbb{R}^{d\times n}$ containing n basic images and matrix $\boldsymbol{H} \in \mathbb{R}^{n\times m}$ are m coefficient vectors encrypting information of a training set.

The *gradient descent* method is used to solve (4). Let $\boldsymbol{V} = [\boldsymbol{v}_1, \boldsymbol{v}_2, ...\boldsymbol{v}_m] \in \mathbb{R}^{d\times m}$ is the initial training set containing m training image represented as a column vector $\boldsymbol{v}_i \in \mathbb{R}^{d\times 1}$. Hoyer [13] proposed the following learning algorithm:

NNSC learning method

Step 1 Randomly initialize matrix $\boldsymbol{W}$ and vector $\boldsymbol{H}$. Normalize vectors in $\boldsymbol{W}$ and $\boldsymbol{H}$ into unit-norm vectors

Step 2 Loop until convergence:
- ○ $\boldsymbol{W} \leftarrow \boldsymbol{W} + \mu(\boldsymbol{V} - \boldsymbol{W}\boldsymbol{H})\boldsymbol{H}^\top$
- ○ Assign negative elements of $\boldsymbol{W}$ by zeros
- ○ Normalize vectors in $\boldsymbol{W}$ into unit-norm vectors
- ○ $\boldsymbol{H} \leftarrow \boldsymbol{H}.*(\boldsymbol{W}'\boldsymbol{V})./(\boldsymbol{W}'\boldsymbol{V}\boldsymbol{H} + \lambda)$
- ○ Repeat Step 2

the two symbols .* and ./ are the element-wise multiplication and division operators. The convergence of (4) is when it less than a predefined threshold.

After training, we archive matrices $\boldsymbol{W}$, which contains basic vectors, and $\boldsymbol{H}$, which are coefficient vectors.

4 New Initialization Methods for the NNSC

In the previous NNSC method, basic vector matrix $\boldsymbol{W}$ is randomly initialized. This leads to a slowly convergence and poor performance because random initialization cannot assure to converge basic vectors into local facial features and also not distinct from each other. So it's not robust enough to occlusions such as sunglasses or scarves. Three initialing methods proposed in this paper guarantee for a faster convergence and more localized facial features. The results are presented in more detail in the experiment section.

First, all m images in the original database are normalized to the same size, n ($n \leq m$) images are picked up randomly for the following three partitioning methods to be studied: (1) grid based, (2) horizontal, and (3) geometric region based partitioning. Note that the random choice of n samples doesn't mean the algorithm just focus on these samples, but to make sure the process more general. The learning phase still utilizes all images in the database. The following sections explain in more detail about the partitioning process.

4.1 Grid-Based Partition

This partitioning method is quite popular such as in [6], [8]. An image is split into a $r \times c$ grid, with $c = r = 3$ in this paper. Grid cells are indexed from left to right, top to bottom, from 1 to 9. A cell from the k-th image $(1 < k < n)$ is chosen to initialize k-th column vector in matrix $\boldsymbol{W}$ so that the final basic vectors could represent all regions over a face and will not be limit in any specific region (Fig. 1).

4.2 Horizontal Partition

Based on a fact that the nature of occlusions (eyes or mouth region) are usually horizontal, partitioning in this way may leave out obscured regions and keep information of the clear ones. An original image is divided into r horizontal regions, $r = 3$ is chosen in experiments. For each k-th image, a region (among three regions) is used to initialize k-th vector in matrix $\boldsymbol{W}$ as illustrated in Fig. 2.

4.3 Geometric Region Based

In [5], authors proposed splitting an image into geometric regions based on International Civil Aviation Organization's (ICAO) suggestion [4]. This paper makes a little modification in referent to the occlusions. An original image is split in to 5 regions: face region excluding ears (region A), face region from forehead down to the mouth (region B), upper and lower half of a face (region C and D, respectively), and eye region (region E). Figure 3 shows 5 regions as discussed. For each image, a region is extract for being the k-th initial basic vectors. Not many research on this kind of geometric features available so far, so this paper would like to contribute in this research topic. Figure 4 presents the initialization using this partitioning method.

5 Face Recognition in Local Feature Space

Basically, recognition steps are similar to [2]. This process is summarized in Fig. 5. Given a database, only one straight face image are stored for each person, one

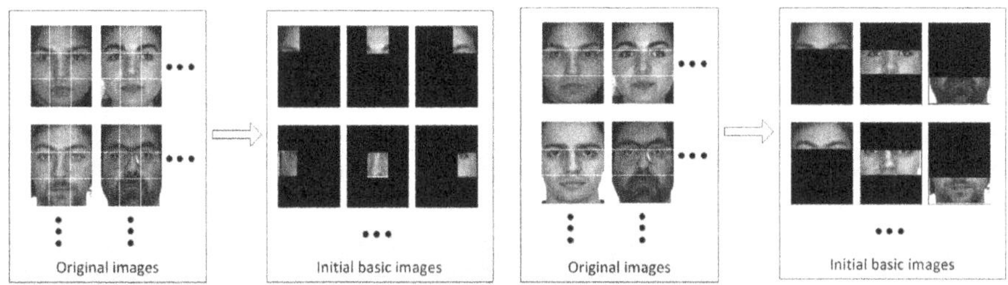

Fig. 1. Grid-based initialization

Fig. 2. Horizontal initialization

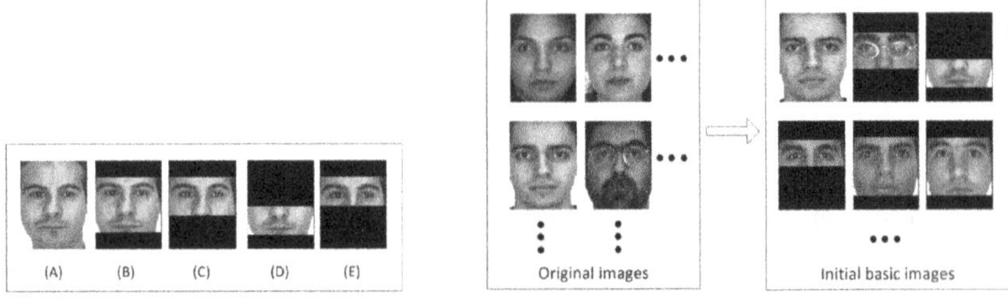

Fig. 3. Five facial geometric regions

Fig. 4. Geometric based partitioning

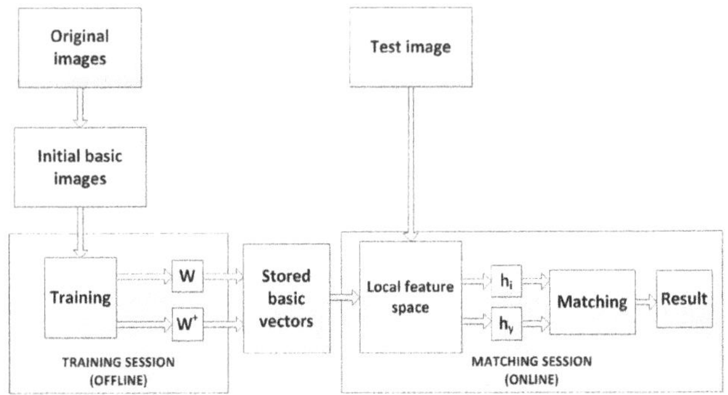

Fig. 5. Training and matching processes

of the three initializing methods are applied to construct an initial basic vectors set. This basic vector set is trained by the NNSC algorithm presented in section 3.2). The final matrix $\boldsymbol{W}$ contains basic vectors forming a local feature space. Then, features are extracted from training and testing set for matching.

5.1 Feature Extraction

Each image represented by a column vector $\boldsymbol{v}_i$ in a training set are projected onto matrix $\boldsymbol{W}^+$ to produce a coefficient set:

$$\boldsymbol{h}_i = \boldsymbol{W}^+ \boldsymbol{v}_i \ . \tag{5}$$

with $\boldsymbol{W}^+ = (\boldsymbol{W}^\top \boldsymbol{W})^{-1} \boldsymbol{W}^\top$.

Vector $\boldsymbol{h}_i$ is a feature vector extracted from i-th image, size of $\boldsymbol{h}_i$ depends on the number of basic vectors chosen during the training phase. Note that number of basic vectors also affects the generalization of learned features and will be further discussed in experiment section.

For a test image $\boldsymbol{y}$, its feature are extracted by the same way:

$$\boldsymbol{h}_y = \boldsymbol{W}^+ \boldsymbol{y} \ . \tag{6}$$

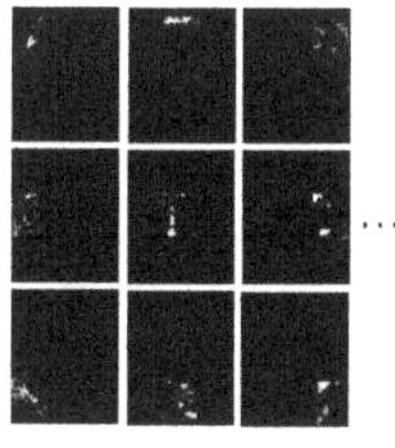

Fig. 6. Some basic vector samples learned from grid-based partitioning

5.2 Feature Matching

Two similarity distances L_1 and L_2 are used for matching.

L_1 similarity measurement. The similarity between two vectors $\boldsymbol{h}_i$ and $\boldsymbol{h}_y$ measured by L_1 distance is computed by sum of absolute difference between pairs of elements in two vector:

$$d(\boldsymbol{h}_i, \boldsymbol{h}_y) = L_1(\boldsymbol{h}_i, \boldsymbol{h}_y) = \sum_{k=1}^{n} |\boldsymbol{h}_i - \boldsymbol{h}_y| \ . \tag{7}$$

L_2 similarity measurement. The L_2 distance between two vector $\boldsymbol{h}_i$ and $\boldsymbol{h}_y$ is measured by square root of sum of squared difference between pairs of elements in two vector:

$$d(\boldsymbol{h}_i, \boldsymbol{h}_y) = L_2(\boldsymbol{h}_i, \boldsymbol{h}_y) = \sqrt{\sum_{k=1}^{n} (\boldsymbol{h}_i - \boldsymbol{h}_y)^2} \ . \tag{8}$$

6 Experiment

6.1 The Aleix-Robert (AR) Database

The AR[7] database includes straight faces of 100 objects, 50 men and 50 women. Each person's photos are taken in two different times (2 weeks apart), each time varies in expression, illumination, occlusion. First, images are normalized into gray scale images with the same size 160 × 125. Subset 8 (sunglasses) and 11 (scarf) are used to evaluate the proposed methods. Each subset contains 200 images of 100 people taken in 2 weeks. Figure 6 is a some learned basic vector samples from this data set.

6.2 Results

Experiment on Sunglasses Data Set. Figure 7a shows classification results for all three initializing methods on sunglasses subset using L_1 distance. Random initialization gave lowest performance in all cases. Grid based method performed best with number of basic vectors varies from 150 to 200. Figure 7b displays results corresponding to L_2 distance. The random method still archived lowest accuracy. Grid based method proved its effectiveness with the high results when the number of basic vectors is between 60 to 90 and got best result with 100 basic vectors.

From experimental results in Fig. 7a Fig. 7b, a conclusion is drawn that the grid based initializing method is the most suitable for recognizing faces wearing sunglasses. The highest score of this method is 90% using L_2 distance, an amount of time for matching is less than 0.1 second comparing to [8] which reported their highest accuracy is 92.5%, but it took 7.5 seconds for one image to be recognized. This is not practical in large scale databases. The proposed method in this paper is dominant when applied to real world applications.

Experiment on Scarf Data Set. Figure 8a and 8b report results on scarf data set. With 100 basic vectors, grid based initialization still get the best result measuring by L_1 distance. It's the same for 50 basic vectors and L_2 distance. So a conclusion for this is grid-based method is the most suitable for faces wearing scarves.

The highest score on this data set is 80% using L_1 measurement, less than 0.1 second for recognizing a test image compared to [8] reporting a result of 90% but consumed 8 second for recognizing one face. So the proposed method is still superior in time with comparable results.

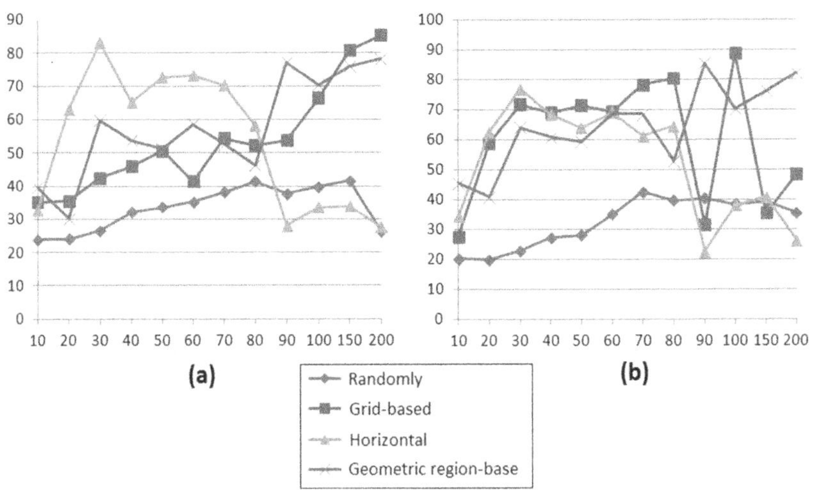

Fig. 7. Results on sunglasses data set using L1 (a) and L2 measurement (b)

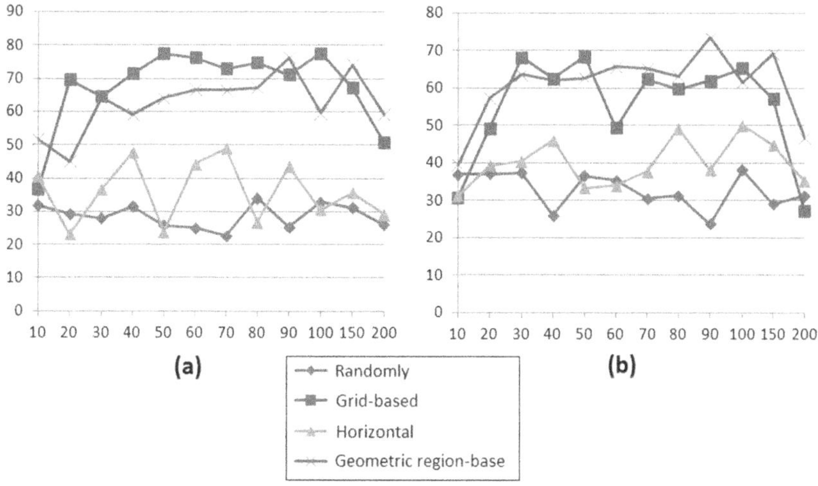

Fig. 8. Results on scarf data set using L1 (a) and L2 (b) measurement

7 Conclusion

In this paper, an improvement was proposed for the disguised face recognition problem (sunglasses or scarf) in case that only one image per person available. The main contribution is boosting the NNSC algorithm with the initialization step so that training converges more quickly, which leads to more distinct basic vectors and thus it's able to represent the whole database by a compact set.

Three initialization methods were tested: grid-based, horizontal and geometric region based partitioning. Experiments on AR database (sunglasses and scarf data sets) proved the proposed methods outperformed compared methods in Sect. 6.2 in time aspect but still retained comparable recognition results. Generally, grid based partitioning archived the best trade-off between time and results of all. This proves a capability to apply in disguised face recognition application with limited databases.

In future works, face recognition with various illumination and emotions will be further studied using the proposed methods. Hopefully, this work will be more completed to be able to apply in real world applications.

References

1. Olshausen, B., Field, D.: Emergence of simple-cell receptive field properties by learning a sparse code for natural images. J. Nature 381, 607–609 (1996)
2. Bhavin, S., Levine, M.: Face recognition using localized features based on non-negative sparse coding. J. Machine Vision and Application 18, 107–122 (2007)
3. Lee, D., Seung, S.: Algorithms for non-negative matrix factorization. In: 13th Advances in Neural Information Processing Systems, pp. 556–562. MIT Press (2000)

4. ISO/IEC JTC1 SC17 WG3/TF1 for ICAO-NTWG: History, interoperability and implementation, Machine readable travel documents. International Civil Aviation Organization (2007)
5. Le, T., Truong, H., Dang, K., Duong, D.: Using Genetic Algorithms to Find Reliable Set of Coefficients for Face Recognition. J. Information Technologies and Communications (JITC) 1(6(26), 124–133 (2011)
6. Wright, J., Yang, A., Ganesh, A., Sastry, S., Ma, Y.: Robust face recognition via sparse representation. In: IEEE Transactions on Pattern Analysis and Machine Intelligence, pp. 210–227. IEEE Press, Washington, DC (2009)
7. Martinez, M., Benavente, R.: The AR face database, `www2.ece.ohio-state.edu/~aleix/ARdatabase.html`
8. Nguyen, M., Le, Q., Pham, V., Tran, T., Le, B.: Multi-scale Sparse Representation for Robust Face Recognition. In: 3rd International Conference on Knowledge and Systems Engineering, Hanoi, Vietnam, pp. 195–199 (2011)
9. Yang, M., Zhang, L., Yang, J., Zhang, D.: Robust sparse coding for face recognition. In: IEEE Conference on Computer Vision and Pattern Recognition, pp. 625–632. IEEE Press (2011)
10. Elhamifar, E., Vidal, R.: Robust Classification using Structured Sparse Representation. In: IEEE Conference on Computer Vision and Pattern Recognition, pp. 1873–1879. IEEE Press (2011)
11. Huang, K., Aviyente, S.: Sparse representation for signal classification. In: Advances in Neural Information Processing Systems (2006)
12. Hoyer, P.: Probabilistic Models of Early Vision. Ph.D. Thesis. Helsinki University of Technology (2002)
13. Hoyer, P.: Non-Negative Sparse Coding. In: 7th IEEE Workshop on Neural Networks for Signal Processing, pp. 557–565. IEEE Press (2002)

Computer Based Psychometric Testing and Well Being Software for Sleep Deprivation Analysis

Vilem Novak[1], Radka Jeziorská[2], and Marek Penhaker[2]

[1] University Hospital Ostrava, Czech Republic
[2] VSB - Technical University of Ostrava,
Faculty of Electrical Engineering and Computer Science,
Ostrava, Czech Republic
vilem.novak@fno.cz,
{radka.jeziorska,marek.penhaker}@vsb.cz

Abstract. In clinical practice we often utilize measurements of psychic functions that help to determine diagnosis, disease stage but also favourable or unfavourable influences of the treatment. So called gnostic and executive psychic functions are important for the quality of life. The process of computer aided evaluation of the psychic functions is valuable especially from the reason of its acceleration and reduction in price. Higher degree of standardization seems to be another benefit of the computer aided evaluation. Within cooperation between the University Hospital Ostrava and the VSB Technical University of Ostrava we have developed software for measurement of the psychometric variables. A part of the research is also standardization of the results by a method of pilot testing within a group of healthy volunteers.

Keywords: Psychometric measurements, cognitive function, sleep laboratory, video-EEG, and software.

1 Introduction

Evaluation of the psychic functions, e.g. the gnostic functions that include selection, maintenance, classification and integration of information, learning and speech [1,2] is a part of the clinical examination in the neurology. The psychometric examination is often utilised within the medical clinical research. In the past, the testing was executed by means of various questionnaires and forms. Their processing was time-consuming. Software for the psychometric measurement accelerates the examination process and reduces its time and price and it gives also more precise results.

The following psychometric variables are measured routinely:

Attention:
- Simple reaction time
- Choice reaction time

Memory:
- Recent memory
- Long-term memory

N.T. Nguyen et al. (Eds.): *Adv. Methods for Comput. Collective Intelligence*, SCI 457, pp. 207–216.
DOI: 10.1007/978-3-642-34300-1_20

Speech functions:
Vocabulary and fluency
Praxis and motorcoordination:
Space memory, figures
Test of figure tracing (dominant and subdominant hand)
Complex functions
Digit symbol substitution test
Vigilance:
Epworth sleepiness scale

Changes of these functions are important indications of diseases as e.g. Alzheimer's disease [3], Parkinson's disease or sleep disorders. However, they might be also influenced by applied medicinal drugs. Most types of drugs affecting the central nervous system and many others such as antihistaminergic agents produce impairment in human cognitive functions [4]. Sleep disturbances and their impact to cognitive functions is another field for psychometric testing.

Psychometric measurements are therefore frequently used in clinical trials with new pharmaceutic agents to monitor their cognitive effects.

There are many computer-based systems measuring the psychic performance of the patient used both in clinical and research settings. One of the most sofisticated systems is the software product of the Cognitive Drug Research Ltd. (CDR) now as a part of the United BioSource Corporation. The system is being developped since 1970s and has more than 50 possible sub-tests localized to most languages. It is composed from a series of brief neuropsychological tests (batteries). Most of the tests are brief in duration (1 to 3 minutes). The computer keyboard is not used in any test. Many responses are made via USB device with YES and NO buttons [4]. The CDR system has become a standard in testing the cognitive effects in pharmaceutical research.

In general, the effect of the interface between computer and patient (buttons device, tablet, mouse, keyboard) is crucial for the accuracy, because of inequal computer skills in different persons.

2 Study Aim

Within the biomedical research programme of two cooperating institutions, a new application for the psychometric measurements was developed. The aim was to develop software performing various psychometric tests. The software might be utilised on PCs as well as on other types of hardware as tablets or smart mobile phones. Another aim was the standardization of tests' results that come from the computer aided testing of a group of healthy volunteers of selected age groups.

The integration of some testing modules to the video-EEG system with a possibility of attention deficit disorders detection and its correlation with epileptiform activity in EEG is recording as the next step.

3 Technical Specification

3.1 Development Environment

The program was created in C# language with utilisation of the development environment of Microsoft Visual C# 2008 Express Edition. This environment is used to develop console applications and graphical user interface applications along with Windows Forms applications, web sites, web applications, and web services in both machine code and managed code for all platforms supported by Microsoft Windows, Windows Mobile, and for some more supported platforms. The built-in tools include a forms designer for building GUI applications, web designer, class designer, and database schema designer. Data are saved to MS Access database by means of ODBC interface.

3.2 Controling the Software

During the psychometric software development we had to solve the problem of various levels in skill to work with a computer mouse. Ability to use the computer mouse might significantly distort inter-individual comparability of the results of some tasks.

Therefore the tablet was selected as a compromise between the mouse control and realisation of a special control element. It consists of a fixed pad with the active rectangular area and from a moving sensing device in a form of the wireless pen. This computer input periphery enables controlling the computer by a similar way as the computer mouse (cursor control).

3.3 Tablet G-Pen F610

The tablet G-Pen F610 was selected for the testing (Pict. 12). It deals with a widescreen tablet that disposes of 150 x 250 mm (6 x 10") working area. It is connected to a computer by means of USB interface from which it is also supplied. For the intrinsic work with the tablet a pressure-sensed pen serves that can distinguish 1024 levels of the pressure for precise sensing. The pen is wireless and disposes of special push buttons that serve as the right and left mouse keys.

When the tablet is used for the testing, the mode of absolute positioning when the pen position on the tablet corresponds to the cursor position on the screen is necessary. The software is also adapted to that mode. Individual elements of the forms can adapt to the screen resolution and therefore screens of various dimensions and of various resolutions might be used.

For the tasks of „Numeral Row Alignment“ and „Test of Reaction Time“ templates were created (Fig. 2) that enable better control of the tasks by tablet. These templates are inserted under the upper foil of the tablet.

Fig. 1. Tablet G-Pen F610

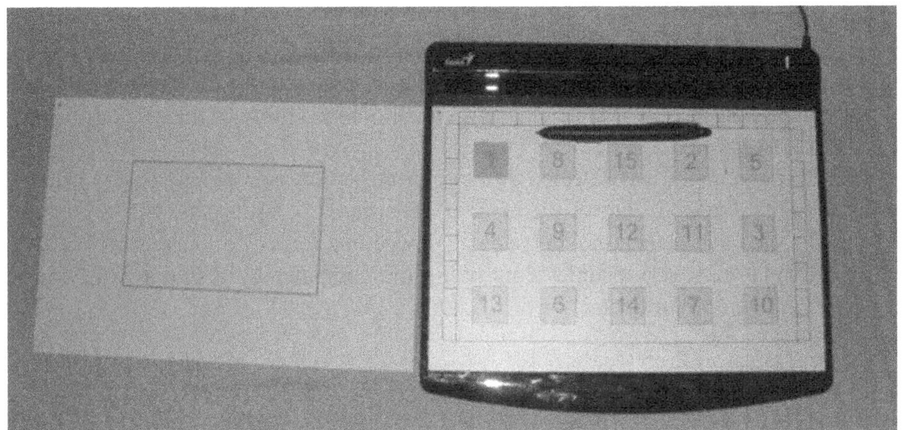

Fig. 2. Tablet and Templates for Test

4 Implemented Tests

4.1 Memory Recall Test

Thirty words are gradually displayed on the screen to the tested subject. The task is to recall as many words as possible, when their order is not important. The words recall might be tested immediately after their presentation (early recall) as well as later (delayed recall). Numbers of correct and incorrect answers are recorded.

4.2 Numeral Row Alignment

This task is to order numbers from 1 up to 15 as quickly as possible by clicking it. A form for this task contains 15 keys with the individual numbers (Fig. 3). These keys are arranged in three rows and five columns and their order is random. After clicking on the correct key it disappears. Not only time, in which the respondent arranges the numeral row in correct order, but also number of incorrect answers/clicks is recorded. This test is to be performed on the tablet.

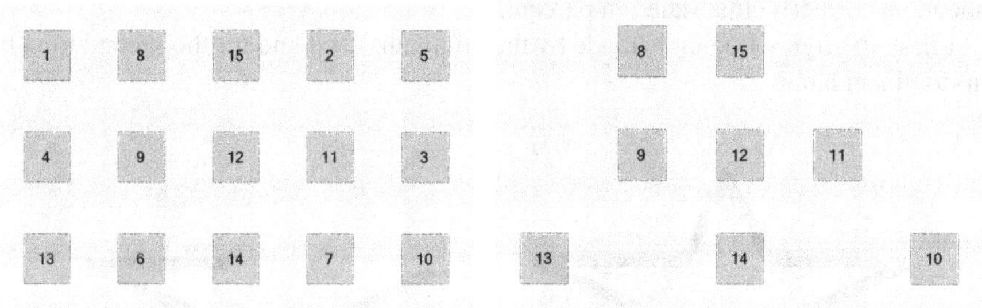

Fig. 3. Ordering of Numeral Row

4.3 Measurement of Simple Reaction Time

This task comes from the psychomotor tests of wakefulness when the respondent reacts by pushing a key when a bulb is switched on. In our implementation a green rectangle appears on the monitor that changes its colour to red in random time intervals (Fig. 4). The tested subject has to click as quickly as possible after the colour was changed from green to red. The colour changes haphazardly within the interval from 2 up to 10 seconds. The total time of this task is 2 minutes. Reaction times are saved in the database. When the task is ended, an average reaction time is calculated. A number of incorrect clicks, i.e. a click when the rectangle is green, is also an important piece of information.

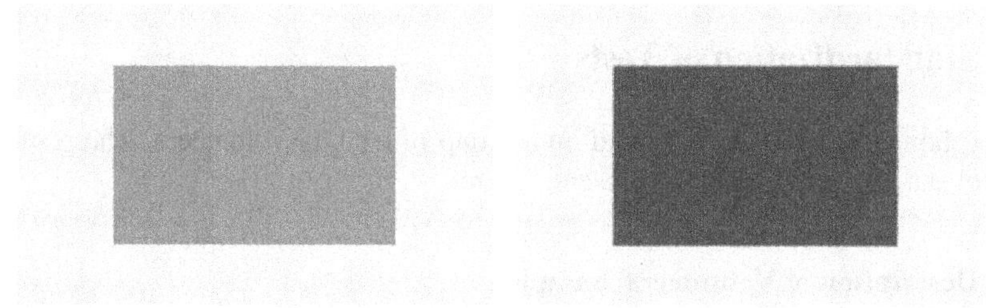

Fig. 4. Measurement of Simple Reaction Time

4.4 Figure Tracing

The task is focused on space orientation and movement coordination. There is a figure in the shape of a star on the screen (Fig. 5). The task is to trace the shape in clockwise direction. The task is complicated by the fact that cursor reaction does not comply with the mouse movement in upwards and downwards direction while the cursor reaction complies with the mouse movement in leftwards and rightwards direction.

The shape traced correctly changes the colour. If the examined person gets off the figure while tracing, the cursor changes its colour to red and it is necessary to return to the point where it got off. The monitored value is a trajectory that is traced by the respondent correctly. It is stated in percent.

At first, the figure tracing is made by the dominant hand and for the second time by non-dominant hand.

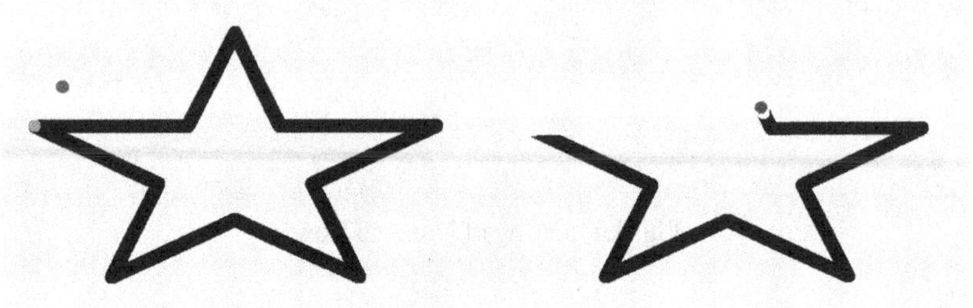

Fig. 5. Figure Tracing Test

4.5 Epworth Sleepiness Scale

It is a simple questionnaire for adults' sleepiness measurement. The patient self-evaluates his/her tendency for falling asleep on a scale of 4 grades (from 0 up to 3 points) in 8 common daily situations. The sum of points might range from 0 up to 24 while 10 points and less are considered the normal value, 11 points and more give evidence of excessive daily sleepiness. The software offers electronic implementation of this widely used questionnaire [4,5].

5 Standardization of Tests

The psychometric tests were applied on a group of healthy volunteers. The results of this pilot statistic processing are presented here.

5.1 Description of Volunteers' Sample

The same number of men (n=30) and women (n=30) took part in the measurement. Number of tested with dominant right hand is n=51 what creates 85% of all respondents. Number of testees with dominant left hand is n=9 what creates 15% of all respondents

The testing was perfomed since the 15th March to 4th April 2011. In total, 65 persons took part in the testing. Their age ranged from 21 to 62 years from which 34 men and 31 women created the sample.

The examined group of persons that was selected for this study is the group of young people of age between 20 to 30 years who, in time of the testing, did not suffer from any hypnophrenosis and who had normal sleep habits. 60 respondents of the age from 21 to 28 years complied with this criterion from which there were 30 men and 30 women. Before the intrinsic test completing the tested were asked not to use any drinks with caffeine content, e.g. tea or coffee, and also not to smoke.

Two versions of the program were created for the testing in order that we obtained the same parameters. The first version of the program needs assistance of a person testing for the task 1 – Words presentation and their consequent recall. The tested subject does not write the recalled words to the database himself/herself but the person testing can see a list of all the words after their presentation and ticks correct words in the list that the tested gradually recalls. The advantage for this version is that numbers of words are directly saved to the database for the individual attempts. The second version does not require assistance of the person testing because the examined person writes the recalled words directly to the database. The action of the person testing is not necessary earlier than at results evaluation when the presented words must be compared manually with the recalled words.

Time demand on the test gave rise to the creation of two versions because its length ranges approximately from 20 up to 30 minutes. Therefore a part of data was collected by means of the first version when the tablet was used for controlling. The rest of data was obtained from the respondents to whom the program was sent by e-mail. The advantage of this version was addressing of a huge number of persons. The response rate was about 50% in this case.

5.2 Results Assessment

Sixty five persons of age from 21 to 62 years took part in the testing. From the point of view of the statistical analysis, the respondents aged between 20 to 30 years were selected. It dealt with 60 tested from whom there were 30 men and 30 women aged from 21 to 28 years with the average value 24.1 ± 1.77 years. A number of the tested with dominant right hand was 51 (85%), a number of the tested with dominant left hand was 9 (15%).

5.3 Memory Recall Test

Within the first attempt from 4 to 17 correct words were recalled with the average value 10.1 ± 3.1 words. From the total number of 60 tested, 44 respondents did not have any wrong word and 45 respondents had maximally 1 duplicate word. Within the second attempt from 7 to 23 correct words were recalled with the average value 16.2 ± 3.8 words. From the total number of 60 tested, 54 respondents did not have any wrong word and 31 respondents had maximally 1 duplicate word. Within the third attempt from 14 to 28 correct words were recalled with the average value 21.0 ± 3.5 words. From the total number of 60 tested, 43 respondents did not have any wrong word and 34 respondents had maximally 1 duplicate word.

5.4 Ordering of Numeral Row

For the second task, the time necessary for ordering of the numeral row ranged from 8.7 to 22.4 seconds. There were some distant measurements caused by excellent skill in work with the mouse in the area of best times and on the contrary the worst times were the cause of insufficient skill in the tablet control. The average time was 13.8 ± 2.4 seconds. A number of wrong clicks ranged from 0 to 3 mistakes when 41 respondents coped with the task without a mistake.

5.5 Simple Reaction Time

Within the third task, the average simple reaction time moved in the range from 0.315 – 0.684 seconds. Two distant measurements occurred here that were the cause of insufficient skill in the tablet control. The average reaction time was 0.438 ± 0.067 seconds. A number of wrong clicks ranged from 0 to 4 mistakes when 40 respondents coped with the task without a mistake. Interesting question for a new measurement and psychological consideration that came out is the relationship between the colour and shape of the signal figures and the reaction time.

5.6 Figure Tracing

A range of the traced trajectory by dominant hand moved from 34% to 100% when the average trajectory was 70.7% ± 22.5%. A range of the traced trajectory by subdominant hand moved from 12% to 100% when the average trajectory was 69.3% ± 26.0%. When the relation between the trajectory length traced by dominant and subdominant hand was evaluated, it was found out that the difference between the dominant and subdominant hand is not statistically significant. This is surprising, because in general the non-dominant hand should be more successful in this type of test (we should perform more testing of this point).

5.7 Epworth Sleepiness Scale

The last task determinated the sleepiness level by means of the Epworth sleepiness scale [5, 6]. Within our study this method is considered as the reference one because its results might be compared with some studies that dealt with this topic. The total score range was from 1 to 16 with the average value 7.1 ± 3.2. One distant measurement was discovered which means that the tested subject probably suffers from hypersomnia.

6 Conclusion

Computer aided testing of psychometric variables seems to be very useful in the medical practice and research. Many testing procedures are well-grounded by a simple algorithm and they are suitable for the implementation in a form of the computer software. The software enables a quick and cheap measurement of the psychometric variables.

The effect of the subject-person/computer interface device is crucial for accuracy and reliability of the psychometric measurement. Keyboard and mouse is not very appropriate because of variable computer skills in population. The button device as used in the CDR system is more useful due to its simplicity. Use of pressure-sensed pen and tablet is another interesting way because of almost general writing skills in population.

In the future, we plan continual measurements of some psychometric parameters as e.g. the reaction time and its integration to video-EEG signal in a form of a polygraphic channel. The psychometric tests are important not only for the clinical practice but they also offer possibilities for some student within training of the programming and mathematic statistics at the Technical University.

Acknowledgment. The work and the contributions were supported by the project SV SP 2012/114 "Biomedical engineering systems VIII" and TACR TA01010632 "SCADA system for control and measurement of process in real time". Also supported by project MSM6198910027 Consuming Computer Simulation and Optimization. The paper has been elaborated in the framework of the IT4Innovations Centre of Excellence project, reg. no. CZ.1.05/1.1.00/02.0070 supported by Operational Programme 'Research and Development for Innovations' funded by Structural Funds of the European Union and state budget of the Czech Republic.

References

1. Boller, F., Grafman, J. (eds.): Handbook of Neuropsychology, vols. 1-11. Elsevier, Amsterdam (1988/1997)
2. Ardila, A.: Houston Conference: Need for More Fundamental Knowledge in Neuropsychology. Neuropsychology Review 12(3) (September 2002)
3. Lewis, L., Mahieux, F., Onen, F., Berr, C., Volteau, et al.: Early Detection of Patients in the Pre Demented Stage of Alzheimers Disease; The Pre-Al Study. The Journal of Nutrition, Health & Aging 13(1), 21 (2009)
4. Wesnes, K.A.: The value of assessing cognitive function in drug development. Dialogues Clin. Neurosci. 2(3), 183–202 (2000)
5. Johns, M.W.: A new method for measuring sleepiness: the Epworth sleepiness scale. Sleep 14, 540–545 (1991)
6. Epworthsleepinessscale.com, The Epworth sleepiness scale (2010), `http://epworthsleepinessscale.com/` (cit. January 02, 2011)
7. Havlík, J., Dvořák, J., Parák, J., Lhotská, L.: Monitoring of Physiological Signs Using Telemonitoring System. In: Böhm, C., Khuri, S., Lhotská, L., Pisanti, N. (eds.) ITBAM 2011. LNCS, vol. 6865, pp. 66–67. Springer, Heidelberg (2011)
8. Hudak, R., Michalikova, M., Toth, T., Hutnikova, L., Zivcak, J.: Basics of bionics and biomechanics: an e-learning course on the ulern platform. Lékař a Technika 36(2), 209–213 (2006) ISSN 0301-5491

9. Kasik, V.: Acceleration of Backtracking Algorithm with FPGA. In: 2010 International Conference on Applied Electronics, Pilsen, Czech Republic, pp. 149–152 (2010) ISBN 978-80-7043-865-7, ISSN 1803-7232
10. Cerny, M.: Movement Monitoring in the HomeCare System. In: Dössel, O., Schlegel, W.C. (eds.) WC 2009. IFMBE Proceedings, vol. 25/V, pp. 356–357. Springer, Heidelberg (2009)
11. Krejcar, O., Jirka, J., Janckulik, D.: Use of Mobile Phone as Intelligent Sensor for Sound Input Analysis and Sleep State Detection. Sensors 11(6), 6037–6055 (2011), doi:10.3390/s110606037, ISSN: 1424-8220
12. Krejcar, O.: Human Computer Interface for Handicapped People Using Virtual Keyboard by Head Motion Detection. In: Katarzyniak, R., Chiu, T.-F., Hong, C.-F., Nguyen, N.T. (eds.) Semantic Methods for Knowledge Management and Communication. SCI, vol. 381, pp. 289–300. Springer, Heidelberg (2011)
13. Katarzyniak, R.P., Chiu, T.-F., Hong, C.-F., Nguyen, N.-T. (eds.): Springer, Heidelberg, doi:10.1007/978-3-642-23418-7_25, ISBN 978-3-642-23417-0, ISSN 1860-949X
14. Krejcar, O., Motalova, L.: Home Care Web Services Evaluation by Stress Testing. In: Yonazi, J.J., Sedoyeka, E., Ariwa, E., El-Qawasmeh, E. (eds.) ICeND 2011, Part 16. CCIS, vol. 171, pp. 238–248. Springer, Heidelberg (2011)
15. Brida, P., Machaj, J., Benikovsky, J., Duha, J.: An ExperimentalEvaluation of AGA Algorithm for RSS Positioning in GSM Networks. Elektronika Ir Elektrotechnika 104(8), 113–118 (2010) ISSN 1392-1215
16. Liou, C.-Y., Cheng, W.-C.: Manifold Construction by Local Neighborhood Preservation. In: Ishikawa, M., Doya, K., Miyamoto, H., Yamakawa, T. (eds.) ICONIP 2007, Part II. LNCS, vol. 4985, pp. 683–692. Springer, Heidelberg (2008)
17. Brad, R.: Satellite Image Enhancement by Controlled Statistical Differentiation. In: Innovations and Advances Techniques in systems, Computing Sciences and Software Engineering, International Conference on Systems, Computing Science and Software Engineering, December 03-12. Electr. Network, pp. 32–36 (2007)
18. Choroś, K.: Further Tests with Click, Block, and Heat Maps Applied to Website Evaluations. In: Jędrzejowicz, P., Nguyen, N.T., Hoang, K. (eds.) ICCCI 2011, Part II. LNCS, vol. 6923, pp. 415–424. Springer, Heidelberg (2011)

Part V
Soft Computing

Dynamic Parameters in GP and LGP

Tomasz Łysek and Mariusz Boryczka

Institute of Computer Science,
University of Silesia, ul.Będzińska 39, Sosnowiec, Poland
{tomasz.lysek,mariusz.boryczka}@us.edu.pl

Abstract. Genetic Programming (GP) is one of Evolutionary Algorithms. There are many theories concerning setting values of main parameters that determine how many individuals will crossover or mutate. In this article we present a method of building dynamic parameter that will improve fitness function. In this way we create hybrid parameters that affect on individual. For testing we use our own dedicated platform. Our investigations of the best range of each parameter we based on our preliminary experiments.

Keywords: Genetic Programming, Linear Genetic Programming, Dynamic Parameters.

1 Introduction

Genetic Programming (GP) is one of Evolutionary Algorithms developed mostly by John Koza[6] and Wolfgang Banzhaf [4]. Few years ago there was a big breakthrough in techniques that provide great solutions for GP, and many researchers have tried to propose new versions of that approach. One of the most interesting is the latest work of Banzhaf and Brameier about Linear Genetic Programming [4]. Genetic Programming is an extension of Genetic Algorithm, and one of the population algorithms based on the genetic operations. The main difference between them is the representation of the structure they manipulate and the meanings of the representation. Genetic Algorithms usually operate on a population of fixed-length binary strings, GP typically operates on a population of parse trees that usually represent computer programs [7]. There are many modifications of GP that operate on other structures. The main factors independent of all population structures are parameters that decide how many individuals will crossover or mutate. It's common that higher value of crossover probability will result in better exploitation and high mutation will improve exploration. In this article we present a new hybrid parameters that will dynamically change values of the main parameters to achieve higher fitness value in shorter time. Our motivation is to create the unified hybrid parameter that will affect all parameters in the same time. For research we use created platform designed to test different GP algorithms. This platform uses authoring language in simple pseudo-code that is converted to C++. That conversion results in better code quality and provides fastest code for testing. Our idea is to improve Tree-based

N.T. Nguyen et al. (Eds.): *Adv. Methods for Comput. Collective Intelligence*, SCI 457, pp. 219–228.
DOI: 10.1007/978-3-642-34300-1_21

GP and Linear GP by using the best-known genetic operation variants for that algorithm and by solving the problem of string evaluation in the manner of fitness function evaluation. This article is organized as follows: first we analyse the related works and ideas of creating the most effective GP in literature. In third section we describe our idea of GP and LGP with build-in functions and libraries. Third section concerns creating dynamic parameter that changes main parameters during calculations. In last section we present experiments with GP and LGP in our platform and compare them to other solutions from literature. We summarize with short conclusions.

2 Related Work

Genetic Programming is strongly developed in different domains. From basic Koza's representation to Linear Genetic Programming it still remains the same algorithm. Brameier introduced population divided into two groups and the leaders of those groups are crossed [4]. Brameier also experimented with a graph version of individual in LGP where each individual is a program written in machine code and instructions are linear. Whole generated code is converted into directed graph. This is a huge leap between the classical approaches with tree-based code. Interesting innovation is to split processing into graphics processing units to improve time-consuming operations using parallel processing [5]. Another solution is to represent genotype as a binary code [1]. This version is based on encoding all set of symbols, compare operations and action symbols into RBGP (Rule-based Genetic Programming). Each individual in this population consists of multiple classifiers (rules) that transform input signals into output signals. The evaluation order of the rules plays absolutely no role except for rules concerning the same output bits. The basic idea behind RBGP approach is to use this knowledge to create a new program representation that retains high Θ-values in order to become more robust in terms of reproduction operations [1]. This solution in mutation and crossing over operators is similar to other genetic algorithms. The last interesting way of representing individual is Gene Expression Programming for Genotype-Phenotype Mappings (GPM) by Ferreira [1], which assumes that individual's structure is a string with head and tail. Head is a list of expressions (functions and symbols) and tail is a list of arguments (Fig. 1). All presented models of population are based on very similar crossover and mutation processes. There were experiments with adaptive parameters (Adaptive Genetic Algorithms) but that works don't include experiments with a predetermined interval of the best fitness function.

3 Genetic Programming and Linear Genetic Programming

Algorithms based on the classical Genetic Programming like LGP, RBGP, etc. have the same structure as a classical one. Because of that we assume that every GP-like algorithm that will be created also will be based on classical one [3]:

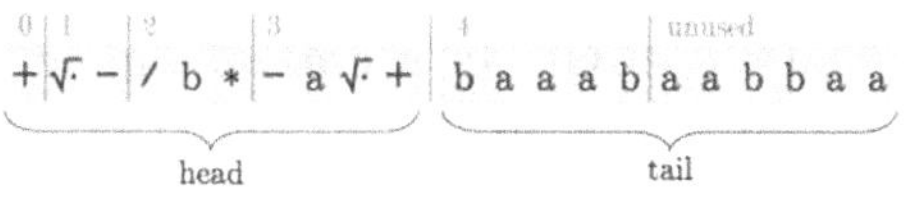

+ √ - / b * - a √ + b a a a b a a b b a a

head tail

Fig. 1. A GPM example for Gene Expression Programming

Algorithm 1. Classical Genetic Programming algorithm

```
Generate population P with random composition of defined functions
while stop criterion is not met do
    parse generated individuals (programs) to set value of fitness function
    copy the best individual
    create new programs using mutation
    create new programs using crossover
Individual with the best value of fitness function is the best solution
```

By the best solution we mean the best function approximation. In this research we make assumption that Genetic Programming may be regarded as prediction models that approximate the objective function:

$$f : I^n \rightarrow O^m$$

where I^n is an input data in the n-dimmensional space, and O^m is an output data in the m-dimensional space [4]. Brameier assumes that collection of input-output vectors describing function which represent the problem can be presented by:

$$T = \{(\boldsymbol{i}, \boldsymbol{o}) | \boldsymbol{i} \in I' \subseteq I^n, \boldsymbol{o} \in O' \subseteq O^m, f(\boldsymbol{i}) = \boldsymbol{o}\}$$

Genotype G in GP is a collection of programs in L language, and phenotype P is a set of functions that:

$$f_{gp} : I^n \rightarrow O^m : f_{gp} \in P$$
$$gp \in G$$

The fitness function will be as follows (V - value):

$$F : P \rightarrow V,$$

and predictable quality sets of phenotype as:

$$f_{gp} \in P.$$

By a reproduction we understand all genetic operations. For mutation there are replacement, insertion and deletion options. Also there is a crossover, permutation, editing and encapsulation. Permutation, like mutation, is used to reproduce only one single tree. The child nodes attached to selected node are shuffled randomly. The main goal is to rearrange the nodes in highly fit sub-trees in order to make them less fragile for other operations [1]. Editing is one of the most interesting genetic operations in GP. It is used to simplify mathematical formulas, e.g. operation 3+5 will be simplified to 7. The idea of the encapsulation operation is to identify potentially useful parts of generated program and to turn them into atomic building block. To improve quality of generated code we build-in Automatically Defined Functions (ADF) [1]. The concept of ADF provides modularity for GP. Finding a way to evolve modules and reusable buildings blocks is one of the key issues in using GP to derive higher-lever abstractions and solutions to more complex problems [1]. This is one of techniques that can cause a reduction of machine code in individual, in LGP algorithm. For node selection in trees and graphs our solution depend on Weise's method [1]. Weise points that in literature selection often speaks of "random selecting" without specifying this process. He propose to define weight of a tree node n to be a total number of nodes in the sub-tree with n as a root, i.e. itself, its children, grandchildren, etc. [1]:

$$nodeWeight(n) = 1 + \sum_{i=0}^{len(n.children)-1} nodeWeight(n.children[i]),$$

where the $nodeWeight$ of the root of a tree is the number of all nodes in the tree and the $nodeWeight$ of each of the leaf is exactly 1. With that modification there are no regions in the trees that have lower selection probabilities than others.

Linear Genetic Programming in our solution is represented by the following structure [5]:

- directed graph instead of classical tree structure or abstract syntax tree,
- linear structure of generated individual program,
- data flow using registers,
- sub-graphs used as functions, variables, etc.,
- algorithm for searching and eliminating ineffective code (expanded reproduction models).

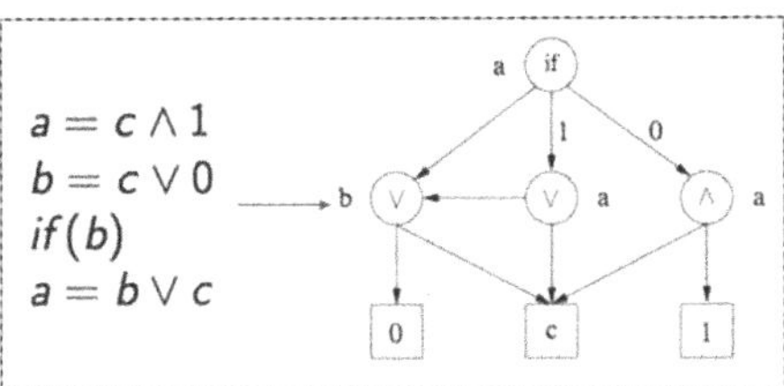

Fig. 2. LGP individual representation in Directed Graph structure

In fig. 2 is presented how the LGP is written as individual (single programm) and how it's image on graph structure. The fitness function depends on the type of a problem, but in all cases this is average of deviation between a solution given by generated program and the expected value. In experiments we use an average value that is average of individuals fitness function value.

4 Dynamic Parameters in Genetic Programming

To build dynamic parameters we use known crossover and mutation operators. There are some versions of crossover but it works slight different in tree-based GP and LGP. Tree-based crossover works by cutting out a part of one tree and exchanging it with a cut-out part of another tree. In LGP crossover is an operator that crosses chosen code-lines in individuals. Mutation in GP algorithms is more complicated, because we have many types of mutation process. Standard mutation affects the function used, e.g. when in tree or code-line we used adding function it can mutate into substitute function. Mutation can react not only with functions but also with terminals. More complicate mutation can perform permutation in tree or code. We also test add mutation, that adds new node in tree or code-line in LGP individual, and cut mutation that is opposite to add mutation. One of the most interesting mutation types is encapsulation that chooses part of tree or code and protects it from future changes. There are also operations dedicated to individual structure. For tree-based GP we test max tree-depth parameter and lifting. Max tree-depth parameter determines how big can be our individuals. Lifting is a special mutation operation that can lift chosen part of tree structure. We loose some part of tree but in effect we can produce shorter individual with same properties and in some case better fitness function. For LGP algorithm we tested max code-lines parameter and max operations in code-line. To test different parameters and determine the appropriate parameters ranges we use our platform and several standard test functions: Mexican Hat, 3 Chains and Schaffer, with tree-based GP and LGP. Parameters for experiments were:

- the size of population $N = 500$,
- the crossover parameter $CR = 0.9$,
- the mutation parameter $F = 0.1$,
- the maximum number of iterations is equal to 1000,
- the maximum program length (LGP) is equal to 200,
- the maximum tree depth (GP) is equal to 20,
- the maximum operator nodes is equal to 200,
- for every testable function the algorithm was run 10 times,
- parameter values for test were between 0.1 and 0.7 with 0.2 progress.

The best result for GP in literature has the fitness function value of 13.7 with depth equal to 16. For LGP fitness function is 5.2 with code length equal to 197. Table 1 presents results for Mexican Hat that were the best adapted in parameter scale. F parameter stand for fitness function value (lower is better). D in tree-based GP is tree depth, and L in LGP is code length.

As the ramification of that experiments we decided to build dynamic parameter that will change other parameters during computations. For each function parameter we chose different interval that can improve exploration or exploitation as needed. Interval was determined by tests of average fitness function value. In experiments we tested parameters with step 0,2. With that experiments we determine parameters values which exploatation are the best. Because of lack of

Table 1. Setting the best ranges for parameters in GP and LGP (part 1)

	function change				terminal change				permutation				inserting			
	GP		LGP		GP		LGP		GP		LGP		GP		LGP	
Parameters value	F	D	F	L	F	D	F	L	F	D	F	L	F	D	F	L
0,1	13,9	19	4,7	192	14,9	20	4,3	200	15,2	20	6,2	200	14,5	17	4,7	189
0,3	13,6	17	3,7	185	13,8	17	3,8	186	14,5	18	4,9	186	14,9	19	5,1	194
0,5	13,5	17	3,7	191	13,7	18	3,7	194	14,3	19	5,1	194	15,2	20	5,3	198
0,7	14,1	20	5,4	200	14,1	20	4,9	200	14,9	20	5,7	200	15,7	20	6,4	200

Table 2. Setting the best ranges for parameters in GP and LGP (part 2)

	cutting				encapsulation				ADF				Lifting	
	GP		LGP		GP		LGP		GP		LGP		GP	
Parameters value	F	D	F	L	F	D	F	L	F	D	F	L	F	D
0,1	14,9	16	4,2	186	14,1	16	4,2	188	14,2	18	4,7	193	13,8	18
0,3	15,1	18	4,9	189	14,4	17	4,6	193	13,8	14	4,2	186	13,6	16
0,5	15,6	18	5,3	194	15,3	19	4,9	197	14	14	5,4	182	13,9	17
0,7	15,9	19	5,4	197	15,8	20	5,1	199	14,5	14,5	9,6	179	14,3	20

space we don't present the entire process of experiments. In consequence hybrid parameters with γ dynamic parameter for tests have form:

- Crossover — range of the best exploration is between 0.8-0.9 for GP and 0.7-0.9 for LGP, then crossover will be $c_{gp} = 0,85 \cdot \gamma$ and $c_{lgp} = 0,8 \cdot \gamma$,
- Mutation — range of best exploration is between 0.1-0.15 for GP and 0.1-0.2 for LGP, then mutation will be $m_{gp} = 0,125 \cdot \gamma$ and $m_{lgp} = 0,15 \cdot \gamma$,
- Function change — range of the best exploration is between 0.3-0.5 for GP, then function change will be $fc = 0,4 \cdot \gamma$,
- Terminal change — range of the best exploration is between 0.4-0.5 for GP and 0.3-0.5 for LGP, then terminal change will be $tc_{gp} = 0,45 \cdot \gamma$ and $m_{lgp} = 0,4 \cdot \gamma$,
- ADF — range of the best exploration is between 0.3-0.5 for GP and 0.1-0.3 for LGP, then ADF will be $adf_{gp} = 0,4 \cdot \gamma$ and $adf_{lgp} = 0,2 \cdot \gamma$,
- Encapsulation — range of the best exploration is between 0.0-0,1 for GP and LGP, then encapsulation will be $enc = 0,05 \cdot 1/\gamma$,
- Cutting — range of the best exploration is between 0.1-0.3 for GP and 0.1-0.5 for LGP, then cutting will be $cut_{gp} = 0,2 \cdot \gamma$ and $cut_{lgp} = 0,3 \cdot \gamma$,
- Permutation — range of the best exploration is between 0.3-0.5 for GP and LGP, then permutation will be $perm = 0,4 \cdot \gamma$,
- Insertion — range of the best exploration is between 0.0-0.1 for GP and 0.0-0.2 for LGP, then insertion will be $ins_{gp} = 0,05 \cdot 1/\gamma$ and $ins_{lgp} = 0,1 \cdot 1/\gamma$,
- Lifting — range of the best exploration is between 0.2-0.3, then lifting will be $lift = 0,25 \cdot \gamma$,

For initial population parameter γ is equal to 1 but after calculating fitness function of second population the parameter is changed in range between 0 and 1. If close to 1 the algorithm is more explorative, but when less than 1 then it is more exploitative.

5 Experiments and Results

The purpose of experiments was to compare effectiveness of GP and LGP in C++ to same algorithms written and compiled in presented platform. For all experiments described in this section we assume the settings as in the previous section. For experiments we use standard Mexican Hat (fig. 3) and 3 chain function (fig. 4) functions. Also we use Schaffer function (fig. 5) for better presentation of results.

For our experiments we use Mexican Hat function [4]:

$$f(x,y) = (1 - \frac{x^2}{4} - \frac{y^2}{4}) \cdot e^{(-\frac{x^2}{8} - \frac{y^2}{8})}$$

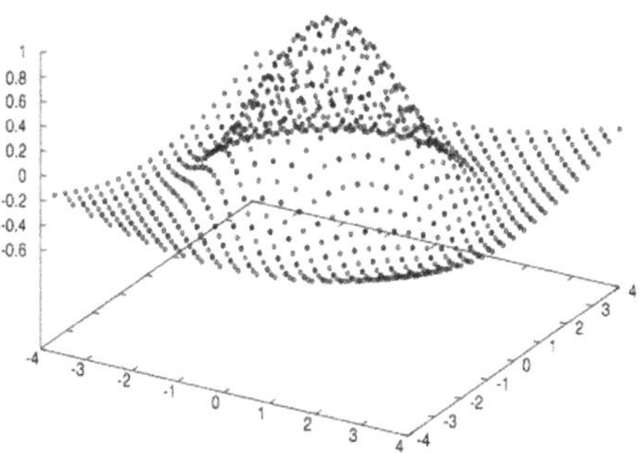

Fig. 3. Mexican Hat function

3 Chains function:

$$f(x_1, y_1, ..., x_n, y_n) = \sqrt{(x_1 - y_1)^2 + ... + (x_n - y_n)^2}$$

and Schaffer function:

$$f(x,y) = 0,5 + \frac{(sin(\sqrt{x^2+y^2}))^2 - 0,5}{(1+0,001*(x^2+y^2))^2}$$

In table 2 and 3 we present results for GP and LGP with established test parameters:

- F — the average fitness function value,
- D — for GP — a tree depth of the best individual,
- L — for LGP — a program length of the best individual.

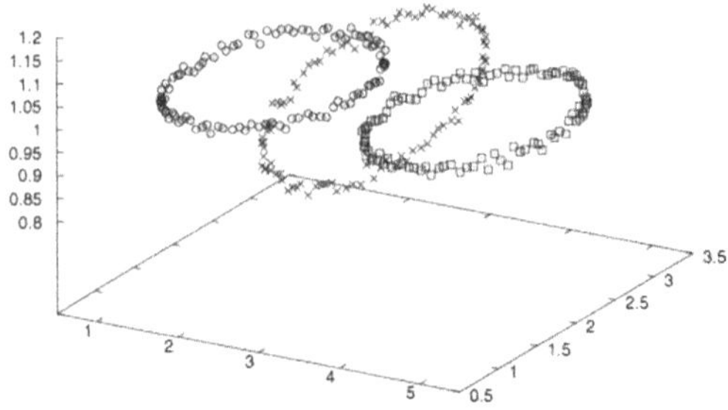

Fig. 4. 3 Chains function

On figures 6 and 7 we present how the fitness function improved when we used dynamic parameters. We conclude that dynamic parameter in all tests gave better results. Interesting fact is that in population size 100 or less, parameter is not so effective in all cases. If population size is bigger than 150 the results obtained are similar as normal algorithm or even better. Same results are on GP and LGP tests. Parameters D and L not correrable because they indicate average largeness of individual. As we conclude out dynamic parameter impoves fitness function value with similar D / L value.

Table 3. Test results for dynamic parameters in GP

	Population 100		Population 300		Population 500	
Name	F	D	F	D	F	D
Mexican Hat	15,9	18	14,8	18	13,5	15
Mexican Hat (γ)	16,4	19	13,7	16	13,4	15
3 chains	18,4	20	16,7	20	14,3	20
3 chains (γ)	18,2	20	14,5	19	14,1	18
Shaffer	23,1	20	20,5	20	18,2	20
Shaffer (γ)	23,4	20	20,3	20	17,8	19

An analysis of performed experiments showed that LGP is not always a better solution for analysed functions. Another interesting conclusion is that using hybrid parameters with dynamic γ parameter can improve algorithm. This is great improvement in GP algorithms. In the few tests presented parameter γ produces better results every test function. Also we showed that dynamic parameter improved our latest experiments. We can assume that if there will be possibility to unify all parameters in one equation with dynamic parameter then it still be good for building populations and don't impair fitness of individuals.

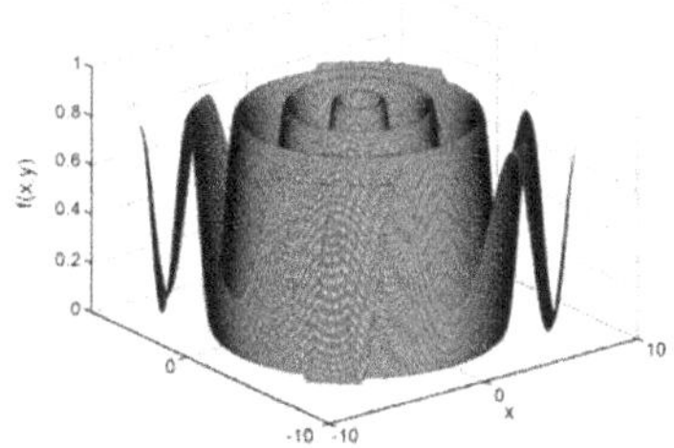

Fig. 5. Scheffer function

Table 4. Test results for dynamic parameters in LGP

	Population 100		Population 300		Population 500	
Name	F	L	F	L	F	L
Mexican Hat	4,7	195	4,1	191	3,5	184
Mexican Hat (γ)	4,5	187	3,8	182	3,2	176
3 chains	10,2	197	9,0	195	7,3	189
3 chains (γ)	10,1	197	7,3	192	5,9	188
Shaffer	16,5	200	9,2	200	8,6	198
Shaffer (γ)	16,2	200	8,4	199	7,1	194

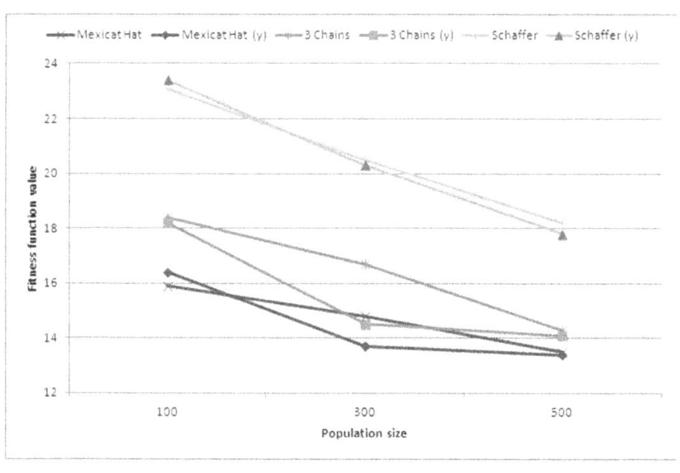

Fig. 6. Dynamic operator in GP

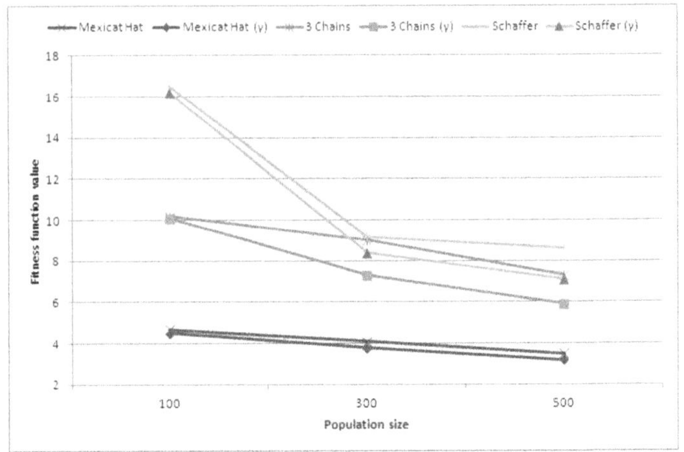

Fig. 7. Dynamic operator in LGP

6 Conclusions and Future Work

This shows that dynamic parameter can be added to parameter without any negative consequences. In fact it can improve fitness of individuals and their quality. In future we will try to unify all parameters to a simple equation for changing each individual in population. Another interesting work to do is to a new mutation function that will affect on all GP-type algorithms with any individual structure. In future research we will create more complex tests and better hybrid solution for dynamic parameter.

References

1. Weise, T.: Global Optimization Algorithms: Theory and Application, pp. 169–174, 191–195, 207–208 (2009)
2. Engelbrecht, A.: Computational Intelligence: An Introduction, 2nd edn., pp. 177–184. John Wiley and Sons Ltd. (2007)
3. Banzhaf, W., Nordin, P., Keller, R., Francone, F.: Genetic Programming - An Introduction, pp. 133–134. Morgan Kaufmann Publishers (1998)
4. Brameier, M., Banzhaf, W.: Linear Genetic Programming, pp. 130, 183–185, 186. Springer (2007)
5. Nedjah, N., Abraham, A., de Macedo Mourelle, L.: Genetic Systems Programming: Theory and Experiences, pp. 16–17. Springer (2006)
6. Riolo, R., Soule, T., Worzel, B.: Genetic Programming Theory and Practice VI, pp. 229–231. Springer (2009)
7. Koza, J.: Genetic Programming: On the Programming of Computers by Means of Natural Selection. MIT Press (1992)
8. Wong, M., Leung, K.: Data Mining Using Grammar Based Genetic Programming And Applications. Kluwer Academic (2002)

GPGPU Implementation of Evolutionary Algorithm for Images Clustering

Dariusz Konieczny, Maciej Marcinkowski, and Paweł B. Myszkowski

Applied Informatics Institute
Wrocław University of Technology
Wyb. Wyspiańskiego 27, 51-370 Wrocław, Poland
{Dariusz.Konieczny,Pawel.Myszkowski}@pwr.wroc.pl,
marcinkowski.maciek@gmail.com
http://www.ii.pwr.wroc.pl/

Abstract. We propose an evolutionary algorithm (EA) usage to image clustering applied to Document Search Engine (DSE). Each document is described by its visual content (including images), preprocessed and clustered by EA. Next, such clusters are core of DSE. However, number of documents and attached images make EA ineffective in such task. Using the natural issue of EA we propose GPGPU (General Purpose Graphic Processing Unit) implementation. The paper describes our proposition, research and results gained in experiments.

Keywords: Document Search Engine, Images Clustering, Evolutionary Algorithm, GPGPU, GPU, CUDA.

1 Introduction

The Document Search Engine can be considered as a problem of finding similar document in a dataset. Problem is connected to a large dataset of documents which can be used in many applications e.g. in libraries to find a relevant document similar to one given in as a query. Such problem can be solved as classification task in data mining stage of knowledge discovery database (KDD) process. However classification based approach needs all documents labeled what is practically impossible according to huge cost/time of labeling process driven by human [6]. Thus our work concerns on clustering task in unsupervised learning mode, where structure of data is not given or we barely know anything about it. Comparing data mining task of classification to clustering, there is the main difference: clustering model uses unlabeled data and its effectiveness is measured by clusters quality. In classification task model quality is measured by accuracy of given classes.

It is worth mentioning that the problem presented in this paper is specific–the document is analyzed only in the visual context (images included). So basically our model bases only on a visual content of document, not a text. The visual content of document, means only included images, figures, tables and schemes [6]. Each of visual element is segmented, preprocessed and described as an image

N.T. Nguyen et al. (Eds.): *Adv. Methods for Comput. Collective Intelligence*, SCI 457, pp. 229–238.
DOI: 10.1007/978-3-642-34300-1_22

features vector. In such approach, our work denotes an alternative to semantic text based document analyses and indeed our approach results can be linked to such method in the complex system. Our approach is based on EA generated clusters. Due to EA computational time we decided to use GPGPU implementation to reduce it.

This paper describes our research for EA GPGPU's implementation for image clustering task in document search engine. The second section shows related works. The third section presents proposed approach, CUDA architecture and details of an implementation. Experiments results are included in section four. The last section concludes and describes directions of further research.

2 Related Work

There are many successful applications of EA in searching solution space problems. EA as one of metaheuristics is based on natural evolution process searches the solution landscape using genetic operators due to optimise fitness function. Nevertheless, curse of dimensionality rises with each new data dimension, and effectiveness of EA decreases significantly. Even if EA is able to find successful solution the computational complexity of EA makes it less attractive in practical usage. There are some techniques of EA effectiveness improvement: the simplest are used when some data (individuals) can be cached. Another approach may change EA paradigm as population steady–state EA, where offspring competes to its parents. There is only one of the whole set of EA evolution time consumption reduction techniques, but another may use not only one processor in computational process. Multipopulation Island EA uses several separated populations and connects them in migration phase.

Another approach, the most important from our point of view, is parallelization of EA basic operations: selection, mutation/crossover and fitness function calculation. Last years gave us very powerful tool: GPGPU. The GPU cards originally designed for 3D graphic acceleration in computer games by parallel processing now can be used not only for metaheuristics computation time reduction. The EA speed-up is connected not only to given problem but also its solution landscape and in literature is given average as 6–70x, but can be found even equals to 700x [3]. In GPGPU implementation of EA can be used several different approaches that differ from other by problem to solve, its representation, used genetic operators but also the parallelism techniques. Parallelism of GPGPU EA implementation may have several degree of grain [1]: it can be some operations (usually because of its most computational requirements), genetic operators or the whole population (island EA model with migration). In paper [9] there is 20 points list of aspects (also technical one) that should be considered in GPGPU EA implementation. Also the paper [4] describes techniques of metaheuristics implementations of GPGPU.

In task of classification GPGPU successful implementation of EA is described in [2][3]. Results are very promising as there is given speedup almost 1000 times. However in other papers there are no such spectacular results, where EA speedup equals from 6 to 100 times [4][7]. Multiple GPU application gives more opportunity to speedup, near to 1280 times [3] comparing to sequential EA implementation. In paper [8] hybrid EA is described very successful implementation on multi GPU, where the computational time is reduced almost 5000 times.

3 GPU Implementation of Evolutionary Algorithms

The aim of our study was to determine the suitability of graphics cards to EA of images clustering task. It was decided to prepare a parallel version of the entire algorithm, but divided into separate kernels with the possibility of substitution of the sequential execution of the host. The most time-consuming operation during solving the selected problem is to calculate the objective function.

Presented EA approach uses standard solution representation – they are composed of floating point values that represent coordinates of cluster centers, length of chromosome i for N dimensions and number of clusters Cl. Such representation allows the classic one–point crossover and mutation usage. Fitness function is based on cluster–image euclidean distance.

The classic measurements for parallel application are not suitable for GPGPU. They assume execution of sequential and parallel version of algorithm on the same processors, which is preposterous for graphics cards. Because of that the observable speedup is used in this paper. The observable speedup S_o is defined as the ratio of the sequential execution time T_s on the selected CPU in relation to the execution time T_g on the graphics card. In the execution time of the graphics card it is also included in the time to copy data between host and device (and vice versa).

3.1 CUDA Environment

The GPU architecture consists of many small simple processors, connected in multithreaded streaming multiprocessors (SMs). The GPU platform is prepared for use of Single Instruction Multiple Thread (*SIMT*) model of parallelism. In the model a group of threads run in parallel the same instruction, while other group can run other instruction (but the same for the other group). CUDA environment extends C language by new instructions to construct, prepare memory and run processes on graphics card. Programs for cards are called kernels. A kernel consists of threads grouped into blocks, and blocks are grouped into grid. Each thread has its own coordinates (threadIdx) in the block, and coordinates of its block (blockIdx) in the grid. This data is used to specify for which the calculation is responsible the thread. One of the challenges of good performance is to use memory. Threads can use the large memory called global memory, or a small but fast shared memory within the block. Every thread has also very small and very fast private memory. There are also two specific memory types: constant and texture memory. Part of the application is executed on the host (CPU) and calls kernel. Often, before the kernel starts, a data have to be moved from the host memory to the device (card) memory, and after processing it have to be moved in the opposite direction. During the kernel execution the card managing system decides which group of 32 threads (called a warp) will be launched on a free multiprocessor (SM). If the warp is waiting for memory access, it is suspended, and another group of threads may be executed on the multiprocessor.

3.2 Details of GPU EA Implementation

There were used EA parameters that are crucial for GPU implementation, as bellows: **Idv** — number of individuals, **Im** — number of images, **Cl** — number of clusters centers in one individual and **Atr** — number of attributes describing image and center of cluster.

We used so called **global** parallelization. The main purpose of this style of parallelization is to parallel all stages of EA independently. Then we had to compare the time of execution with the sequential ones. We decided, after many researches, to compare 3 different algorithms: sequential (with time T_S), mixed - parallel with sequential mutation and selection (T_M) and fully parallelized (T_P). There is another sequential algorithm which is fully optimized and was ran on the best available hardware. It is much faster than sequential algorithm used in researches and will be compared with parallel ones. Every parallel code was using global memory in graphic card. There can be defined 7 stages of executing implemented EA. There are 1) counting distances between images and centers of clusters in individuals, 2) finding minimum distance between centers of clusters and images, 3) summation of minimum distances for one individual, 4) evaluation of every individual, 5) selection, 6) mutation, 7) crossover. All these stages were parallelized independently as follows:

1. The block size is equal to **Im**. Threads in each block are divided on two dimensions the dimension X is equal **Idv** and the dimension Y is equal **Cl**. We think it provides the best distribution of work because the block size is always the biggest and on the other hand **Im** is the biggest value in every test set.
2. The block size is equal to **Im** and there are **Idv** threads in it. Every thread is looking for minimum value by comparing **Cl** values from a distance matrix obtained in step 1.
3. Every minimum value for every individual is summed. It is made by one block of **Idv** threads. The calculated value is called Ek.
4. Two tasks are made in this step. All of them are made by **Idv** blocks with only one thread in them. The first task is to calculate maximum distance between clusters centers in one individual. After that each thread is calculating the evaluation of individual.
5. Selection is distributed on **Idv** blocks with only one thread. Every thread is selecting 5 individuals and choosing the one with the best evaluation.
6. Mutation is made by **Idv** x **Cl** blocks with **Atr** threads. Every thread is changing only one value in an individual cluster center. It checks the mutation condition and if it is true, it makes the conversion. That kind of distribution provides that there will be no need to do more than one change by any thread.

7. Crossover is made by **Idv** x **Cl** blocks of **Atr** threads. The principle of the operation is similar to the mutation. Every thread is moving to the new individual only one value from the center of cluster. It checks the intersection and moves the value on the proper position in the new individual (descendant).

4 Experiments

4.1 Used GPU/CPU Configurations

We used two different graphic cards to test the parallel versions of algorithm and one computer to test the sequential one. We used:

1. nVidia Tesla T10 with clock 1140 MHz and 4294 MB of global memory. It has 240 CUDA cores, 30 multiprocessors and maximum 512 threads per block and 65535 block per grid. Its compute capability is 1.3. The computer with such card is run by Intel(R) Xeon(R) CPU E5540 @ 2.53GHz with clock 2533 MHz and 8192 cache memory, RAM 16465 MB.
2. nVidia GeForce GTX 470 with clock 1215 MHz and 1341 MB of global memory. It has 448 CUDA cores, 14 multiprocessors and maximum 1024 threads per block and 65535 blocks per grid. Its compute capability is 2.0. The computer configuration is: Intel(R) Core(TM) i7 CPU 930 @ 2.80GHz with clock of 1596 MHz and 8192 KB of cache memory. It can execute more than 5667 million operations per second. The RAM size is 24739 MB.

The nVidia Tesla card was invented to proceed parallel calculations and isn't able to display graphics. GeForce GTX470 is a graphic card for domestic use but is still able to proceed parallel calculations.

4.2 Used Datasets

We provided 7 test data sets to evaluate the parallel EA effectiveness. They can be described by parameters: **Idv**, **Im**, **Cl**, **Atr** and number of arithmetic operations **Op** in millions (MM). The last attribute was added to show the real size of problem. The issues of data sets are provided in Table 1.

The first dataset was originally provided in SYNAT project [6]. The rest was created for our experiments validation. The last dataset contains the largest set of images, individuals and attributes used in project. The rest datasets were created to evenly fill the gap between these two datasets. Also every dataset has different attributes to check their impact on the speedup. We tried to provide even growth of every attribute in dataset.

The memory usage during computation for intermediate results can be calculated using formula: **Img** * **Idv** * **Cl** * **4**. For datasets 6 and 7 the needed memory is bigger than memory available on GeForce GTX 470 and that is the reason why the tests for these sets aren't conducted.

Table 1. Description of generated data sets

	Img	Idv	Cl	Atr	Op	Mem usage
Dataset 1	23685	20	10	36	351 MM	19 MB
Dataset 2	60000	40	4	40	789 MM	38 MB
Dataset 3	60000	60	4	40	1184 MM	58 MB
Dataset 4	60000	60	4	60	3520 MM	58 MB
Dataset 5	60000	100	30	60	21969 MM	720 MB
Dataset 6	120000	100	30	60	43935 MM	1440 MB
Dataset 7	170000	100	30	60	62240 MM	2040 MB

4.3 Experimental Results

The experiments were provided for every dataset. Each dataset is described by set of images and set of individuals. All versions of the algorithm were run 10 times and timings were averaged. We also counted the timings of every stage of EA independently. The CUDA events were used to measure time because they take into account the time of a synchronization between graphics card and host computer. The counting for the parallel version starts with copying data from host to device and ends with the final copying data from device to host. Tests were performed for 1, 25, 50 and 100 generations. In this paper we present experimental results for 100 generations, because we suppose our application will be mainly used for these number of generations.

The first study helps to compare parallel versions of algorithm. We counted the speedup of every algorithm (S_M for mixed version and S_P for parallel one) in a comparison with the sequential version run on I7 CPU. The results for 100 generations are provided in Table 2.

Table 2. Speedups of different versions of algorithms

	S_M Tesla	S_M GTX470	S_P Tesla	S_P GTX470
Dataset 1	24,72	6,86	14,32	5,72
Dataset 2	26,96	10,79	19,47	12,79
Dataset 3	40,81	11,12	27,74	19,57
Dataset 4	64,48	32,70	48,02	48,49
Dataset 5	65,13	46,55	50,38	45,85
Dataset 6	111,17	nd	87,86	nd
Dataset 7	148,13	nd	118,55	nd

As we can see the speedup of every algorithm increase with the growth of dataset (Fig. 1). It is in line with the expectations because there are more calculations which can be done parallel. We also can observe that the mixed, parallel — sequential version has better speedup than the parallel one.

The second study helps us compare sequential and parallel execution times of every stage of EA. It is useful for finding stages of EA which shouldn't be

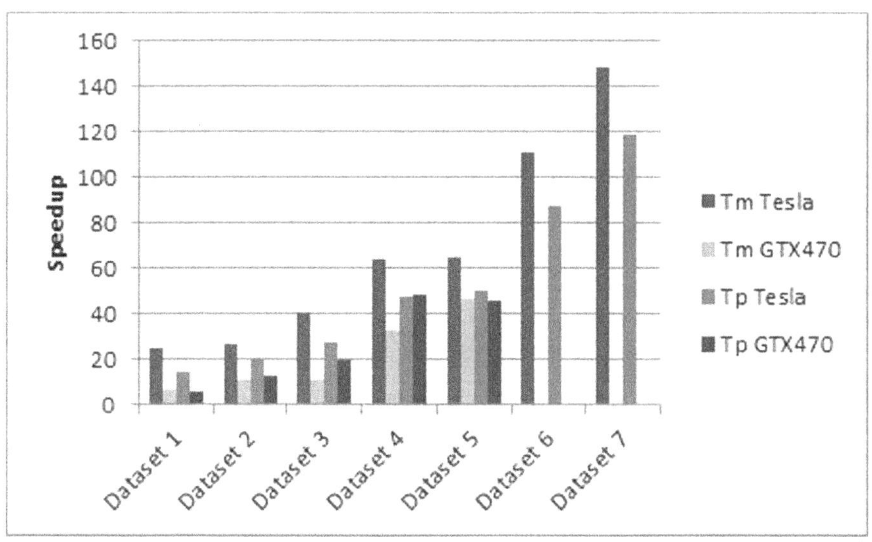

Fig. 1. Speedups of different versions of algorithms for 100 generations

parallelized for this size datasets. The timings for Dataset 1 and 7 are provided in Table 3.

As it can be seen in tables there are differences between the parallel and the sequential timings. Distances counting and minimum finding are much more faster in parallel but the other stages have the speedup less than 1 what means the parallel version is slower than the sequential one. So there is no point in doing these stages in parallel? The answer is no because if we want to do distance counting and minimum finding in parallel we need to do many transfers between host and device with a lot of data which will waste all time saving. Also it proves that for quite small datasets the parallelization can do more harm than good.

The third study provides us an information about growth of speedup between datasets with only different number of individuals. Datasets 3 and 4 are different only in terms of individuals number. It allows us to predict speedup growth for a bigger data sets with more individuals. Gained speedup values can be found in Table 4.

As we can see the growth of speedup for every algorithm is increasing but for fully parallelized version speedup is increasing more rapidly (Fig. 2). We predicts that for datasets with more individuals the fully parallelized version of algorithm can reach better results than the mixed parallel — sequential version.

There were researches on an optimized sequential version performed independently. This version is the fastest available sequential algorithm run on a better workstation than mentioned in subsection 4.1. The speedup results of parallel algorithms compared to this algorithm are shown in Fig 3.

As it can be seen the Tesla version of the parallel algorithm is still faster than the sequential one. The maximum speedup is near 35 times which is very good

Table 3. Timings [ms] of the sequential Ts and the parallel Tp versions of algorithm for datasets 1 and 7

	Dataset 1			**Dataset 7**		
	T_S **I7**	T_P **Tesla**	**Speedup**	T_S **I7**	T_P **Tesla**	**Speedup**
Distances counting	4174,17	127,19	32,82	319383,09	1825,96	174,91
Minimum finding	67,32	3,40	19,47	2227,85	274,89	8,10
Ek counting	13,88	58,50	0,24	99,93	166,60	0,60
I-Index counting	0,24	0,50	0,48	60,18	53,00	1,14
Selection	0,00	4,35	0,00	0,00	4,97	0,00
Mutation	0,20	10,70	0,02	4,65	219,35	0,02
Crossover	0,10	4,90	0,02	1,51	5,79	0,26
Host-device transfer	nd	4,78	nd	nd	19,10	nd
Device-host transfer	nd	0,10	nd	nd	0,45	nd

Table 4. Speedup rate for different individual numbers

	S_M Tesla	S_M GTX470	S_P Tesla	S_P GTX470
Dataset 3	40,81	11,12	27,74	19,57
Dataset 4	64,48	32,70	48,02	48,49

result. Unfortunately the GTX version has speedup less than 1 for dataset 1 what means it is slower than the sequential one. But for bigger datasets speedup increasing and usage GTX for parallelization is still cost–effective.

4.4 Summary

The two graphic cards used to tests were created for different tasks. The nVidia Tesla T10 card is designed to parallel calculations and GeForce GTX 470 can be used to this task but its main task is graphics display. We can feel that looking on the speedup results. Tesla is better and it can do calculations on bigger sets because of its global memory size. But still simple graphic card reaches results which aren't negligible. It still can be used to parallelizing calculations for smaller datasets. If we want to reach better speedup for bigger datasets we will need to use card designed to this task.

We developed more parallel versions of algorithm but we present only two of them - the fully parallelized and the fastest. The other versions of algorithm have different division on parallel and sequential stages and also sometimes different threads distribution. There is a version with only distances counting and Minimum finding done in parallel and the version with all genetic operators calculated sequentially.

The mixed sequential–parallel algorithm reaches the best results for all datasets but for bigger ones we can conclude that the differences in speedup between the mixed and the fully parallel version will decrease.

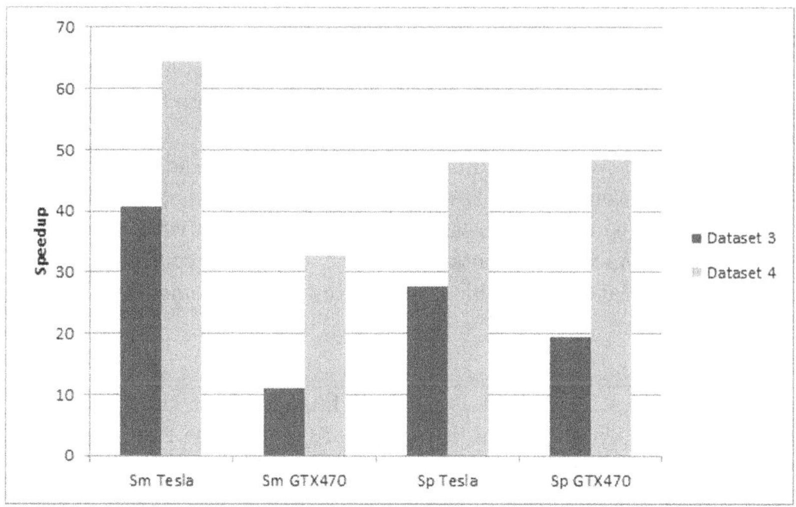

Fig. 2. Speedup rate for different individual numbers

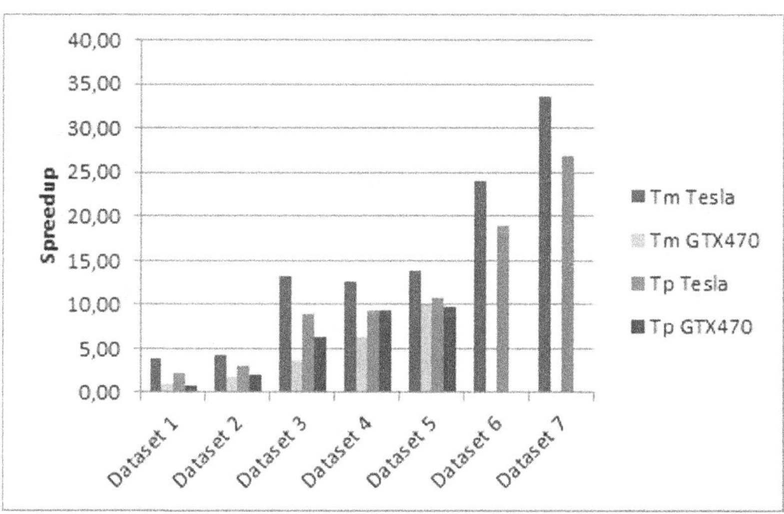

Fig. 3. Comparison speedup to the best sequential EA (for 100 generations)

5 Conclusions

A parallelization of EA (GPU implementation included) is one of the strongest trends of the EA research, next to works connected to specialized representations, still problematic theoretical EA aspects and benchmarks developing [5]. In summary it is worth mentioning about GPU usage costs. The cost measured in money is not problematic (few hundreds of dollars) however the labor cost of a

GPU implementation cannot be ignored: a GPU implementation strongly need specific realization using given card configuration and a problem specification. GPU card changing or even problem size may cause completely different results, not always suitable. Thus, such a specialization is connected to not guarantee a success. Another disadvantage is a specific GPUs SIMT architecture usage, where there is used a completely different programming paradigm. Nevertheless, the main advantage remains the EA computational process speedup, where the EA running time is almost in a real time. That's why we wanted to use GPU architecture to do some experiments series. The results of experiments presented in this paper show that it is possible to get fastest EA implementation in various ways using GPGPU.

Acknowledgements. This work is partially financed from the Ministry of Science and Higher Education Republic of Poland resources in 2010–2013 years as a research SYNAT project (System Nauki i Techniki) in INFINITI-PASSIM.

References

1. Arenas, M.G., Mora, A.M., Romero, G., Castillo, P.A.: GPU Computation in Bioinspired Algorithms: A Review. In: Cabestany, J., Rojas, I., Joya, G. (eds.) IWANN 2011, Part I. LNCS, vol. 6691, pp. 433–440. Springer, Heidelberg (2011)
2. Banzhaf, W., Harding, S., Langdon, W.B., Wilson, G.: Accelerating Genetic Programming through Graphics Processing Units. In: Genetic Programming Theory and Practice VI. Genetic and Evolutionary Computation, pp. 1–19. Springer (2009)
3. Cano, A., Zafra, A., Ventura, S.: Speeding up the evaluation phase of GP classification algorithms on GPUs. Soft Computing - A Fusion of Foundations, Methodologies and Applications 16(2), 187–202 (2011)
4. Langdon, W.B.: Graphics processing units and genetic programming: an overview. Soft. Comp. 15, 1657–1669 (2011)
5. O'Neill, M., Vanneschi, L., Gustafson, S., Banzhaf, W.: Open issues in genetic programming. Genet. Program. Evolvable Mach. 11, 339–363 (2010)
6. Myszkowski, P.B., Buczek, B.: Growing Hierarchical Self-Organizing Map for searching documents using visual content. In: 6th International Symposium Advances in Artificial Intelligence and Applications (2011)
7. Robilliard, D., Marion-Poty, V., Fonlupt, C.: Genetic programming on graphics processing unit. Genet. Program. Evolvable Mach. 10, 447–471 (2009)
8. Sharma, D., Collet, P.: GPGPU-Compatible Archive Based Stochastic Ranking Evolutionary Algorithm (G-ASREA) for Multi-Objective Optimization. In: Schaefer, R., Cotta, C., Kołodziej, J., Rudolph, G. (eds.) PPSN XI, Part II. LNCS, vol. 6239, pp. 111–120. Springer, Heidelberg (2010)
9. Wahib, M., Munawar, A., Munetomo, M., Akama, K.: Optimization of Parallel Genetic Algorithms for nVidia GPUs. In: IEEE Congr. on Ev. Comp. (CEC), pp. 803–811 (2011)

Recombination Operators in Genetic Algorithm – Based Crawler: Study and Experimental Appraisal

Huynh Thi Thanh Binh, Ha Minh Long, and Tran Duc Khanh

School of Information and Communication Technology
Hanoi University of Science and Technology, Hanoi, Vietnam
{binhht,khanhtd}@soict.hut.edu.vn,
minhlong293@gmail.com

Abstract. A focused crawler traverses the web selecting out relevant pages according to a predefined topic. While browsing the internet it is difficult to identify relevant pages and predict which links lead to high quality pages. This paper proposes a topical crawler for Vietnamese web pages using greedy heuristic and genetic algorithms. Our crawler based on genetic algorithms uses different recombination operators in the genetic algorithms to improve the crawling performance. We tested our algorithms on Vietnamese newspaper VnExpress websites. Experimental results show the efficiency and the viability of our approach.

Keywords: Genetic Algorithms, Focused Crawler, Keyword, Vietnamese Word Segmentation.

1 Introduction

Nowadays, demand for information searching of internet users is increasing. However, very few Vietnamese search engines are truly successful and they can hardly compete with popular search engine around the world such as: Google, Yahoo... Thus, utilizing the knowledge of Vietnamese language and constructing a domain-specific search engine is a greatly promising research field.

Crawler is one of the most important components in a search engine. It browses the web by following the hyperlinks and storing downloaded documents in a large database of the search engine. A general-purpose crawler downloads as many pages as possible without contents concernment. Nevertheless, the growth of information on the World Wide Web is so tremendous that crawling all documents becomes very difficult and time-consuming. One of the most popular methods to cope with this problem is to use focused crawlers or topical crawlers. A focused crawler tries to find the shortest path leading to relevant documents and neglect irrelevant areas. They usually use some greedy search methods called the crawling strategies to find which hyperlink to follow next. Better crawling strategies result in higher precision of document retrieval.

In [23], Nhan et.al built a topical Vietnamese web pages crawler based on genetic algorithms. Starting with an initial set of keywords, the crawler expands the set by

N.T. Nguyen et al. (Eds.): *Adv. Methods for Comput. Collective Intelligence*, SCI 457, pp. 239–248.
DOI: 10.1007/978-3-642-34300-1_23

adding the most suitable terms that is intelligently selected during the crawling process by genetic algorithm. Multi-parents recombination operators are used in genetic algorithms to crawl web pages. The idea is to create only one child from n parents by selecting (n-1) crossover point. For creating a segment of children, the corresponding segment from the parents is found. Different number of parents in recombination operators are used.

The result of genetic algorithms depends on many parameters, such as: recombination operator, mutation operator, size of population and model of genetic algorithms... To our best knowledge, there are several researches on genetic algorithms in focused crawler, but there is no research on the difference between recombination operators. The result of genetic algorithms using multi-parent recombination operators depends not only on the method of choosing the parents but also on the way to recombine the offspring.

In this paper, we introduce different recombination operators in genetic algorithms to improve the crawling performance. We consider four different methods for choosing parents: the first one is based on roulette wheel selection, the second one uses rank selection, the third one uses random selection, the last one uses Levenshtein distance between the parents (minimum and maximum distance). We also experiment each method with five different methods to recombine to the child: one-point crossover, uniform crossover and multi-parent recombination operator with the different number of parents (2, 4, 6). We concentrate on analyzing the recombination operators in genetic algorithms to crawl topical Vietnamese web pages. We also propose a greedy heuristic algorithm to crawl topical Vietnamese web pages and compare the results with [23] and the results found by our proposed genetic algorithms.

As in [23], we use a Vietnamese text segmentation tool which implements automata and part of speech (POS) tagging techniques.

The rest of this paper is organized as follows: in section 2, we briefly overview works done in focused crawler, genetic algorithms in focused crawler and the motivation for our work. Section 3 describes in detail our algorithm. Section 4 discusses the results of our experiments and comparisons with other crawlers. The paper concludes with section 5, where we describe some possible directions for future works.

2 Related Works

2.1 Focused Crawler

Focused Crawler was firstly introduced by Chakrabarti et al. in 1999 [7]. They built a model using a document taxonomy and seed documents for classifying retrieved pages into categories (corresponding to nodes in the taxonomy). In [6], Filippo Menczer et al. proposed three different methods to evaluate crawling strategies and compared their crawler with other ones. Bing Liu et al. combined the crawling strategy with clustering concepts [14]. For each topic, their crawler first takes a specific number of top weighted retrieved pages from Google for that topic and then extracts some other keywords from them. As a result, a greater numbers of pages are returned and with greater precision. Anshika Pal et al. [5] utilized Content and Link Structure Analysis to improve the quality of web navigation. They also presented a method for traversing the irrelevant pages to boost the coverage of a specific topic. In [10], Aggarwal introduced another intelligent crawler that can adapt online the queue link-extraction strategy using a self-learning. Alessandro Micarelli and Fabiano Gasparetti [3] synthesize all existing algorithms used in focused crawler and giving evaluating method for these systems. Angkawattanawit et al. [12] enhance re-crawling performance by utilizing several data such as seed URLs, topic keywords and URL relevance predictors. These data are constructed from previous crawling logs and used to improve harvest rate.

2.2 Genetic Algorithms in Focused Crawler

Gordon [16] was the first to propose a new approach in document descriptors. Documents are represented by an array of keywords and they will evolutes through natural selection and genetic operators. The last result will be the best string describing the document. Yang [17] also employs this technique to improve the weight of associated keywords in documents. Petry et al. applied genetic algorithms to increase efficiency of collecting information from documents marked with indexes by changing weights of query term [8]. In [1], Chen compares Best-First search and genetic algorithm search crawler to gather the most closely related homepages based on the links and keyword indexing. Through the experiments, genetic algorithm crawler reaches higher recall value whereas precision values of both crawlers are nearly equal. Menczer and Belew proposed InfoSpiders [2], a collection of autonomous goal-driven crawlers without global control or state, in the style of genetic algorithms. In [11], Hsinchum Chen puts forward an individuality spider based on web page context. Jialun Qin et al. use genetic algorithm model to build a Domain-Specific Collections [19].

This paper presents an extension of the ViCrawler, which crawls topical Vietnamese web pages, named ViCrawler as in [23]. ViCrawler is a system which applies genetic algorithms in collecting information in Vietnamese and creating a Crawler with high adaptability. In contrast with the experiment in [23], we consider four different methods for choosing parents and each method of choosing parents we experiment with five methods to recombine to the child. We also implemented a greedy heuristic algorithm to crawl topical Vietnamese web pages.

3 Proposed Algorithm

3.1 Genetic Algorithm

Chromosome representation

We use real encoding method as in [23]. From seed URLs, ViCrawler downloads a set of documents.

There are many weighting schemes to calculate the weight of word in a document: Lnu.ltu weighting scheme, Simple weight, Binary weight, Term frequency weight, Inverse document frequency. Lnu.ltu weighting scheme often produces the best clustering results compared to other weighting schemes. So, we use Lnt.ltu weighting scheme for calculating the weight of word in our system as [18], [23].

Fitness Function

All chromosomes in the population have to be evaluated to identify the relevance of the document according to initial set of keywords. There are many similarity functions such as Jaccard, Dice and cosine. We choose the cosine function as fitness function of the chromosomes in our implementation as in [23].

Parent Selection

In a recombination operator, choosing parents and recombine them are the most important steps. We use five methods to select chromosomes to be parents in crossover operators:

- *Roulette wheel selection (rw-method):* parents are selected according to their fitness. The better the chromosome are, the more chances they have to be selected. Chromosome with bigger fitness will be selected more frequently.
- *Rank selection (rk-method):* Roulette wheel selection will have problems when the fitness value differs significantly. For example, if the best chromosome fitness is 90% of all, then the other chromosomes will have very few chances to be selected if the roulette wheel approach is used. Rank selection first ranks the population from highest to lowest fitness, then every chromosome receives order of this ranking. After this all the chromosomes have a chance to be selected.
- *Random selection (rm-method):* parents are selected randomly from the population.
- *Levenshtein distance (lmin-method* and *lmax-method):* this method is based on Levenshtein distance between the parents.

We use two different Levenshtein distance methods:

- *lmin-method:* chooses a set of chromosomes from a ranking table based on the minimum Levenshtein distance between them.
- *lmax-method:* similar to the previous method but based on the maximum value.

Crossover operator

To create the offspring, we use three following ways:

- *One-point crossover* (*op-crossover*): creates two children from two parents by selecting one crossover point in the parents.
- *Multi-parents crossover (mp-crossover):* creates only one child from n parents by selecting (n-1) crossover points in the parents randomly. The child is inherited n segments from n parents, one segment per parent.

Uniform crossover (*uf-crossover*): It uses a fixed mixing ratio between two parents. If the mixing ratio is 0.5, the offspring has approximately half of the genes from first parent and the other half from second parent, although crossover point can be randomly chosen.

Mutation operator

Mutation operator is used to avoid local optimization and further vary the population. We apply the single point mutation. ViCrawler replaces the value of the chosen gene with a random value between 0 and 1.

After GA process finish, ViCrawler finds the chromosome with highest fitness. Then it extends the initial set of keywords by adding a new gene of this chromosome to the initial keywords.

ViCrawler continues to collect data from web with extended set of keywords by using GAs.

3.2 Greedy Algorithm

In this paper, we propose Greedy Algorithm to crawl topical Vietnamese web pages:

- A candidate set: we select a set of new words from downloaded documents. For example, we download 100 documents, they have 1000 words. If 50 of 1000 words are already exists in set of keywords, we select 950 remaining words as a candidate set.
- A selection function: return a set of new words which have highest weight.
- Add new words to set of keywords.

The pseudo code for Greedy Algorithm is shown below:

```
Procedure GreedyAlgorithm
Begin
//D all of new words from downloaded documents
//K set of keywords
D, K ← initial()
//C candidate set
C ← null
For (i = 1 to length(D)) do begin
if (D[i] ⊄ K) then C.add(D[i])
Endfor
sortByWeight(C);
For (j=1 to number_of_NewKeywords) do begin
K.add(C[j])
Endfor
End Procedure
```

4 Experimental Results

4.1 Problems Instances

The crawlers operated in Vietnamese top online newspapers: www.vnexpress.net. Target topic is Estate. The crawler started with one URL, and crawled five thousands pages.

4.2 Experiment Setup

We implement our system with two main components: Vietnamese word segmentation and genetic algorithm crawler. Our system is written in Java and run on a server with Intel Pentium E2180, 1GB of RAM.
We compare the results of our genetic algorithms, greedy heuristic algorithm and GA in [23].

In the experiments, the population size is 50, crossover rate is 0.8, and mutation rate is 0.05. The number of generation is 5000.

In GA which uses uf-crossover method, we experiment with 2, 4, 6 parents in recombination operator, denoted mp(2), mp(4), mp(6) corresponding (see Table 1).

Table 1. Experimental genetic algorithms

Selections / Crossover	rw	rk	rm	lmin	lmax
op-crossover	GA1	GA6	GA11	GA16	GA21
uf-crossover	GA2	GA7	GA12	GA17	GA22
mp(2)	GA3	GA8	GA13	GA18	GA23
mp(4)	GA4	GA9	GA14	GA19	GA24
mp(6)	GA5	GA10	GA15	GA20	GA25

GAx: the genetic algorithm used in ViCrawler. Columns represent the method to select parents to recombine for the child. Rows represent the way to recombine. GA26: the genetic algorithm used in [23] with rank selection and 3-parents crossover.

In Greedy Algorithm: the number of new keywords is 5, the number of documents is 500. Every 500 downloaded documents, we choose 5 new keywords which have highest weight to add to set of keywords.

4.3 Results and Discussions

We use the following metrics for evaluating the crawler performance.

- Precision metric

$$precision\_rate = \frac{relevance\_pages}{pages\_downloaded}$$

relevance_pages: the number of pages related to the desired topic
pages_downloaded: the number of pages downloaded by ViCrawler.

- Convergence rate: the numbers of generations where the fitness is max. The max fitness does not improve until the number of generations increase.

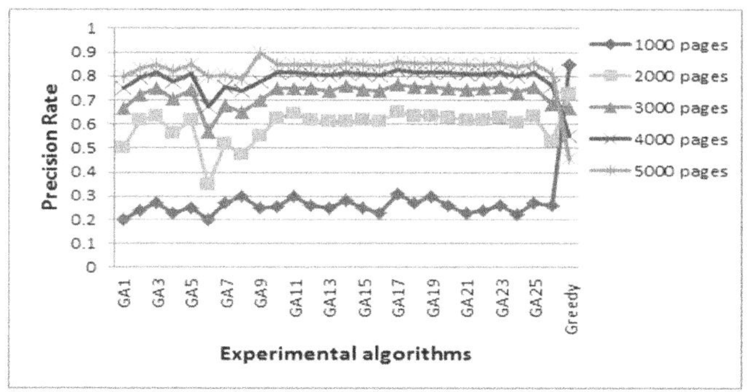

Fig. 1. Precision rate after crawl 5000 pages

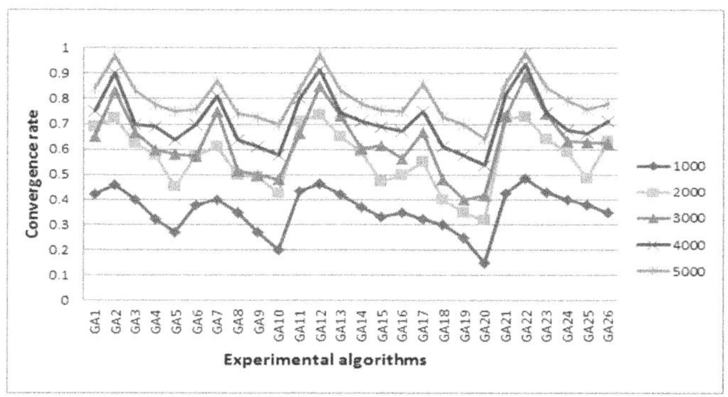

Fig. 2. Convergence rate after 5000 generations

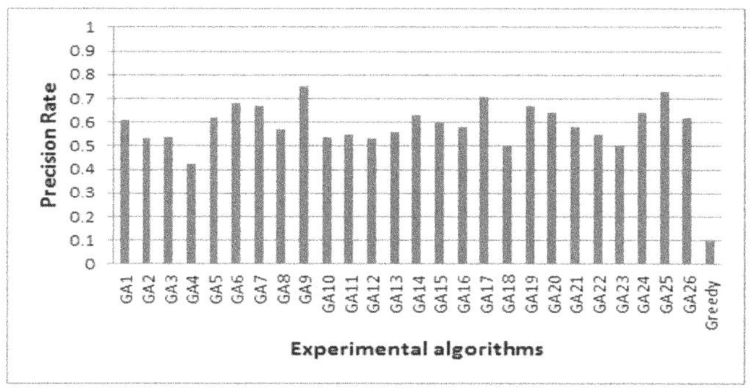

Fig. 3. Precision rate of new keywords added by ViCrawler

The results in Fig. 1 show that:

- The number of crawled pages is increased; the precision of ViCrawler is increased as well.
- For the genetic algorithm which uses roulette wheel selection, the precision rate is highest with mp(2).
- For the genetic algorithm which uses rank selection, the precision rate is highest with mp(6).
- For the genetic algorithm which uses uniform crossover, the precision is the worst with rank selection and is best with lmin selection.
- The precision rate of the greedy algorithm is highest on small number of crawled pages. When number of crawled pages is increased; the precision rate of greedy algorithm is decreased.

The results in Fig. 2 show that:

- With the same method for selection the parents: The convergence rate of GA which uses uniform crossover is slowest and multi-parent recombination operator is fastest.
- With the same method to recombine to the child: The convergence rate of GA which uses lmax selection is slowest and lmin selection is fastest.

The results in Fig. 3 show that:

- The precision rate of new keywords added by ViCrawler which uses rank selection and 4 parents; lmax selection and 6 parents are the highest.
- The precision rate of new keywords added by ViCrawler which uses roulette wheel selection and 4 parents is the lowest.
- The precision rate of new keywords added by Greedy Algorithm is lower than ViCrawler which uses Genetic Algorithm.

5 Conclusion and Future Work

We have studied the use of recombination operators in genetic algorithms to crawl Vietnamese web pages. We have experiment our approach with different ways of selecting parents in the recombination operator and different ways to create the child. The experimental results show some improvements with respect to Nhan et al. [23].

There are some lines of future work. It would be important to perform more extensive tests with larger volume of web pages and with more specific topics. Also it would be interesting to experiment with different genetic operators in order to find the most appropriate ones for the interested instances.

Acknowledgement. This work was partially supported by the project "Constructing an intelligent system for crawling, classifying and searching Vietnamese web pages vertically" funded by the Ministry of Science and Technology, Vietnam under grant number B2010-01-364. The Vietnam Institute for Advanced Study in Mathematics provided part of the support funding for this work.

References

1. Chen, H., Chung, Y., Ramsey, M., Yang, C.: A smart Itsy Bitsy Spider for the Web. Journal of the American Society for Information Science 49(7), 604–618 (1998)
2. Menczer, F., Belew, R.K.: Adaptive retrieval agents: Internalizing local context and scaling up to the Web. Machine Learning 29(2/3), 203–242 (2000); Longer version available as Technical Report CS98-579, University of California, San Diego
3. Micarelli, A., Gasparetti, F.: Adaptive Focused Crawling. In: Brusilovsky, P., Kobsa, A., Nejdl, W. (eds.) Adaptive Web 2007. LNCS, vol. 4321, pp. 231–262. Springer, Heidelberg (2007)
4. Shokouhi, M., Chubak, P., Raeesy, Z.: Enhancing Focused Crawling with Genetic Algorithms. In: Proceedings of the International Conference on Information Technology: Coding and Computing (ITCC 2005), pp. 503–508 (2005)
5. Pal, A., Tomar, D.S., Shrivastava, S.C.: C Shrivastava, Effective Focused Crawling Based on Content and Link Structure Analysis (IJCSIS) International Journal of Computer Science and Information Security 2(1) (June 2009)
6. Menczer, F., Pant, G., Srinivasan, P., Ruiz, M.: Evaluating Topic-Driven Web Crawlers. In: Proceedings of the 24th Annual International ACM/SIGIR Conference, New Orleans, USA, pp. 241–249 (2001)
7. Chakrabarti, S., van den Berg, M., Domc, B.: Focused crawling: a new approach to topic-specific Web resource discovery. In: Proceedings of the 8th International World Wild Web Conference, Toronto, Canada, pp. 1623–1640 (1999)
8. Petry, F., Buckles, B., Prabhu, D., Kraft, D.: Fuzzy Information Retrieval Using Genetic Algorithms and Relevance Feedback. In: Bonzi, S. (ed.) Proceedings of the Fifty-Sixth Annual Meeting of the American Society for Information Science Annual Meeting, Silver Spring, MD, vol. 30, pp. 122–125 (1993)
9. David, E.: Goldberg, Genetic Algorithms in Search, Optimization, Machine Learning. Addison Wesley (1989)
10. Aggarwal, C., Al-Garawi, F., Yu, P.: Intelligent Crawling on the World Wide Web with Arbitrary Predicates. In: Proc. 10th Int. World Wide Web Conf., Hong Kong, pp. 96–105 (2001)
11. Hsinchum, C., Chen, Y.M., Ramsey, M., Yang, C.C., Ma, P.C., Yen, J.: Intelligent spider for Internet searching. In: Proceedings of the Thirtieth Hawaii International Conference on System Sciences, Maui, Hawaii, January 4-7, pp. 178–188 (1997)
12. Angkawattanawit, N., Rungsawang, A.: Learnable Crawling: An Efficient Ap-proach to Topic-specific Web Resource Discovery. Journal of Network and Computer Applications, 97–114 (April 2005)
13. Chen, H.: Machine learning for information retrieval: Neural networks, symbolic learning, and genetic algorithms. Journal of the American Society for Information Science, 194–216 (1995)
14. Liu, B., Chin, C.W., Ng, H.T.: Mining Topic-Specific Concepts and Definitions on the web. In: Proceedings of the 12th International World Wild Web Conference (www 2003), Budapest, Hungary, pp. 251–260 (May 2003)
15. Raghavan, V., Aggarwal, B.: Optimal Determination of User-Oriented Clusters: An Application for the Reproductive Plan. In: Proceedings of the Second International Conference on Genetic Algorithms and Their Applications, Cambridge, MA, pp. 241–246 (1987)
16. Gordon, M.: Probabilistic and Genetic Algorithms for Document Retrieval. Communications of ACM 31(2), 152–169 (1988)

17. Yang, J., Korfhage, R., Rasmussen, E.: Query Improvement in Information Retrieval Using Genetic Algorithms: A Report on the Experiments of the TREC Project. In: Harman, D.K. (ed.) Proceedings of the First Text Retrieval Conference, pp. 31–58. National Institute of Standards and Technology (NIST) Special Publication 500-207, Washington, DC (1993)
18. Reed, J.W., Jiao, Y., Potok, T.E., Klump, B.A., Elmore, M.T., Hurson, A.R.: TF-ICF: A New Term Weighting Scheme for Clustering Dynamic Data Streams. In: Proceedings of the 5th International Conference on Machine Learning and Applications, pp. 258–263 (2006)
19. Qin, J., Chen, H.: Using Genetic Algorithm in Building Domain-Specific Collections: An Experiment in the Nanotechnology Domain. In: Proceedings of the 38th Hawaii International Conference on System Sciences, vol. 102 (2005)
20. Hông Phuong, L.ê., Thi Minh Huyên, N., Roussanaly, A., Vinh, H.T.: A Hybrid Approach to Word Segmentation of Vietnamese Texts. In: Martín-Vide, C., Otto, F., Fernau, H. (eds.) LATA 2008. LNCS, vol. 5196, pp. 240–249. Springer, Heidelberg (2008)
21. Daciuk, Jan, Watson, B.W., Watson, R.E.: Incremental construction of minimal acyclic finite state automata and transducers. In: Proceedings of the International Workshop on Finite State Methods in Natural Language Processing, Ankara, Turkey, June 30-July 1, vol. 1, pp. 48–56 (1998)
22. Maurel, D.: Electronic Dictionaries and Acyclic Finite-State Automata: A State of The Art. In: Published in Grammars and Automata for String Processing, Ankara, Turkey, June 30-July 1, vol. 1, Part 3, pp. 177–188 (1998)
23. Nhan, N.D., Son, V.T., Binh, H.T.T., Khanh, T.D.: Crawl Topical Vietnamese Web Pages using Genetic Algorithm. In: Proceedings of Second International on Knowledge and System Engineering, pp. 217–223 (2010)

Meta-evolution Modelling: Beyond Selection/Mutation-Based Models

Mariusz Nowostawski

Information Science Department,
University of Otago, Dunedin, P.O. BOX 56, New Zealand
mariusz@nowostawski.org

Abstract. In this position article we argue the need for integrative approach to evolutionary modelling and point out some of the limitations of the traditional selection/mutation-based models. We argue a shift towards fine-grained detailed and integrated evolutionary modelling. Selection/mutation-based models are limited and do not provide a sufficient depth to provide reductionists insights into the emergence of (biological) evolutionary mechanisms. We propose that selection/mutation should be augmented with explicit hierarchical evolutionary models. We discuss limitations of the selection/mutation models, and we argue the need for detailed integrated modelling approach that goes beyond selection/mutation. We propose our own research framework based on computational meta-evolutionary approach, called Evolvable Virtual Machines (EVM) to address some of the challenges.

1 Motivation

Frontiers of evolutionary research (in all related disciplines, including biologiy and artificial life) are overlapping and they cross-reference research programmes from various disciplines. There are many common themes in biological and artificial evolutionary modelling. Despite large overlaps, there is a visible fragmentation and difficulty in combining research results from various fields. Progress in this area may require a more integrative approach. Models are needed to bring dispersed results together in a unified research framework. Answering fundamental questions requires insights and understanding on interrelations of biological organisation on various levels. Selection/mutation-based models are sufficient and widely used for some of the artificial and biological systems. However, many of the challenges that researchers of biological evolution and artificial life face can only be addressed through frameworks that can model fine-grained evolutionary dynamics and complex hierarchical evolutionary organisation. Unfortunately, formal analytical models are often intractable. We argue the need for complex dynamical and hierarchical approach towards modelling evolutionary dynamics on multiple levels of abstraction. We review the current state of computational evolutionary systems and propose a broader, more generic computational evolutionary framework, called Evolvable Virtual Machines, as a direction for making progress within the theoretical studies of evolution, meta-evolution and properties of living systems. Expanding and going beyond pure selection/mutation-based models is the main focus of this article.

N.T. Nguyen et al. (Eds.): *Adv. Methods for Comput. Collective Intelligence*, SCI 457, pp. 249–258.
DOI 10.1007/978-3-642-34300-1_24

Using a somewhat simplified outlook on the field, the study of evolution proceeds on two basic levels:

(a) collecting, analysing and modelling (mathematically and computationally) naturally occurring (biological) evolutionary phenomena;
(b) formulating mathematical and computational generalisations of the evolutionary theories, based on and derived from results of the activities of the previous point (a).

The latter point (b) is (generally) the main focus of research efforts within the artificial life and others (multi-agent systems, complex adaptive systems, evolutionary computations, etc), whereas the former (a) is of main concern to evolutionary biology and related fields (among others population dynamics, molecular biology, genomics, rnomics [7], etc). These two levels influence each other and, ultimately, work on the same goals. Various research activities often use quite different modelling techniques and tools, which makes an integrative approach cumbersome or impossible. It is appropriate and fruitful to use coarse-grained models (such as selection/mutation) for analysis of certain statistical properties and selected evolutionary trajectories (for a). On the other hand, it is extremely difficult, and often not possible (nor appropriate) to use such models for fine-grained modelling, integrative and generalisation studies (b).

Selection/mutation, due to its inherent limitations (discussed later) are not appropriate to be used for integrative modelling and generalisations of evolutionary dynamics. We propose that different models for evolutionary dynamics are needed to augment (or replace) selection/mutation paradigm. We argue that new research frameworks based on complex hierarchical interacting systems are more suited for modelling and gaining insights into the evolutionary dynamics. Even though such models suffer from intractability in a traditional mathematical sense, they will nevertheless provide a better platform for researching and discussing complex aspects of evolutionary processes. This is most notably argued by Wolfram [27]. Such approach will facilitate better understanding and ways to obtain insights into the areas that are particularly challenging, such as the genetic architectures[1], genotype-phenotype dynamics[2], evolution of adaptation and hierarchical organisation [10].

2 Common Objectives

The 1998 report *Frontiers in population biology* (National Science Foundation) [10] identified six essential research areas in the field:

[1] Genetic architecture refers broadly to the set of the properties of genes and their interactions, including epigenetic effects, that determine the evolutionary dynamics of a given system and, effectively, phenotype of an organism.

[2] Phenotypic characteristics can span variety of phenomena, from gene expression profiles, through sensory signalling and performance within a given environment. Integrated system modelling approach is necessary, instead of isolated traits studies.

- Linkages to genomics and molecular genetics
  - Genetic architectures and evolutionary trajectories
  - Divergence of genetic architectures
  - Evolution of complex phenotypes
- Linkages to environmental issues
  - Invasive species
  - Community genetics
  - Evolution in human-dominated systems

In 2005 panel of experts met again to review and identify the main challenges and frontiers in evolutionary research for the next decade. The panel produced a new report, *Frontiers in Evolutionary Biology* (National Science Foundation) [11] that identified the following research themes, with various linkages and inter-dependencies between them:

- The evolution of genome structure and function
- Genetics/evolution of adaptation
- Population divergence and speciation
- Microevolution of development
- Evolution of integrated phenotypes
- Microbial systems
- Evolution of conflict and cooperation
- Large-scale patterns of diversity in time and space
- Applied evolution

It is implicit in both of the reports that the main challanges in the field are all in the area of modelling evolutionary dynamics in a holistic and integrated way. The stress is on integrated systems and fine-grained models. These are all well reflected by the research programmes and research themes pursued independently by many researchers. To proceed with the integrating research platform discussed in Section $1a$, there is a need to integrate various research results into a common framework. In other words, there is a need for a good platform to bridge the gap and to provide such integrating facilities and research programmes.

3 How to Integrate

To facilitate the integration process, a common set of assumptions and base principles should be established. At the moment, this common platform is based on the notions derived from Darwinian (or more accurately Mendelian) models based on selection and mutation [25]. Despite their huge popularity and abundance in evolutionary research, selection/mutation-based models suffer inherent limitations. Some of these limitations are of philosophical importance and some of practical nature. In general, we can classify these limitations into four classes:

(a) The mathematical models based on selection/mutation rely on certain properties, that are (usually) not met in the phenomena that is being modelled. That leads to discrepancy that is blurred or purposefully neglected by the researchers using these models. Note also, apart simple cases the mathematical models of evolutionary dynamics are intractable (that defeats the purpose of using them in the first place).

(b) The selection/mutation models often rely on the notion of fitness. Fitness expressed as an external influences (or bias) on the dynamics of the evolutionary processes is useful and provides the researchers the ability to focus on certain aspects of the evolutionary dynamics alone, neglecting other aspects and dependencies. Fitness simplifies the analysis of certain phenomena. However, the use of fitness itself presents a number of issues, among others: 1) it makes it harder to investigate and gain insights into the evolutionary trajectory derived exclusively from the dynamics of the interacting components by abstracting some (and usually unknown and intractable portion) of that dynamics into the given external bias (fitness); and 2) it implicitly assumes that global information about the state of the system is readily available to all of the components within the dynamical system itself (this violates a useful property of locality of interactions and local information spreading).

(c) The interpretative capabilities of models based on selection/mutation for general evolutionary dynamics are limited. The interplay of sampling effects (genetic drift), statistical tendencies of selection/mutation and the choice of the external bias in the form of fitness, can be adjusted arbitrarily in such a way as to provide a full spectrum of possible dynamical trajectories. There is no significance between various concrete experimental setups, thus applicability and interpretative capabilities of such models are limited.

(d) Selection/mutation-based models are inhibitely difficult to be used for modelling complex evolutionary systems, e.g. adaptability and symbiogenesis.

In the following sections we will briefly discuss the limitations of selection/mutation-based models, and identify what is needed to augment the selection/mutation for modelling various complex phenomena. Then, we introduce our approach and briefly discuss how we envision the Evolvable Virtual Machines (EVM) model to help gaining insights into the evolutionary trajectories, architectures and symbiosis research [14,12].

4 Selection/Mutation Models

Selection/mutation modelling has proved itself fruitful in many areas of evolutionary research. For practical reasons, the notion of selection/mutation and fitness provide useful tools with predictive capabilities. However, the interpretative values of selection/mutation models are questionable (in terms of causality). Selection/mutation models rely heavily on the notion of fitness. In simple terms, fitness represents/captures an external, influence or bias, that makes a random heuristic search with heuristic function G ([25], 3.2, p.9) to converge to the observed in biological system attractor A. In the model, it is G (and in special case of SGA the fitness/selection/mutation/mixing decomposition of G) that guides the process towards a particular attractor. However, in the biological system, it is more appropriate to think that complex (unknown) evolutionary interactions between all of the interacting entities produce a dynamics with a given attractor. G in this view, is just the mathematical result, the final effect, not the cause of the dynamics.

In [11], page 3, we read: "Adaptation requires both that organisms differ in fitness and that those fitness differences be heritable." Fitness, a simplified modelling approximation and tool for studies of certain coarse-grained statistical regularities, becomes in the statement above a first class causal agent. It may be useful, in certain studies to treat fitness and selection/mutation as causal forces, but one should not forget that these are ultimately just pure statistical constructs, that are derived from deterministic interactions of large collections of components on different level of evolutionary organisation. Selection/mutation/fitness are observable effects of the underlying evolutionary dynamics of the complex biological systems. The explanatory capabilities of selection/mutation are not applicable or particularly useful for studies of these underlying dependencies and dynamics.

Consider the case of two simple computational self-replicators, R_1 and R_2, such as the time to replicate R_2 is twice as long as R_1. The actual physical cause of this difference may be the length of the program, or inefficiency of the self-replication algorithm used (or any other structural or dynamical relationship). Now, when an initial population consisting of 50% of R_1 and 50% of R_2 is observed, a simple selection model can be inferred, such as $S =< 2/3, 1/3 >$. The explanatory power of such a model is however limited to global statistical effects of the underlying (detailed) dynamics. To understand why the population converges to R_1, or in other words, why R_1 is much more efficient self-replicator will not be possible with this simple selection-based model. A detailed model of all the interacting entities with their inderdependencies is required. Given such a model, selection-based model becomes irrelevant. It is our belief, that only through such fine grained models a progress can be made into understanding of long standing evolutionary challenges (Section 2).

Another problem occurs due to finite population sampling, referred as genetic drift. If there are two replicators as before, R_1 and R_2, and the replication rate for both is the same, then selection/mutation model is simply unable to capture such a system. Any selection model due to the sampling effect, will eventually converge to a single fixed point. The final population will be uniform (either R_1 or R_2). Note, that sampling effects are inherent in the modelling based on selection/mutation-based models. Any actual model with self-replicating entities will not suffer such a limitation. Note also, that the relevance of statistical sampling may or may not have any relevance to the actual phenomena being modelled. It may or may not be just the artifact of the model itself.

5 Meta-evolution: Going beyond Traditional Models

In [11] page 3 we read:

> Genomes have specific architectures and components that vary dynamically both within and among species. We are beginning to understand that single-gene approaches to understanding genome structure and function are insufficient. Individual gene products are embedded in large-scale interaction networks that represent integrated functional units at the molecular genetic level. Our understanding of the evolutionary dynamics of diversification in genome structures and their associated molecular genetic networks, however, remains limited. [...] we lack theoretical analyses to inform our understanding of how genomes and their associated genetic networks evolve. The revolution in genomics technologies and resources, including whole genome sequences, low-cost sequencing, microarray technologies and computational power, now

> allow us to begin to address questions surrounding the evolution of genome structure and function.

This statement suggest the need for theoretical and experimental frameworks suitable for conducting interpretative studies of complex systems based on aggregate network of interacting entities. This is what our EVM model (discussed later) tries to address. Explicit modelling of dynamics of complex networks of interacting entities is necessary, to enable research and analysis of complex dynamical genetic networks. A good indication of this is already placed in the mentioned above report:

> Complex phenotypic characteristics are determined by interaction of multiple causes. Such characters represent a challenge to developmental evolutionary biologists because there are typically no readily apparent intermediate phenotypes that would be likely to be maintained by selection. The classic example is the vertebrate eye.

Using selection models is often not appropriate and not fruitful way to pursue the research. Especially for complex phenomena discussed above it seems that there is a need to develop proper detailed hierarchical and network-based models of many interacting entities, that augment simple statistical models and selection/mutation.

6 Symbiosis and Symbiogenesis

Two of the main challenges that have been identified in the Section 2 are the evolution of adaptability and symbiosis. Adaptation has strong linkages with all other areas of evolutionary research and general properties cannot be studied in isolation. We believe, that adaptation should be studied in the context of complex evolutionary architectures (and without the notion of fitness, see discussion in Section 5). We believe, that symbiosis and symbiotic linking on various levels of organisation is crucial for the concept of adaptation. Report [10] identified symbiosis as one of a central themes:

> To date, the best-studied interactions have been infectious diseases, which have traditionally been studied from biomedical and agricultural perspectives. Most of this work has focused on elucidating the molecular bases of pathogenicity and susceptibility. But the evolution of pathogenicity and susceptibility are not well understood and such a perspective will provide important new insights for understanding and managing host- pathogen interactions. Moreover, pathogenic interactions are only one class among many symbioses, which take a variety of forms and are involved in such important yet poorly understood evolutionary phenomena as the generation of biodiversity and the emergence of biocomplexity. New discoveries have shown, for example, that both inter and intra genomic interactions can promote population divergence and speciation.

Symbiosis is defined as the interaction between two organisms living together. At least one member benefits from the relationship. The other member (the host) may be positively or negatively affected. Proponents of symbiogenesis argue that symbiosis is a primary source of biological variation and that acquisition and accumulation of random mutations alone are not sufficient to develop high levels of complexity [4,5].

K. Mereschkowsky [8] and I. Wallin [26] were the first to propose that independent organisms merge (spontaneously) to form composites. According to Margulis [6], "Life did not take over the globe by combat, but by networking". To model and study symbiotic relationships we have advised a generic evolutionary modelling architecture called Evolvable Virtual Machines (EVM). We have also conducted a number of experimental studies on symbiogenesis [13]. In the following section we will briefly discuss the main elements of the architecture.

7 Computational Modelling

When designing our computational modelling framework, our first objective was to provide a robust and flexible virtual machine, which can manipulate its own underlying implementation in order to adapt itself to the needs of a particular running program. Existing computational architectures provide little support for such an adaptation of the underlying virtual machine.

The second objective was to investigate different properties that influence the evolvability and possible meta-evolutionary[3] computational models. We are interested in properties that increase evolvability of a particular computing language. Most notably, we used some of the ideas from the Tierra [17,18] and Avida [1,15] systems, the ADATE system [16], Grammatical Evolution [19], the Push programming language [24] and the OOPS system (bias-optimal search) [23].

The third objective is related to the second. The earliest attempts at multi-level evolutionary computational designs were undertaken relatively recently and we were able to trace these ideas back to Juergen Schmidhuber's diploma work. Schmidhuber himself reports [21]:

> Pages 7-13 of Schmidhuber's (sic) diploma thesis [21] are devoted to a more ambitious self-improving GP [Genetic Programming] approach that recursively applies metalevel GP (first introduced here) to the task of finding better program-modifying programs on lower levels – the goal was to use GP for improving GP.

This line of research has also been taken up by Olsson's ADATE system [16]. Meta-level evolutionary systems provide a sound and important contribution to the understanding and synthesis of artificial and biological evolutionary processes. Their theoretical frameworks are difficult to evaluate empirically, however, due to high processing requirements. It is important to note that only two-level (meta-level and base-level) approaches have ever been studied empirically in automatic program generation (among others, by [16]). Only the work of Schmidhuber expands this notion into the meta-meta-. . .-level architectures. Theoretical work based on Levin's search [3,2,20] is valuable and provides a background for further exploration. The Optimal Ordered Problem Solver implementation [22] (OOPS) has been one of the inspirations for the EVM architecture implementation.

[3] By meta-evolution we mean the process of evolving evolvable structures, that is, structures that are easily evolvable and adaptable.

The original OOPS system has been criticised as not having much practical value due to the extremely high computational costs when applied to even "simple" computational problems. Such a critique, though valid for complex meta-programming paradigms in general, is missing an important point: for simple problems it does not make sense to use program generator techniques, or to use Levin's search for an optimal algorithm. However, beyond a certain complexity threshold, the exploration of program spaces may be a better, more efficient way to find a solution to a given task. The main point of meta-level automated program generators is to investigate ways for more efficient and effective program generators. One of the meta-programming aspect is related to the modelling of tendencies and algorithm generators. The second aspect is to make sufficient effort, through implementation techniques and through incorporating expert knowledge, such that the final system is efficient and can be effectively employed to tackle real-life problems.

The EVMA departs from the pure deterministic search processes of existing meta-search systems, such as those based on Levin search [3]. For example, the OOPS system [22], conducts bias-optimal exhaustive search in the program space. Not only is the search exhaustive, but also all the partial results are assumed to be stored indefinitely in an unlimited storage. Those properties make OOPS non-applicable to most real-life computational problems. The EVMA on the other hand does not assume the following:

- deterministic exhaustive search
- infinite storage for partial solutions

This means, that a) our EVM model employs stochastic methods to explore (only) part of the total search space. This may be taken arbitrarily close to exhaustive search by regulating the actual search space that a given method explores; b) EVM needs to dispose of and disregard some of the previously constructed structures due to space limitations. As in case a), for many problems that can incorporate all the partial solutions within the storage, this can be treated as effectively unlimited storage capabilities. Based on these two main points, EVM can be treated as a generalisation of generic bias-optimal search techniques, such as OOPS for example. In fact, our framework can use various stochastic search techniques to explore an unknown search space. The main reason for this is unsuitability of exhaustive search for large search spaces (due to high computational costs). In some circumstances it pays to make a good guess instead of covering the entire search space. This is especially true for infinite rugged search spaces like in evolutionary computation. The EVM provides a framework for program space exploration appropriate for multi-task distributed environments.

8 Conclusion

In this article, we have argued for meta-evolutionary modelling and provided brief introduction to our computational meta-evolutionary framework called EVM. We have focused on one of the aspects that is commonly re-occurring in many disciplines conducting studies of evolution, namely, inappropriate use of selection/mutation models for generalisations and fine-grained simulation of complex evolutionary trajectories. Research efforts based on Darwinian selection/mutation models are abundant and dominate the overall research landscape. They are well-suited for coarse statistical analysis of certain evolutionary phenomena. Even though they are popular and widely used, these models are insufficient for generalisations and fine-grained information-centric evolutionary modelling. To provide insights into and detailed fine analysis of aspects of the evolutionary dynamics, different models must be used to augment the popular selection/mutation paradigm. We have discussed some of the limitations of selection/mutation models and presented a general abstract architecture called EVM that can be used for modelling complex evolutionary phenomena to the arbitrary level of precision. The EVM architecture represents a generalised framework suitable for meta-evolutionary computational modelling.

In the future, we will investigating the ways EVM can be used to model hypercyclic dependencies in complex causal networks. We work on demonstration of emergence and analysis of the mechanisms of symbiogenesis. We also work on formalisation of EVM through pi-calculus and process algebra that will allow formal analysis of emerging computational properties through graph metrics a similar fashion to recent work by [9].

References

1. Adami, C.: Introduction to Artificial Life, 1st edn. Springer (July 30, 1999)
2. Hutter, M.: The fastest and shortest algorithm for all well-defined problems. International Journal of Foundations of Computer Science 13(3), 431–443 (2002), `http://citeseer.ist.psu.edu/hutter02fastest.html`, `http://arxiv.org/abs/cs.CC/0206022`
3. Levin, L.A.: Universal sequential search problems. Problems of Information Transmission 9(3), 265–266 (1973)
4. Margulis, L.: Origin of Eukaryotic Cells. University Press, New Haven (1970)
5. Margulis, L.: Symbiosis in Cell Evolution. Freeman & Co., San Francisco (1981)
6. Margulis, L., Sagan, D.: Microcosmos: Four Billion Years of Evolution from Our Microbial Ancestors. Summit Books, New York (1986)
7. Mattick, J.S., Gagen, M.J.: The evolution of controlled multitasked gene networks: The role of introns and other noncoding rnas in the development of complex organisms. Molecular Biology and Evolution 18(9), 1611–1630 (2001)
8. Mereschkowsky, K.S.: Ueber ber natur und ursprung der chromatophoren im pflanzenreiche. Biol. Zentralbl. 25, 593–604 (1905)
9. Moncion, T., Amar, P., Hutzler, G.: Automatic characterization of emergent phenomena in complex system. Journal of Biological Physics and Chemistry 10 (2010)
10. Frontiers in population biology. Tech. Rep. biorpt1098, National Science Foundation (NSF), USA, Rutgers University (October 1998), `http://www.nsf.gov/pubs/reports/frontiers_population_bio_disclaimer.pdf`
11. Frontiers in evolutionary research. Tech. Rep. biorpt080706, National Science Foundation, Rutgers University (March 2005), `http://www.nsf.gov/pubs/reports/frontiers_evolution_bio.pdf`
12. Nowostawski, M.: Evolvable Virtual Machines. Ph.D. thesis, Information Science Department, University of Otago, Dunedin, New Zealand (December 2008)
13. Nowostawski, M., Epiney, L., Purvis, M.: Self-adaptation and Dynamic Environment Experiments with Evolvable Virtual Machines. In: Brueckner, S.A., Di Marzo Serugendo, G., Hales, D., Zambonelli, F. (eds.) ESOA 2005. LNCS (LNAI), vol. 3910, pp. 46–60. Springer, Heidelberg (2006)

14. Nowostawski, M., Purvis, M.K.: Engineering Self-Organising Systems. In: Brueckner, S.A., Hassas, S., Jelasity, M., Yamins, D. (eds.) ESOA 2006. LNCS (LNAI), vol. 4335, pp. 176–191. Springer, Heidelberg (2007)
15. Ofria, C., Wilke, C.O.: Avida: A software platform for research in computational evolutionary biology. Artificial Life 10, 191–229 (2004)
16. Olsson, R.: Inductive functional programming using incremental program transformation. Artificial Intelligence 74(1), 55–81 (1995), http://www.ia-stud.hiof.no/~rolando/art_int_paper_74.ps
17. Ray, T.S.: An approach to the synthesis of life. In: Langton, C., Taylor, C., Farmer, J.D., Rasmussen, S. (eds.) Artificial Life II, Santa Fe Institute Studies in the Sciences of Complexity, vol. XI, pp. 371–408. Addison-Wesley, Redwood City (1991)
18. Ray, T.S.: Is it alive, or is it GA? In: Belew, R.K., Booker, L.B. (eds.) Proceedings of the 1991 International Conference on Genetic Algorithms, pp. 527–534. Morgan Kaufmann, San Mateo (1991)
19. Ryan, C., Collins, J.J., Neill, M.O.: Grammatical Evolution: Evolving Programs for an Arbitrary Language. In: Banzhaf, W., Poli, R., Schoenauer, M., Fogarty, T.C. (eds.) EuroGP 1998. LNCS, vol. 1391, pp. 83–96. Springer, Heidelberg (1998), http://link.springer.de/link/service/series/0558/papers/1391/13910083.pdf
20. Schmidhuber, J.: The speed prior: a new simplicity measure yielding near-optimal computable predictions (2002), http://citeseer.ist.psu.edu/schmidhuber02speed.html
21. Schmidhuber, J.: Self-referential learning, or on learning how to learn: The meta-meta-... hook. Diploma thesis, Institut fuer Informatik, Technische Universitaet Muenchen (1987), http://www.idsia.ch/~juergen/diploma.html
22. Schmidhuber, J.: Optimal ordered problem solver. Tech. Rep. IDSIA-12-02, IDSIA (July 31, 2002), ftp://ftp.idsia.ch/pub/juergen/oops.ps.gz
23. Schmidhuber, J.: Optimal ordered problem solver. Machine Learning 54, 211–254 (2004)
24. Spector, L., Robinson, A.: Genetic programming and autoconstructive evolution with the Push programming language. Genetic Programming and Evolvable Machines 3(1), 7–40 (2002)
25. Vose, M.D.: The Simple Genetic Algorithm: Foundations and Theory. MIT Press, Cambridge (1999)
26. Wallin, I.: Symbionticism and the Origin of Species. Williams & Wilkins, Baltimore (1927)
27. Wolfram, S.: A New Kind of Science, 1st edn. Wolfram Media, Inc. (May 2002)

Probabilistic Splicing Systems

Sherzod Turaev[1], Mathuri Selvarajoo[2], Mohd Hasan Selamat[3], Nor Haniza Sarmin[2,4], and Wan Heng Fong[4]

[1] Department of Computer Science
Kulliyah of Information and Communication Technology
International Islamic University Malaysia
50728 Kuala Lumpur, Malaysia
turaev@sherzod.info

[2] Department of Mathematical Sciences, Faculty of Science
Universiti Teknologi Malaysia
81310 UTM Johor Bahru, Johor, Malaysia
nhs@utm.my

[3] Faculty of Computer Science and Information Technology
Universiti Putra Malaysia
43400 UPM Serdang, Selangor, Malaysia
hasan@fsktm.upm.edu.my

[4] Ibnu Sina Institute for Fundamental Science Studies
Universiti Teknologi Malaysia
81310 UTM Johor Bahru, Johor, Malaysia
fwh@ibnusina.utm.my

Abstract. In this paper we introduce splicing systems with probabilities, i.e., *probabilistic splicing systems*, and establish basic properties of language families generated by this type of splicing systems. We show that a simple extension of splicing systems with probabilities may increase the computational power of splicing systems with finite components.

1 Introduction

DNA molecules are double stranded helicoidal structures composed of four nucleotides A (*adenine*), C (*cytosine*), G (*guanine*), and T (*thymine*), paired A-T, C-G according to the so-called *Watson-Crick complementary*. Watson-Crick complementary and *massive parallelism*, the other fundamental and distinctive feature of DNA molecules, are taken as the main characteristics of *DNA computing*.

Adleman's [1] famous biological experiment, which could solve Hamiltonian Path Problem using these two features, indeed gave a high hope for the future of DNA computing. Since there have been obtained a number exciting results showing the power of DNA computing, for instance, Lipton [2] showed that how to use DNA to solve the problem to find satisfying assignments for arbitrary contact networks. Boneh et al. [3] showed that DNA based computers can be used to solve the satisfiability problem for Boolean circuits.

N.T. Nguyen et al. (Eds.): *Adv. Methods for Comput. Collective Intelligence*, SCI 457, pp. 259–268.
DOI: 10.1007/978-3-642-34300-1_25

One of the early theoretical proposals for DNA based computation was made by Head [4] who used the *splicing operation* – a formal model of the cutting and recombination of DNA molecules under the influence of restriction enzymes. This process works as follows: two DNA molecules are cut at specific subsequences and the first part of one molecule is connected to the second part of the other molecule, and vice verse. This process can be formalized as an operation on strings, described by a so-called *splicing rule* of the form $(u_1, u_2; v_1, v_2)$ where u_1u_2 and v_1v_2 are the subsequences in question and the cuts are located between u_1 and u_2 as well as v_1 and v_2. These rules are the basis of a computational model (language generating device) called a *splicing system* or also *H system*. A system starts from a given set of strings (*axioms*) and produce a *language* by iterated splicing according to a given set of splicing rules.

Since splicing systems with finite sets of axioms and rules generate only regular languages (see [5]), several restrictions in the use of rules have been considered (see [6]), which increase the computational power up to the recursively enumerable languages. This is important from the point of view of DNA computing: splicing systems with restrictions can be considered as theoretical models of *universal programmable DNA based computers.*

Different problems appearing computer science and related areas motivates to consider suitable models for the solution of these problems. For instance, in order to develop accurate tools for natural and programming language processing, the probabilistic models have been widely used. In fact, adding probabilities to grammars allows eliminating ambiguity and leads more efficient parsing and tagging algorithms for the language processing. The study of probabilistic grammars (defined by assigning a probability distribution to the productions) and probabilistic automata (defined by associating probabilities with the transitions) started in the 1960s (for instance, see [7–11]). The recent results on probabilistic grammars and automata can be found, for instance, in [12–14].

In general, the probability of a generated (accepted) string is computed by the multiplication of the probabilities of those rules (transitions) which participated in the derivation (acceptance) of the string (though, in [11], the computation of probabilities is defined slightly different). Different threshold probabilistic languages can also be defined by using different thresholds (numbers, sets, etc.) and their modes (order relations, a membership to a threshold set, etc.)

The interesting and natural fact that the probabilistic concepts in formal language and automata theories can also be adapted in DNA computing theory, i.e., we can define probabilistic splicing and sticker systems as well as probabilistic Watson-Crick automata.

In this paper we introduce *probabilistic splicing systems*. In such systems, probabilities are associated with the axioms (not with the rules), and the probability of the generated string from two strings is calculated by multiplication of their probabilities. In order to overcome the ambiguity (the same string may have different probabilities), we can use another operation such as addition. We also define different threshold probabilistic languages using as a threshold segments, sets, real numbers and as a mode the order and equality relations, a membership to a threshold set.

This paper is organized as follows. Section 2 contains some necessary definitions and results from the theories of formal languages and splicing systems. Section 3 introduces the concept of a probabilistic splicing systems and threshold probabilistic languages, explains the specific features of probabilistic splicing systems in two examples and establishes some basic results concerning to the generative power of probabilistic splicing systems. It shows that probabilistic splicing systems with finite components generate not only regular languages but also context-free and context-sensitive languages. Section 4 discusses the obtained results, cites some open problems and indicates possible topics for future research in this direction.

2 Preliminaries

In this section we recall some prerequisites, by giving basic notions and notations of the theories formal languages, and splicing systems, which are used in sequel. The reader is referred to [6, 15, 16] for detailed information.

Throughout the paper we use the following general notations. The symbol $\in$ denotes the membership of an element to a set while the negation of set membership is denoted by $\notin$. The inclusion is denoted by $\subseteq$ and the strict (proper) inclusion is denoted by $\subset$. $\emptyset$ denotes the empty set. The sets of integers, positive rational numbers and real numbers are denoted by $\mathbb{Z}$, $\mathbb{Q}_+$ and $\mathbb{R}$, respectively. The cardinality of a set X is denoted by $|X|$.

The families of recursively enumerable, context-sensitive, context-free, linear, regular and finite languages are denoted by **RE**, **CS**, **CF**, **LIN**, **REG**, **FIN**, respectively. For these language families, the next strict inclusions, named *Chomsky hierarchy*, hold

$$\mathbf{FIN} \subset \mathbf{REG} \subset \mathbf{LIN} \subset \mathbf{CF} \subset \mathbf{CS} \subset \mathbf{RE}.$$

Further, we recall some basic notations and notations of (iterative) splicing systems.

Let V be an alphabet, and $\#, \$ \notin V$ two special symbols. A *splicing rule* over V is a string of the form

$$r = u_1\#u_2\$u_3\#u_4, \text{ where } u_i \in V^*, 1 \leq i \leq 4.$$

For such a rule r and strings $x, y, z \in V^*$, we write

$$(x, y) \vdash_r z \text{ iff } x = x_1u_1u_2x_2,\ y = y_1u_3u_4y_2, \\ \text{and } z = x_1u_1u_4y_2,$$

for some $x_1, x_2, y_1, y_2 \in V^*$.

We say that z is obtained by splicing x, y, as indicated by the rule r; u_1u_2 and u_3u_4 are called the *sites* of the splicing. We call x the *first term* and y the *second term* of the splicing operation. When understood from the context, we omit the specification of r and we write $\vdash$ instead of $\vdash_r$.

An *H scheme* is a pair $\sigma = (V, R)$, where V is an alphabet and $R \subseteq V^*\#V^*\$V^*\#V^*$ is a set of splicing rules. For a given H scheme $\sigma = (V, R)$ and a language $L \subseteq V^*$, we define

$$\begin{aligned}\sigma(L) =& \{z \in V^* \mid (x, y) \vdash_r z, \\ & \qquad \text{for some } x, y \in L, r \in R\}, \\ \sigma^0(L) =& L, \\ \sigma^{i+1}(L) =& \sigma^i(L) \cup \sigma(\sigma^i(L)), i \geq 0, \\ \sigma^*(L) =& \bigcup_{i \geq 0} \sigma^i(L).\end{aligned}$$

An *extended H system* is a construct $\gamma = (V, T, A, R)$, where V is an alphabet, $T \subseteq V$ is the *terminal* alphabet, $A \subseteq V^*$ is the set of *axioms*, and $R \subseteq V^*\#V^*\$V^*\#V^*$ is the set of *splicing rules.* When $T = V$, the system is said to be *non-extended.* The language generated by γ is defined by

$$L(\gamma) = \sigma^*(A) \cap T^*.$$

$\mathbf{EH}(F_1, F_2)$ denotes the family of languages generated by extended H systems $\gamma = (V, T, A, R)$ with $A \in F_1$ and $R \in F_2$ where

$$F_1, F_2 \in \{\mathbf{FIN}, \mathbf{REG}, \mathbf{CF}, \mathbf{LIN}, \mathbf{CS}, \mathbf{RE}\}.$$

Theorem 1 ([6]). *The relations in the following table hold, where at the intersection of the row marked with F_1 with the column marked with F_2 there appear either the family $\mathbf{EH}(F_1, F_2)$ or two families F_3, F_4 such that $F_3 \subset \mathbf{EH}(F_1, F_2) \subseteq F_4$.*

	FIN	**REG**	**LIN**	**CF**	**CS**	**RE**
FIN	**REG**	**RE**	**RE**	**RE**	**RE**	**RE**
REG	**REG**	**RE**	**RE**	**RE**	**RE**	**RE**
LIN	**LIN, CF**	**RE**	**RE**	**RE**	**RE**	**RE**
CF	**CF**	**RE**	**RE**	**RE**	**RE**	**RE**
CS	**RE**	**RE**	**RE**	**RE**	**RE**	**RE**
RE	**RE**	**RE**	**RE**	**RE**	**RE**	**RE**

3 Definitions, Examples and Results

In this section we define probabilistic splicing systems which is specified with probabilities assigned to each string generated by the splicing system and the multiplication operation over the probabilities. Moreover, we define threshold probabilistic splicing languages and show that probabilistic splicing systems with finite components can also generate context-free and context-sensitive languages.

Definition 2. *A probabilistic H (splicing) system is a 5-tuple* $\gamma = (V, T, A, R, p)$ *where* V, T, R *are defined as for a usual extended H system,* $p : V^* \to [0, 1]$ *is a probability function, and* A *is a finite subset of* $V^+ \times [0, 1]$ *such that*

$$\sum_{(x, p(x)) \in A} p(x) = 1.$$

We define a probabilistic splicing operation as follows:

Definition 3. *For strings* $(x, p(x)), (y, p(y)), (z, p(z)) \in V^* \times [0, 1]$ *and* $r \in R$,

$$[(x, p(x)), (y, p(y))] \vdash_r (z, p(z))$$

if and only if $(x, y) \vdash_r z$ *and* $p(z) = p(x)p(y)$.

Thus, the probability of the string $z \in V^*$ obtained by splicing operation on two strings $x, y \in V^*$ is computed by multiplying their probabilities.

Definition 4. *Then the language generated by the splicing system* γ *is defined as*

$$L_p(\gamma) = \{z \in T^* \mid (z, p(z)) \in \sigma^*(A)\}.$$

Remark 5. We should mention that splicing operations may result in the same string with different probabilities. Since, in this paper, we focus on strings whose probabilities satisfy some *threshold* requirements, i.e., the probabilities are merely used for the selection of some strings, this "ambiguity" does not effect on the selection. When we investigate the properties connected with the probabilities of the strings, we can define another operation together with the multiplication, for instance, the addition over the probabilities of the same strings, which removes the ambiguity problem.

Let $L_p(\gamma)$ be the language generated by a probabilistic splicing system $\gamma = (V, T, A, R, p)$. We consider as thresholds (cut-points) subsegments and discrete subsets of $[0, 1]$ as well as real numbers in $[0, 1]$. We define the following two types of *threshold languages* with respect to thresholds $\Omega \subseteq [0, 1]$ and $\omega \in [0, 1]$:

$$L_p(\gamma, *\omega) = \{z \in T^* \mid (z, p(z)) \in \sigma^*(A) \wedge p(z) * \omega\},$$
$$L_p(\gamma, \star\Omega) = \{z \in T^* \mid (z, p(z)) \in \sigma^*(A) \wedge p(z) \star \Omega\},$$

where $* \in \{=, \geq, >, \leq, <\}$ and $\star \in \{\in, \notin\}$ are called *threshold modes.*

We denote the family of languages generated by probabilistic splicing systems of type (F_1, F_2) by $p\mathbf{EH}(F_1, F_2)$, where

$$F_1, F_2 \in \{\mathbf{FIN}, \mathbf{REG}, \mathbf{CF}, \mathbf{LIN}, \mathbf{CS}, \mathbf{RE}\}.$$

Remark 6. In this paper we focus on probabilistic splicing systems with the finite set of axioms, since we consider a finite initial distribution of probabilities over the set of axioms. Moreover, it is natural in practical point of view: only splicing systems with finite components can be chosen as a theoretical models for DNA based computation devices. Thus, we use the simplified notation $p\mathbf{EH}(F)$ of the language family generated by probabilistic splicing systems with finite set of axioms instead of $p\mathbf{EH}(F_1, F_2)$, where $F \in \{\mathbf{FIN}, \mathbf{REG}, \mathbf{CF}, \mathbf{LIN}, \mathbf{CS}, \mathbf{RE}\}$ shows the family of languages for splicing rules.

From the definition the next lemma follows immediately.

Lemma 7

$$\mathbf{EH}(FIN, F) \subseteq p\mathbf{EH}(F),$$

for all families $F \in \{\mathbf{FIN}, \mathbf{REG}, \mathbf{CF}, \mathbf{LIN}, \mathbf{CS}, \mathbf{RE}\}$.

Proof. Let $\gamma = (V, T, A, R)$ be an extended splicing system generating the language $L(\gamma) \in \mathbf{EH}(\mathbf{FIN}, F)$ where $F \in \{\mathbf{FIN}, \mathbf{REG}, \mathbf{CF}, \mathbf{LIN}, \mathbf{CS}, \mathbf{RE}\}$.

Let $A = \{x_1, x_2, \ldots, x_n\}$, $n \geq 1$. We define a probabilistic splicing system $\gamma' = (V, T, A', R, p)$ where the set of axioms is defined by

$$A' = \{(x_i, p(x_i)) \mid x_i \in A, 1 \leq i \leq n\}$$

where $p(x_i) = 1/n$ for all $1 \leq i \leq n$, then $\sum_{i=1}^{n} p(x_i) = 1$. We define the threshold language generated by γ' as $L_p(\gamma', > 0)$, then it is not difficult to see that $L(\gamma) = L_p(\gamma', > 0)$. □

Example 8. Let us consider the system

$$\gamma_1 = (\{a, b, c, d\}, \{a, b, c\}, \{(cad, 2/7), (dbc, 5/7)\}, p_1, R_1)$$

where

$$R_1 = \{r_1 = a\#d\$c\#ad,\ r_2 = db\#c\$d\#b,\ r_3 = a\#d\$d\#b\}. \tag{1}$$

It is not difficult to see that the first rule in (1) can only be applied to the string cad, and the second rule in (1) to the string dbc. For instance,

$$[(cad, 2/7), (cad, 2/7)] \vdash_{r_1} (caad, (2/7)^2),$$

and

$$[(dbc, 5/7), (dbc, 5/7)] \vdash_{r_2} (dbbc, (5/7)^2).$$

In general, for any $k \geq 1$ and $m \geq 1$,

$$[(ca^k d, (2/7)^k), (cad, 2/7)] \vdash_{r_1} (ca^{k+1} d, (2/7)^{k+1}),$$

and

$$[(db^m c, (5/7)^m), (dbc, 5/7)] \vdash_{r_2} (db^{m+1} c, (5/7)^{m+1}).$$

From the strings $ca^k d$, $k \geq 1$, and $db^m c$, $m \geq 1$, by the rule r_3,

$$[(ca^k d, (2/7)^k), (db^m c, (5/7)^m)] \vdash_{r_3} (ca^k b^m c, (2/7)^k (5/7)^m).$$

Thus,

$$L_p(\gamma_1) = \{(ca^k b^m c, (2/7)^k (5/7)^m) \mid k \geq 1, m \geq 1\}.$$

We consider the threshold languages generated by this probabilistic splicing systems with different thresholds and modes:

(1) $L_p(\gamma_1, = 0) = \emptyset$;

(2) $L_p(\gamma_1, > 0) = L(\gamma_1')$;

(3) $L_p(\gamma_1, > \nu^i) = \{ca^k b^m c \mid 1 \leq k, m \leq i\}$;

(4) $L_p(\gamma_1, \in \{\nu^n \mid n \geq 1\}) = \{ca^n b^n c \mid n \geq 1\}$;

(5) $L_p(\gamma_1, \notin \{\nu^n \mid n \geq 1\}) = \{ca^k b^m c \mid k > m \geq 1\} \cup \{ca^k b^m c \mid m > k \geq 1\}$

where $\nu = 10/49$, $i \geq 1$ is a fixed integer, and γ_1' is the "crisp" variant of the splicing system γ_1, i.e., γ_1 without probabilities. We can see that the second language is **regular**, the third language is **finite**, the fourth and fifth languages are not regular but **context-free**. □

Example 9. Consider the probabilistic splicing system

$$\gamma_2 = (\{a, b, c, w, x, y, z\}, \{a, b, c, w, z\}, A_2, R_2, p_2)$$

where $A_2 = \{(wax, 3/19), (xby, 5/19), (ycz, 11/19)\}$ and

$$R_2 = \{r_1 = wa\#x\$w\#a, r_2 = xb\#y\$x\#b, r_3 = yc\#z\$y\#c, \\ r_4 = a\#x\$x\#b, r_5 = b\#y\$y\#c\}.$$

Using the first axiom and rule r_1, we obtain strings

$$(wa^k x, (3/19)^k), k \geq 1,$$

the second axiom and rule r_2,

$$(xb^m y, (5/19)^m), m \geq 1,$$

the third axiom and rule r_3,

$$(yc^n z, (11/19)^n), n \geq 1.$$

The nonterminals x and y from these strings are eliminated by rules r_4 and r_5, i.e.,

$$[(wa^k x, (3/19)^k), (xb^m y, (5/19)^m] \vdash_{r_4} \\ (wa^k b^m y, (3/19)^k (5/19)^m),$$

and

$$[(wa^k b^m y, (3/19)^k (5/19)^m), (yc^n z, (11/19)^n] \vdash_{r_5} \\ (wa^k b^m c^n z, (3/19)^k (5/19)^m (11/19)^n).$$

Then the language generated by the probabilistic splicing system γ_2

$$L_p(\gamma_2) = \{(wa^k b^m c^n z, \tau_1^k \tau_2^m \tau_3^n) \mid k, m, n \geq 1\}$$

where $\tau_1 = 3/19$, $\tau_2 = 5/19$ and $\tau_3 = 11/19$.

Further, we consider the following threshold languages:

$$L_p(\gamma_2, > 0) = L(\gamma_2') \in \mathbf{REG}$$

where γ_2' is the "crisp" variant of the splicing system γ_2.

$$L_p(\gamma_2, > \tau^i) = \{wa^k b^m c^n z \mid 1 \leq k, m, n \leq i\} \in \mathbf{FIN}$$

where $\tau = 165/6859$, and $i \geq 1$ is a fixed positive integer.

Now, let $\Omega = \{(165/6859)^n \mid n \geq 1\}$, then

$$L_p(\gamma_2, \in \Omega) = \{wa^n b^n c^n z \mid n \geq 1\} \in \mathbf{CS} - \mathbf{CF}$$

and

$$L_p(\gamma_2, \notin \Omega) = \{wa^k b^m c^n z \mid k, m, n \geq 1 \wedge k \neq m, m \neq n, k \neq n\} \in \mathbf{CS} - \mathbf{CF}. \quad \square$$

The examples above illustrate that the use of thresholds with probabilistic splicing systems increase generative power of splicing systems with finite components. We should also mention two simple but interesting facts of probabilistic splicing systems. First as Proposition 3 and second as Proposition 4, stated in the following.

Proposition 10. *For any probabilistic splicing system γ, the threshold language $L_p(\gamma, = 0)$ is the empty set, i.e., $L_p(\gamma, = 0) = \emptyset$.*

Proposition 11. *If for each splicing rule r in a probabilistic splicing system γ, $p(r) < 1$, then every threshold language $L_p(\gamma, > \eta)$ with $\eta > 0$ is finite.*

From Theorem 1, Lemma 7 and Examples 8,9, we obtain the following two theorems.

Theorem 12

$$\mathbf{REG} \subset p\mathbf{EH}(\mathbf{FIN}) \subseteq p\mathbf{EH}(F) = \mathbf{RE}$$

where $F \in \{\mathbf{REG}, \mathbf{CF}, \mathbf{LIN}, \mathbf{CS}, \mathbf{RE}\}$.

Theorem 13

$$p\mathbf{EN}(\mathbf{FIN}) - \mathbf{CF} \neq \emptyset.$$

4 Conclusions

In this paper we introduced probabilistic splicing systems by associating probabilities with strings and also established some basic but important facts. We showed that an extension of splicing systems with probabilities increases the generative power of splicing systems with finite components, in particular cases, probabilistic splicing systems can generate non-context-free languages.

The problem of strictness of the second inclusion in Theorem 12 and the incomparability of the family of context-free languages with the family of languages generated by probabilistic splicing systems with finite components (the inverse inequality of that in Theorem 13) remain open.

We should mention the possible and interesting topics for the study in probabilistic DNA computing in general and in probabilistic splicing systems in particular:

- Since extended splicing systems with finite set of splicing rules can generate languages in different language families in Chomsky hierarchy (see Theorem 1), it would be interesting to consider splicing systems with an infinite probability distribution over the set of axioms, and investigate the properties of the generated languages.
- It is also interesting to define probabilistic sticker systems and probabilistic Watson-Crick automata with different thresholds and modes.
- Another interesting topic in this direction is to consider splicing systems extended with fuzzy characteristic functions and weights, which can also be defined in the same way as probabilistic splicing systems.

Acknowledgements. This work has been supported by Ministry of Higher Education Fundamental Research Grant Scheme FRGS /1/11/SG/UPM/01/1 and University Putra Malaysia via RUGS 05-01-10-0896RU/F1.

References

1. Adleman, L.: Molecular computation of solutions to combinatorial problems. Science, 1021–1024 (November 11, 1994)
2. Lipton, R.J.: Using DNA to solve NP–complete problems. Science, 542–545 (April 28, 1995)
3. Boneh, D., Dunworth, C., Lipton, R., Sgall, J.: On the computational power of DNA. Discrete Applied Mathematics, Special Issue on Computational Molecular Biology, 79–94 (1996)
4. Head, T.: Formal language theory and DNA: An analysis of the generative capacity of specific recombination behaviors. Bull. Math. Biology, 737–759 (1987)
5. Pixton, D.: Regularity of splicing languages. Discrete Applied Mathematics, 101–124 (1996)
6. Păun, G., Rozenberg, G., Salomaa, A.: DNA computing. New computing paradigms. Springer (1998)
7. Booth, T., Thompson, R.: Applying probability measures to abstract languages. IEEE Transactions on Computers 22, 442–450 (1973)

8. Ellis, C.A.: Probabilistic Languages and Automata. PhD thesis, Department of Computer Science, University of Illinois (1969)
9. Fu, K.S., Li, T.: On stochastic automata and languages. Inter. J. Information Sci. (1969)
10. Rabin, M.: Probabilistic automata. Information and Control 6, 230–245 (1963)
11. Salomaa, A.: Probabilistic and weighted grammars. Information and Control 15, 529–544 (1969)
12. Smith, N., Johnson, M.: Weighted and probabilistic context-free grammars are equally expressive. IEEE Transactions on Computers 33, 477–491 (2007)
13. Kornai, A.: Probabilistic grammars and languages. Journal of Logic, Language and Information 20, 317–328 (2011)
14. Fowler, T.: The Generative Power of Probabilistic and Weighted Context-Free Grammars. In: Kanazawa, M., Kornai, A., Kracht, M., Seki, H. (eds.) MOL 12. LNCS, vol. 6878, pp. 57–71. Springer, Heidelberg (2011)
15. Rozenberg, G., Salomaa, A.: Handbook of formal languages, vol. 1-3. Springer (1997)
16. Dassow, J., Păun, G.: Regulated rewriting in formal language theory. Springer, Berlin (1989)

How to Model Repricable-Rate, Non-maturity Products in a Bank: Theoretical and Practical Replicating Portfolio

Pascal Damel and Nadège Ribau-Peltre

Assistant Professor at the University of Lorraine, France
damel.pascal@wanadoo.fr, nadege.peltre@neuf.fr

Abstract. Managing value has taken on considerable importance as a financial topic at both the theoretical and practical level. Value management enables a capital's economic value and commercial margins to be calculated. The actuarial process to calculate this value is hampered by repricable-rate banking products without maturity (as non maturity deposits). This is a significant issue as this type of product figures widely on balance sheets. The aim of this article is to present and test several methods for calculating this value.

Keywords: value, ALM "asset liability management", replicating portfolio, repriceable rate, non-maturity product.

1 Introduction

The aim of this article is to propose a theoretical and practical approach on banking products that pose a double challenge of repriceable rates without any explicit revision formulation or maturity dates. For example, a repriceable-rate current account with an undefined maturity date would disrupt traditional bank management. The absence of maturity obstructs the structuring of liquidity and rate gaps as well as making it difficult to use rate simulations to calculate the value evolution of banks. Value is indispensible for asset/liability management services that determine economic value, duration and convexity of equity capital. In the context of value management as a principal of utmost importance, banks lack the adequate tools to calculate the value of non-maturity products, and are incapable of properly managing the value in their balance sheets.

The lack of maturity creates an additional concern in the domain of financial controlling. Without maturity date, it is difficult to implement traditional techniques to calculate the margin for non-maturity products (the method for fund transfer pricing using market rates. A stable margin is important for widely-used products such as current accounts and sight deposits. The margin on this type of product is a major organizational consideration for banks to legitimize the existence of a distribution network and bear its costs.

Our article is divided into two parts. The first part will provide an overview of the subject. The second part will deal with repriceable rate modeling. In this second part, we will present to the scientific options for modeling value either via a profit-and-loss-based static replicating portfolio known as the optimal value method, or a

N.T. Nguyen et al. (Eds.): *Adv. Methods for Comput. Collective Intelligence*, SCI 457, pp. 269–278.
DOI: 10.1007/978-3-642-34300-1_26

balance-sheet-based dynamic replicating portfolio method. To illustrate our research, we will use a non-maturity product (the savings account) from a major European bank. We carried out this study with market and accounting data taken daily over an extended period of 30 years.

2 Modeling Non-maturity Products: An Overview

There are several methods to calculate value and value modifications of non-maturity deposits. The first ones rely on modeling based on short-term rates.

2.1 The Problem to Solve

In Finance, a perceived euro today is not the same value as a euro seen tomorrow. The sum of cash flow perceived at any time, we should take into account the price of the time (discount rate)

The cash flow discounted call current value V_0

$$V_0 = \sum_{t=1}^{n} \frac{CF_n}{(1+x)^n} \tag{1}$$

With

FC: cash flow

n: time

x: the discount rate (the time price)

For an undefined maturity product with repricable rate, we have two problems. The first is that we don't know the time. The client of the Bank can take his money at any time.

We don't know to the forward rate that the bank will pay to the customers. The interest rate depends on the situation of bank's competition and financial markets. It's the second problem.

In the formula of the current value see below, we have two unknown variables CF and n.

2.2 Replicating Portfolios

These methods aim to reproduce the evolution of the historical value of non-maturity products with financial instruments from the market such as bonds and options.

Static or optimal value replicating portfolios by Wilson [7] based on the Profit and Loss statement. This method was developed by Smithson in 1990, Wilson in 1994 and Damel [2] in 2001, and is based on the "building" or "Lego" approach. The goal of a replicating portfolio is to transform the volume and the retail rate of a repriceable-rate, non-maturity product into a bond portfolio of the same volume with a preferably stable margin.

In this method, the stable component of a non-maturity product is thought to be dependent on standard bond rate contracts with known maturity dates. In this problematic context, stability can be defined from the daily or monthly volume of retail products. The equation (2) uses "time series model" for optimal representation. This model defines a stable component (trend identified from an econometric model) and a volatile component, which is the statistical residual of the stability equation. This latter component is represented by a short-term bond with a short-term monetary reference rate (overnight rate or one-month rate, for example).

The stable component, or the trend, is made up with linked bonds. These linked bonds have several maturity terms from 3 months to 30 years. These linked bonds define a global strategy in refinancing founded on the money market and the primary bond market.

This stable component is made up of interest rate products. The advantage of this approach is that it takes into account both anticipated risk redemption and the eventual repricing of the product. We compute with SAS software an econometric link between the volum of a savings account multiplied by the retail rate with a roll-over bond volume multiplied by the market rate. The difference between the retail rate and replicating bond rates in a portfolio enables the margin to be established. A bond-based replicating portfolio helps define indispensible tools such as duration and convexity. This approach is comprised of two equations:

(2) is the equation of the historical volume, and

(3) the historical borrowing rate paid by the bank on an undefined maturity product.

$$\text{Volum}_t = \text{Stab}_t + \varepsilon_t \tag{2}$$

$$\text{Volum}_t \cdot \text{retail rate}_t = \varepsilon_t \cdot 1\,\text{month rate}_t + \sum_{i=3\,\text{mois}}^{15\,\text{ans}} \alpha_i\ \text{Bond}_i \cdot \text{rate}_{it} + \text{margin}_t + \delta_t \tag{3}$$

With the following constraints:

$$\sum_{i=3\,\text{months}}^{15\,\text{years}} \alpha_i = 1 \tag{4}$$

- Volumt represents the accountig position at a given moment t.
- Stabt represents the stable component or volum trend at a given moment t.
- Retail ratet is the rate of a product at a given moment t and 1 month rate, 2 month rate, etc. represent the market reference rate at a 1 month maturity, 2 month maturity, etc.
- εt represents the statistical residual or the volatile component of accounting positions or retail contracts that abruptly crop up and disappear.
- Bondi•rateit represents the interest for a zero bond coupon of maturity i and a market reference i.

- αi represents the stable component replicated by the position Bondi•tauxi.
- $\Sigma\, \alpha i$ is equal to 1 (i= from 3 months to 15 years).
- margint is the statistical intercept or the constant respresents the rate component of a sight account that is not contingent on market rates.

Technically, equation (3) is obtained thanks to non-linear statistical optimization. A replicating bond portfolio is obtained while respecting statistical criteria.

This approach is used as it takes into account the multi-periodicity of maturity dates and in particular allows a stable margin to be calculated through the difference between the borrowing rate and the market replicating portfolio rate. The drawback is that the replicating portfolio remains static and constant over that period.

Balance-sheet-based dynamic replicating portfolio: Elkenbracht and Nauta's approach [5]. Jarrow and Van Deventer [6] were the first to build a model based on the replication of non-maturity products from a dynamic portfolio. As opposed to a static portfolio, the composition of a dynamic portfolio is not stable. This modeling is based on the book value and not on the interest paid (Profit and Loss) as proposed by static replication in equation (3).

The authors have taken the equations presented below, with PV(c(t)) defining the value of the non-maturity product:

$$PV(c(t)) = V(t_0) + \Sigma_{i=0}^{N-1} d(t_{i+1})[\Delta V(t_{i+1}) - V(t_i)c(t_i) + r(t_i)V(t_i)] - d(t_{N+1})V(t_N) \quad (5)$$

$$PV(c(t)) = \sum_{i=0}^{N-1} d(t_{i+1})[-V(t_i)c(t_i) + r(t_i)V(t_i)] \quad (6)$$

$$PV(c(t)) = \sum_{i=0}^{N-1} d(t_{i+1})[-V(t_i)(c(t_i) + TS) + r(t_i)V(t_i)] = 0 \quad (7)$$

$$PV(c(t)) = \sum_{i=0}^{N-1} d(t_{i+1})[-V(t_i)(c(t_i) + TS) + r(t_i)V(t_i)] + P\&Lportfolio = 0 \quad (8)$$

The authors responsible for ALM modeling at ABN AMRO propose an approach that ties the non-maturity retail product to a reinvestment portfolio so as to hedge its book value. The reinvestment strategy presumes that the profits/losses on the portfolio will be compensated by the opposing profits /losses on the non-maturity product (including the margin).

This study was carried out over a period of 22 years. The study shows a high stability of the margin over the period of the analysis. The rate model used is the classic forward model.

3 Methods for Modeling a Non-maturity, Repriceable Rate Savings Account

The aforementioned methods show that the ones which are the most complete come from replicating portfolios.

We thus propose to model repriceable rates differently and then build a replicating portfolio using static and dynamic methods.

3.1 The Data

We used a time series daily volume rendering of savings accounts from 1981 to 2007 in this study. The trend started with an amount of 6 billion euros and stabilized at the end of 2007 with an outstanding amount of roughly 25 billion euros. Growth occurred during periods of stability. For reasons of confidentiality, the most recent periods have not been included.

3.2 Modeling Retail Rate Behavior

Modeling retail rate behavior is necessary for calculating the value of a savings account. The scope is to understand how rate revision works. Indeed, rate changes on savings accounts in European banks are decided by the bank without referring explicitly to reference rates[1]. Moreover, the effect from the timelines between the decisions made on the two rate revisions must be taken into account. To model this type of rate, we used the approach by Damel, which incorporates logistic regression in addition to static margins, dynamic margins and time as explicative variables [1].

Numerous tests have been carried out to calibrate the scientific approach and bring about a stronger reasoning behind the history and the best backtesting methods from 1981 to 2007. We've put forward several methods.

We tested several variants of static and dynamic margins. The variants consisted of absolute, relative (percentage) and averaged calculations of the margins. A two-step process was used to select the explicative reference variables.

The first step consisted of pre-selecting the explicative variables by testing various methods (simple correlations - canonical correlations - variation coefficients). The most effective method in this study was simple correlation.

The obvious aim of this research was to obtain a model with the best forecasting ability. Forecasting ability was measured by the backtesting made possible by weekly data on the retail rate, the market rate and the volumes from 1981-2007. Several levels of tests were carried out. In spite of some excellent econometric adjustments, the tests on historical data initially showed a weak forecasting capacity. To get around this problem, we created models over shorter periods to forecast each rate revision and to optimize the backtesting over short periods (a few years) and to anticipate subsequent repricing. We noted certain recurrent explicative variables in the short-term

[1] As Euribor + spread.

forecasts between 1981 and 2007. We used these margin-based variables to build a long-term forecasting model. We also observed that the rising trends in repriceable rates were different from the falling trends.

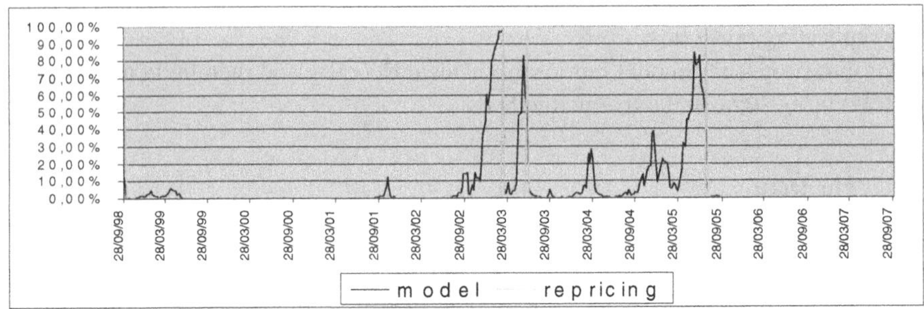

Fig. 1. Model from 1981 - 1998 / forecast 1999 - 2007

3.3 Wilson's Static Replicating Portfolio

Scientific justification: for the static replicating we use time series model as forecasting methods and works of Pr. Granger (Nobel Laureate). The software used is Eviews for time series model and SAS to estimate the system of Wilson's equation (model as linear regression with constraints). The model will be correct if the scientific criteria associated with these methods are met.

Wilson's optimal value static replication is founded on historical data: modeling the optimal portfolio and commercial margin from the P & L statement. The time series of sight deposit accounts is stationary with two differentiations using the 0% probability Augmented Dickey-Fuller test.

What is obtained is and ARMA with a Garch (3.0) term error. Econometric criteria used are correlogram, test of DW, White, Arch, etc… It should be reiterated that this model can identify the stable component or the part that can be modelized in the series. The statistical residual is the short-term, volatile component.

The tests we carried out incorporated the following facts. In the Wilson model, the margin (margint) is an intercept in the equation system. The margin is the interest differential obtained by making the absolute difference between the repriceable rate and the refinancing market rate, multiplied by the volume of the savings account.

Lastly, we propose improving the equation specification to take into account this strong rate and volume effect on the commercial margin. The intercept (the margin) of Wilson's equation is associated with an index that follows the evolution of loan interest paid (retail rate * volume) by the bank to its savings account holders. The stable margin is thus a stable proportion of paid loan interest. This econometric specification enables both the rate and volume effect to be taken into account in the historical modeling. The model obtained is the model at its optimum (R2 equal to 98.7%, respecting linear constraints on a replicating portfolio, and minimizing the variance of the margin estimator). This replicating portfolio can thus calculate the duration (here 3.10 years) and even the convexity. The commercial margin using Wilson's method declines on the whole to reach 174 BPS at the end of the timeline.

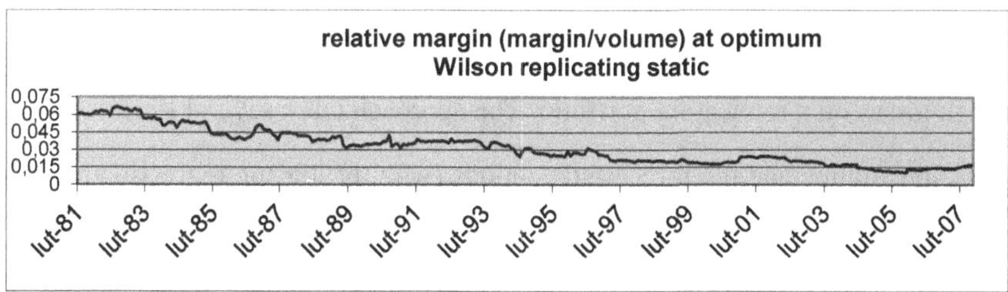

Fig. 2. Commercial margin in billions euro and relative margin in BPs

Static replication has the advantage of being able to transform a non-maturity product in a portfolio made up of bonds with a limited number of restrictive hypotheses. On the other hand, there are drawbacks as well, as Mr. Elkenbracht points out. Indeed, the bond refinancing that enables the margin to be calculated is stable on the timeline. Even if the the econometric adjustment is correct, it can't be said that this rule of replicating refinancing will be applied in the future. Likewise, a replicating portfolio by nature cannot be applied to new products. This is why we covered the dynamic replicating portfolio approach in our research; it breaks away from historical interest/volume logic.

3.4 Dynamic Replication

Optimal value dynamic replication based on historical data: optimal portfolio modeling and commercial margin from a profit and loss statement. Scientific justification: the static replicating is an iterative system programmed in VBA. The quality of replication is effective if the convexity of the value have the same value between the banking product and replicating portfolio.

Dynamic replication is a method that requires a runoff technique or the amortization of outstanding loans. We tested several economic forecasting models. The Pindyck-Rubinfeld method is best adapted for runoff as it uses double smoothing and the stochastic properties of time series models.

This method enables the volume of a savings account to be distributed over a 30 year period[2]. The final maturity of this amortization profile depends on the confidence interval used in Pindyck-Rubinfeld's model. This amortization profile will be used in the actuarial calculations of a dynamic replicating portfolio.

Dynamic Replication. We shall revert to Jarrow & Van Deventer's paradigm, revised by Mr. Elkenbracht (cf above).

$$PV(c(t)) = \sum_{i=0}^{N-1} d(t_{i+1})[-V(t_i)(c(t_i)+TS) + r(t_i)V(t_i)] + P\&Lportfolio = 0 \quad (8)$$

To solve this equation, we had at our disposal an interest rate distribution model, an equation for the retail rate that relies on the process of interest rate distribution, and an amortization model.

[2] This period corresponds to a strict condidence interval from a liquidity stress scenario.

To hedge the value of a savings account, it suffices to add the commercial margin to the loan rate paid to customers. Our approach creates an investment strategy in such a way that earnings/losses on the value of a savings account are inverse to the earnings/losses on a replicating portfolio. The approach takes several steps:

- In t_0, the volume and the forecasted coupons (with thr logistic regression model) of savings accounts are respectively amortized. The current value of these flows must be identical to the book value in t_0. The discount factors are increasing with a commercial spread (SWAP zero coupon + Spread). We implicitly obtain this margin and the uniformity of actuarial values. For technical purposes, an iteration was made.
- Then, the partial sensitivity of the value of the savings account for each point on the curve was calculated by shifting to roughly 10 points on either side of the discount factor. A variation of the value for each point on the curve was obtained.
- This variation in value is replicated in a bond portfolio[3] in such a way as to obtain the same sensitivity by curve point. The difference in the nominal between the replicating portfolio and the volume of the savings account is invested/borrowed at an overnight monetary rate.
- In tn, market rate variations create variations in value both for the savings account and the replicating bond portfolio. Because the partial sensitivities are identical, there isn't a big gap between the gains and losses. Nevertheless, differences remain as the profile of cash flows is not purely identical; in other words, there are different convexities, hence differences in creating value.
- On the whole, the value of a savings account increased more than that of a replicating portfolio[4] in the rate environment of our test. This difference in value, or convexity, poses a recurring problem for the ALM of major banks.
- The new production, or the difference between the initial amortization and the true volume of a given period adheres to the same rules of construction, with a marginal markup that can be different.
- The partial sensitivities of a savings account are recalculated. The portfolio is adjusted so as to maintain uniformity among the partial sensitivities.

We thus have a dynamic portfolio thanks to this interactive process.

- If there are still problems in convexity, our technique also allows the commercial margin of the former production to be changed (volume to be amortized). We can too integrate non-linear market derivatives such as options.

The dynamic replicating portfolio gives us a real tool to hedge value. There is no denying that the calculated commercial margin is 30 base points under the Wilson static method, yet is much more stable (tests carried out below).

[3] Classic plain vanilla at a fixed rate.

[4] Positive histograms of the difference in value. When value declines, the decrease in debt value (savings account) is on the whole less than the replicating portfolio. Inversely, when value increases, the debt rises on the whole more than the replicating portfolio.

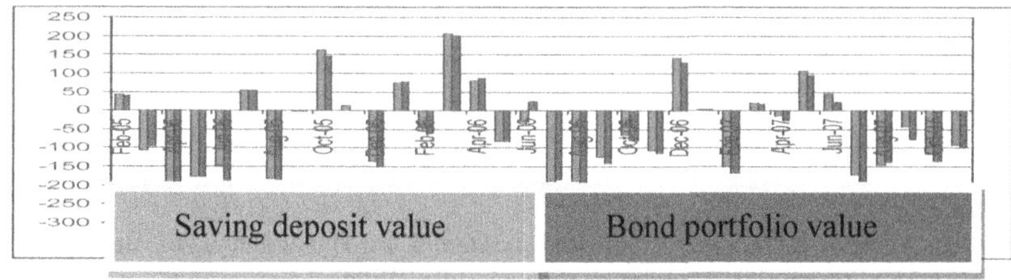

Fig. 3. Differences of value between the replicating and the non maturity product

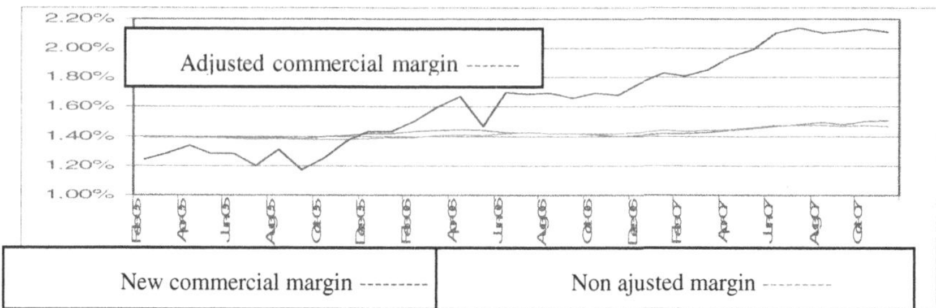

Fig. 4. Commercial margins

The adjusted commercial margin both integrates the margin of the new production and modifies the net present value of the former production by adjusting the commercial margin.

4 Conclusion

Managing value is at the core of modern financial concerns. The classic tools of calculating and hedging value are not suitable for non maturity product with repricable rate. The replicating portfolio is both a theoretical and concrete response to modeling and hedging this value. The shared limits of financial theory tied to the prediction of economic variables apply to our methodology as well. This research, which warrants reflection of the repriceable rate and on the amortization profile, draws a comparison between a static replicating portfolio and a dynamic replicating portfolio. The various tests presented show the obvious advantages of the dynamic replicating technique. The commercial margin is much more stable, and an exact hedging of value is made possible.

References

1. Damel, P.: La modélisation des contrats bancaires à taux révisable: une approche utilisant les corrélations canoniques (Modeling repriceable rate banking contracts: an approach using canonic correlations). Banque et Marchés (39) (1999)

2. Damel, P.: L'apport des méthodes de 'replicating portfolio' ou portefeuille répliqué en A.L.M.: méthode contrat par contrat ou par la valeur optimal (The contribution of replicating portfolio methods in ALM: a contract-by-contract or optimal value method). Banque et Marchés (51) (2001)
3. Dewachter, H., Lyrio, M., Maes, K.: La modélisation de la valeur des produits sans maturité (Modeling the value of non-maturity products). Working Paper of The National Bank of Belgium (2005)
4. Esch, L., Kieffer, R., Lopez, T., assisted by Damel, P., Hannosset, J.-F., Berbé, C.: Asset and Risk Management: Risk-Oriented Finance. John Wiley & Sons (2006)
5. Elkenbracht, M., Nauta, B.J.: Managing interest rate risk for non-maturity deposits. Risk (November 2006)
6. Jarrow, R., Van Deventer, D.: The arbitrage-free valuation and hedging of non-maturity deposits and credit card loan. Journal of Banking and Finance (22), 249–272 (1998)
7. Wilson, T.: Optimal Value: Portfolio Theory. Balance Sheet 3(3) (1994)

The Use of Artificial Neural Networks in Company Valuation Process

Zofia Wilimowska[1] and Tomasz Krzysztoszek[2]

[1] Industrial Engineering and Management Institute, Faculty of Computer Science and Management, Wroclaw University of Technology, Poland
[2] School of Higher Vocational Education in Nysa, Poland
{zofia.wilomowska,tomasz.krzysztoszek}@pwr.wroc.pl

Abstract. An increase in a company value is the main goal of the firm activity that creates opportunities for long term functioning and its development. The goal realization makes the investors see the company better and the company can find the capital easier. Many factors (external and internal) cause the company value. The paper presents the factors that should be taken into consideration in the process of company valuation; moreover, the articles presents a method of drivers value forecasting. The authors proposed the method based on artificial neural networks. The structure and simulation of the model is conducted. The implementation of the model is presented on the example of "Hama-Bis" company.

Keywords: Computer simulations, artificial neural networks, company valuation, mixed methods of valuation, value drivers.

1 Introduction

The idea of Value Based Management needs determining the factors whose influence firm value and make it possible to control them (value drivers). Managers can increase the firm's value by creating factors value at an efficient level. Uncertainty and risk which are associated with business activity require a special instrument supporting a managerial decision that creates the firm's value. The choice of the company valuation method – methods based on real assets or based on future cash flows – depends on the accepted goals of the firm and valuation process. The income methods treat the firm as a cash flow generator from today to the infinity – under the assumption of the going concern:

$$V = \sum_{t=1}^{\infty} \frac{E(CF)_t}{(1+r)^t}, \tag{1}$$

where : $E(CF)_t$ – expected cash flow in period t, r – discount rate.

Changeability and risk connected with economic activity create the necessity of the firm environmental changing reaction. The firm's valuation methods used up to now

N.T. Nguyen et al. (Eds.): *Adv. Methods for Comput. Collective Intelligence*, SCI 457, pp. 279–288.
DOI: 10.1007/978-3-642-34300-1_27

are not good enough, they do not value active firm managing in changing environment [1]. The factors that determine company value which exist in dynamically changing environment and influencing the forecasted cash flow level and discount rate are not deterministic. They follow the unpredicted environment changes and internal changes of the company. So the valuation process should use modern technologies to describe factors influencing company value.

2 Methods of Company Valuation

There is no doubt that company valuation should use value drivers. Depending on the valuation method, the researcher considers drivers which affect assets, income or both of them. Whichever approach is chosen, the value drives should be carefully analyzed relative to how much they influence and create the defined type of value.

For the verification of the usefulness of value drivers in companies being in business on the Polish market questionnaire research was carried. The research included twenty two organizations. The first and the second questions concerned the usefulness of value drivers on the strategic and operative level. Questionnaire results are presented in Table 1.

At the strategic level the most useful factors which were recognized were the growth of sales ratio (68%) and operative profit margin and free cash flows (50%). However, on the operative level as the most useful the respondents defined operative

Table 1. Usefulness of value drivers on strategic and operative rung (Source: [5])

No	Value driver	Usefulness				
		none	low	I don't know	average	high
		Strategic level				
1.	Growth of sales ratio	9%	0%	0%	23%	68%
2.	Operative profit margin	5%	5%	9%	32%	50%
3.	Investments in working capital	5%	5%	27%	50%	14%
4.	Investments in assets	5%	14%	50%	23%	9%
5.	Capital structure	5%	14%	50%	18%	14%
6.	Cost of capital	5%	5%	36%	36%	18%
7.	Period of growth of value	0%	0%	45%	32%	23%
8.	Cash tax rate	5%	9%	59%	27%	0%
9.	Free cash flows	5%	9%	18%	18%	50%
10.	Management of client loyalty	0%	0%	36%	45%	18%
11.	Operative lever	0%	5%	50%	23%	23%
		Operative level				
1.	Growth of sales ratio	0%	0%	9%	50%	41%
2.	Operative profit margin	0%	0%	14%	36%	50%
3.	Investments in working capital	0%	14%	45%	36%	5%
4.	Investments in assets	5%	0%	64%	23%	9%
5.	Management loyalty of client	5%	14%	32%	36%	14%

profit margin (50%) and the growth of rate ratio. Considering external factors (macro-economic drivers), the inflation was regarded by 18% of respondents, 36% of the asked recognized it as rather deciding. The rates of interest and gross domestic product were regarded as rather deciding on enterprise value properly by 68% and 50% of responders, respectively.

The main idea of mixed methods is that the value of company depends on its assets and its goodwill. Mixed methods are a kind of compromise. They deal with static assets valuation and with future profits gathering. Regardless of their classification, the main idea of the mixed methods of valuation is based on the following formula:

$$V = V_A + goodwill \quad \text{or} \quad V = V_A + f(V_I), \tag{2}$$

where: V- value of the company, V_A- company value based on assets, V_I- company value based on income.

The interaction of V_A, V_I and their influence on the final value of the company is described by weights which stand by asset and goodwill elements:

$$V = a \cdot V_A + b \cdot V_I\,, \tag{3}$$

where: a -weight attributed to company value based on assets, b-weight attributed to company value based on income. Conditions:

$$\cdot V_A \leq \cdot V_I\,, a, b \in (0,1) \tag{4}$$

Components V_A and V_I can be easily swapped with more sophisticated equivalents like real options etc. The researchers have developed several mixed methods which differ in a sense of weights settlement and treatment of V_A and V_I.

3 The Idea of Neural Networks Used for Company Valuation

Depending on the valuation method, the researchers consider value drivers which affect assets, income or both of them. The mixed method of valuation considers mainly two groups of drivers: asset and income ones. Describing a simple model of company value, the following relation can be noticed:

$$V = F(V_A) + F(V_I) = F(x_1, x_2, x_3, \ldots x_n) + F(y_1, y_2, y_3, \ldots y_m) \tag{5}$$

Variables x_1, x_2, x_3,..,x_n and y_1,y_2,y_3,...,y_n are value drivers, influencing in sequence the explained variables V_A and V_I where V_A, will be find as Net Assets Value and V_I as Discounted Cash Flows. To use the mixed method valuation at the beginning the prognosis of assets value and income value has to be conducted. Fig. 1 illustrates possible connections and relations between several groups of drivers. The main value drivers of V_A and V_I are final value factors and they are also divided into other numerous drivers.

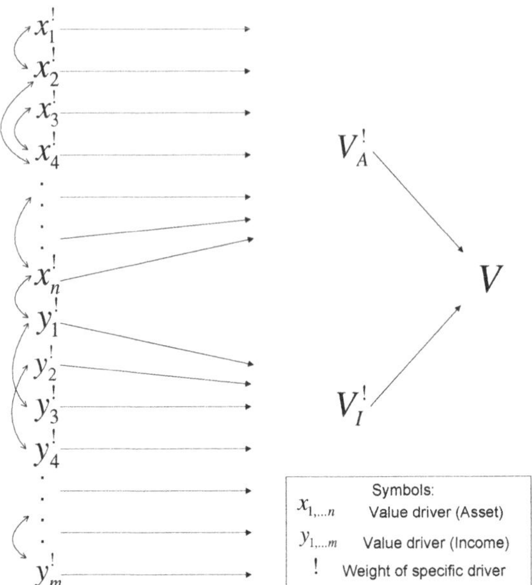

Fig. 1. Value drivers structure against the background of firm value

Artificial neural networks will be used as a tool to build and perform the model of company valuation.

The main difference between artificial neural networks and traditional models is that ANNs have an ability to learn [2. 6]. In short, neural networks, basing on exemplary data, can detect and memorize regularity and patterns occurring in the delivered data set. After the training process is performed, the researcher decides whether the network works properly or not. The correct results of training lead to test phase, where the neural system generates solutions to the defined problems.

The task being solved in this paper is a prediction type of problem. To sum up, the possibility of accomplishing defined tasks, proper conduction of simulation and gaining satisfying results using ANNs demand considering several issues: building the network, training the network, testing the network.

Generally, the simpler structure of artificial neural network is, the easier it is to manipulate and modify it. Building neural network form the start is very time-consuming and often difficult, even for advanced programmers.

4 Simulation Results of Neural Networks Used for Hama-Bis Valuation

"Hama-Bis" is a civil company. It is a hatchery and poultry distribution company situated in Stary Widzim near Wolsztyn in Wielkopolska province in Poland. Founded in 1997, the hatchery can presently produce 50 million day-old chickens per year, which equals to 10% of all national production. The final product of "Hama-Bis" activity is day-old chicken, defined as a meat consumption broiler bird.

Raw material in the form of hatching eggs is delivered by Polish and foreign suppliers. After being hatched, chickens are sold and distributed to broiler farms. 70% of "Hama-Bis" hatchery production covers the Polish market; the rest goes to Eastern Europe costumers mainly in Ukraine, Russia, Belarus and Lithuania.

4.1 Value Drivers Used in the Valuation Process

The construction of neural network was preceded by the aggregation of value drivers. They are described by [3, 4]:

- Warsaw stock exchange index WIG: 188 sessions covering 47 months were considered.
- Hatchability: Generally, average hatchability between years 2003 and 2005 was at the level of 82%. Important information is that 2% lower hatchability results in 1 million less chicken production per year. In 2007 this factor dropped down to 78%.
- Economic situation: the economic situation of poultry market depends on wild fowl epidemic diseases and its side effects. In 2006 many companies were afflicted by bird flu disease. As a result, the liquidity of most of them drastically dropped below the safe level.
- The average contract price of hatching eggs: The cost of hatching egg amounts to around 70% of chicken sales price.
- Average sales price: This factor is established by market and it constantly changes. This is also an example of external environment driver.
- Export: At the end of 2006, 32% of total "Hama-Bis" production was assigned for export. Average sales prices in the eastern part of Europe are higher than domestic, therefore maintaining other factors stable, export is more profitable.
- Consumption per capita: In the last quarter of 2006, statistical consumption in Poland equaled 21 kilos per person. At present, this value driver is mainly determined by the market price.
- Expenditure on marketing: Calculated as relation between expenditure on marketing in total capital expenditure.
- Share in the market: Polish poultry market is well-developed. There are over 150 hatcheries across the country. In June 2007, "Hama-Bis" had a 10% share of one day old chicken production in the Polish market.
- Risk of bird flu: Although risk is included in company valuation, particularly in DCF methods, but it is being chosen arbitrarily.
- ROE: Return on equity is a value driver easy to estimate, which should be used during company valuation.
- Sales ratio for 10 largest costumers: After revising "Hama-Bis" financial reports, the researcher noticed that forefront clients generate major company profit. The company aims to increase strategic clients instead of the minor ones.

As stated before, the above-mentioned drivers can be treated as input data. The goal of the first step research is to measure how strong they influence outputs, which are drivers as well, but the direct ones. There were defined two outputs, used in most common income and asset methods of valuation. These factors are also the main components in mixed approach to company valuation:

- Cash flow: Free Cash Flow to Equity (FCFE). The source of this driver is cash flow statement.
- Net assets: The choice of this factor was made because of its application in both in asset and mixed approach to valuation.

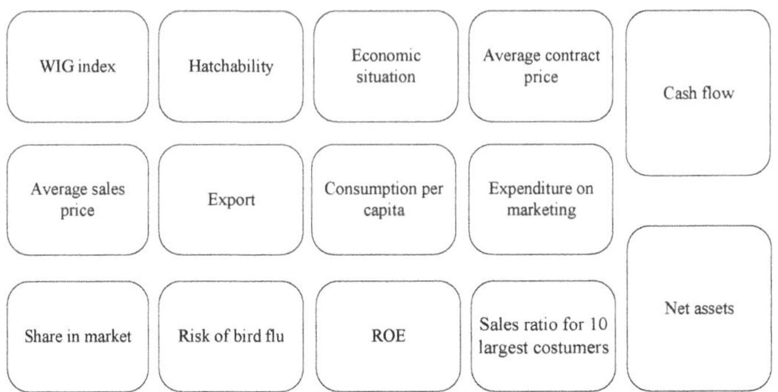

Fig. 2. Exemplary value drivers for "Hama-Bis" company [3]

The Presented value drivers illustrated in Fig. 2 should be taken into consideration and treated as model variables. Such description can facilitate their usage in computer modeling which will be applied in the mixed method of company valuation.

4.2 Model Construction

The collected data was fed to spreadsheet. The structure of Neural Networks used to company valuation is shown in Fig. 3.

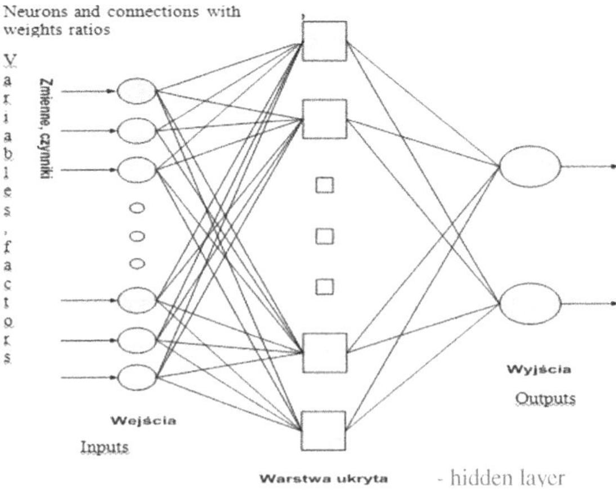

Fig. 3. The example of the ANN used in the valuation process

Input data are value drivers and output date are the company values. The software used to create neural network and conduct simulations was NeuroSolutions version 4.32 by NeuroDimension. This comprehensive tool lets the user build artificial neural network with the help of 4 modules: NeuralBuilder, NeuralExpert, NeuroSolutions for Excel, Custom Solution Wizard. NeuroSolutions for Excel was used in the following research. This module works with Microsoft Excel as an add-in it can be found in Excel's tool bar. In order to solve the defined problem, NeuroSolutions for Excel offers 12 modules, which help the user build, train, test network and analyze its results, Fig. 4.

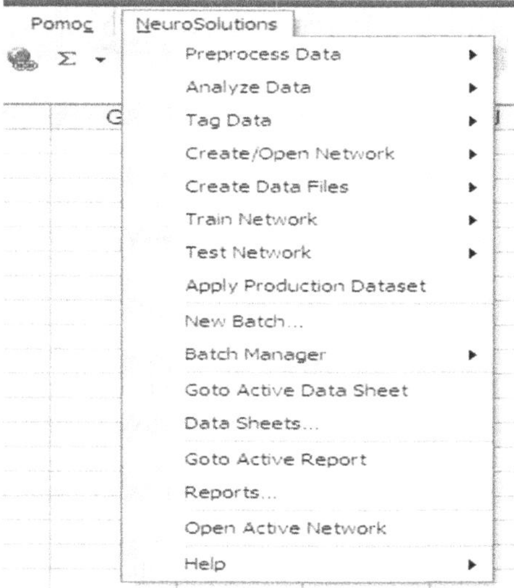

Fig. 4. Functions of NeuroSolutions for Excel

The software requires data to be contained in separate columns. Thus, the number of columns equals the number of total inputs and outputs. The labels of data sets must be allocated in the first row of spreadsheet. The following rows contain the values of data samples. There are 12 inputs, which represent explanatory variables and 2 outputs, which are explained variables. ANNs requires numerous data sets. Total data set covers 188 samples for each variable. It gives a matrix of 14 columns and 188 rows: 14x188=2632 relations, excluding feedback connections. Value drivers were gathered from financial reports, balance and cash flow sheets. Tagging is also used for training and testing purpose. It is necessary to mark the number of data samples designated to train and test the neural system. After the data is tagged, a neural network can be built. NeuroSolutions for Excel offers several neural networks, which are ready to use. They are automatically offered after the delivered data is checked. In the presented research, multilayer perception network was created. This type of network is relatively simple to modify and is often used for forecasting issues.

The created network has one with four processing elements. The transfer function of neurons in this layer is TanhAxon. The learning rule for designed network is back propagation with momentum algorithm. Supervised learning control is enabled. After the neural network was created it had been trained. The learning process was firstly conducted with 1000, 8000, 10000, 30000 epochs with randomized initial weights. Training data set covered 60 samples. The best training results were obtained in simulation with 10000 epochs as it had the lowest Mean Squared Error (MSE) and its learning curve (MSE versus epochs) looked like L letter. To verify correct network performance, a test was preceded. If the results of test are satisfying, the training part can be approved as well as the prognosis of outputs value.

There were used two types of errors measures in forecasting :

$$MSE = \sqrt{\frac{1}{T^*}\sum_{t=1}^{T^*}\left(y_t^* - y_t\right)^2} \text{ and } RASE = \sqrt{\frac{1}{T^*}\sum_{t=1}^{T^*}\left(\frac{y_t^* - y_t}{y_t}\right)^2} \times 100 \quad (6)$$

where: y_t*- forecasted value, required net output, y_t- real value, T*- horizon of forecast.

Two tests were conducted. In testing, the learning is turned off and the chosen data set is fed through the network. Test data covered 127 samples. The best network weights achieved from training were used. The first test included Cash Flow output stated earlier as (income value driver). The results of the desired output and actual network output were shown on network statistics. Linear correlation coefficient (r) for this network equaled 0,726. The second test included Net Assets output (-value of assets driver). This test gave better results, mainly by higher linear correlation value. It is due to the fact that Net Assets behaves in a more stable way than Cash Flow. The network forecast for Net Assets did not generate such high oscillation as its Cash flow equivalent. The conclusion is that the utilized value drivers which influence asset value and goodwill generate more goodwill part than company material property. Anyhow, the results of training network for both outputs did not achieve its satisfying level, especially testing for Cash Flow output. Because of that, a new training of neural network was started.

This time, the data set to train covered 80 samples, which are 20 more than in the previous learning process. The results of training with 1200 epochs are illustrated in Fig. 5.

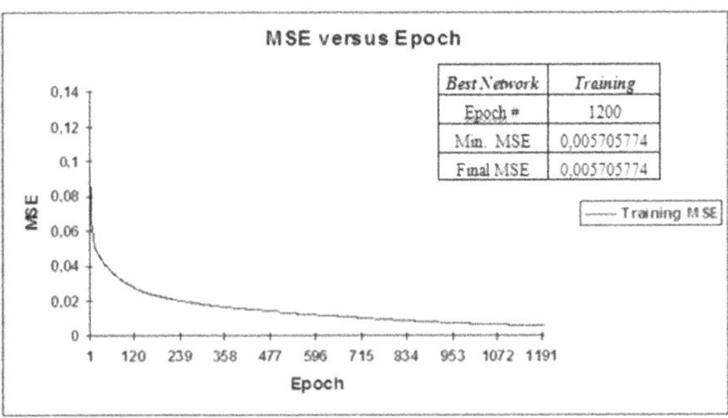

Fig. 5. Learning curve and statistics for 1200 epochs

Because the training sample increased, the testing samples decreased to 108 testing samples. It is a kind of compromise. Like in the earlier example, two tests were conducted [3]. The first one checks network behavior with Cash Flow output, the other one with Net Assets (see Fig. 6 and Fig. 7).

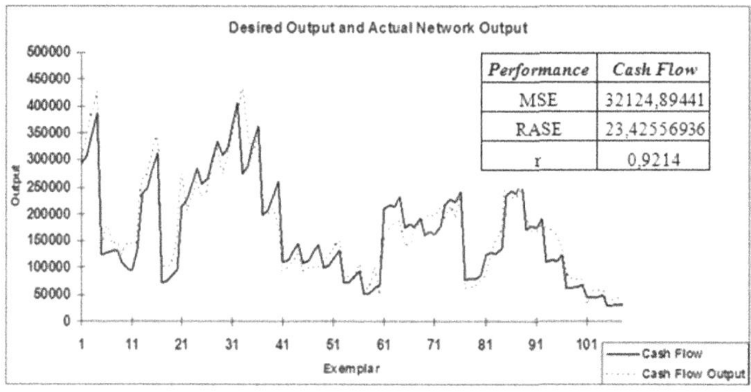

Fig. 6. Desired and actual network output for Cash Flow

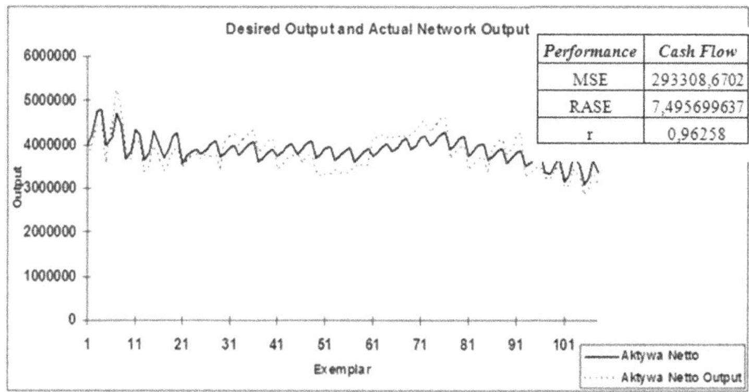

Fig. 7. Required and actual network output for Net Assets

In the research it is assumed that the forecasted values of the factors determine asset value and income value of the company. RASE for Cash Flow is about 23,42%, it means that RASE for *goodwill* $f(V_I)$, will be the same. RASE for the forecasted assets value is 7,49%, so for V_A will be the same.

5 Summary

Each economic organization needs to be analyzed. On the one side, the analysis allows checking now if firm activities meet the expected results; on the other side, it allows to define directions and ways of further improvement activities.

Dynamic environment changing and company internal factors changing influence company drivers value and company value. The classical methods are not good enough to describe the valuation process. Using IT technologies, neural networks technologies make dynamic valuation of the parameters possible.

Value based management requires specific tools that are able to describe not only factors in the static sense but dynamic too. It makes interactive valuation possible. Using unconventional methods for modeling these specific and complicated processes, managers will be able to create, monitor and control the process of valuation.

References

1. Łukaniuk, M.: Methods of Company Valuation and Real Option. Doctor thesis, Wroclaw University of Technology, Wroclaw (2003) (in Polish)
2. Tadeusiewicz, R.: Elementary Introduction to Neural Networks with Samples Programs. PWN, Warsaw (1998) (in Polish)
3. Wilimowska, Z., Krzysztoszek, T.: Computer method of weights settlement in mixed methods of company valuation. In: Wilimowska, Z., et al. (eds.) Information Systems Architecture and Technology. Application of Information Technologies in Management Systems, pp. 67–79. Printing House of Wroclaw University of Technology, Wroclaw (2007)
4. Wilimowska, Z., Krzysztoszek, T.: Value drivers in mixed methods of company valuation. In: Świątek, J., et al. (eds.) Information Systems Architecture and Technology. Decision Making Models, pp. 23–33. Printing House of Wroclaw University of Technology, Wroclaw (2007)
5. Wilimowska, Z., Buła, B.: Value drivers. In: Borzemski, L., et al. (eds.) Information Systems Architecture and Technology. Information Models, Concepts, Tools and Applications, pp. 271–278. Printing House of Wroclaw University of Technology, Wroclaw (2006)
6. Witkowska, D.: Artificial Neural Networks and Statistical Methods. Selected Financial Issues. Publishing House C.H. Beckm, Warsaw (2002) (in Polish)

Circadian Rhythm Evaluation Using Fuzzy Logic

Martin Cerny and Miroslav Pokorny

VSB – Technical University of Ostrava, Ostrava, Czech Republic
{martin.cerny,miroslav.pokorny}vsb.cz

Abstract. Useful information about person's behavior and its changes provides the measurement of the physical activity of the monitored person in flat. The identified changes are cyclic with a period of approximately 24 hours – this is Circadian Rhythm of Activity, CAR. In the event that we correlate CAR with information about the type of room and activities envisaged in this room, we can define the circadian rhythm (CR). The CR evaluation is made by different mathematical and statistical procedures now. Such systems do not mostly have predictive character.

This work uses for classification and prediction of significant circadian rhythms deviations diagnostic method - fuzzy expert system. This methodology allows quick and effective decision-making and it shows predictive ability to detect deviations of circadian rhythm.

Keywords: Circadian rhythm, Fuzzy logic, Remote home care.

1 Introduction

The remote home care systems are currently the subject of great interest of research institutes and companies around the world. There has already been designed and implemented several remote home care systems. Most of these systems meet the general scheme and they are intended primarily for monitoring elderly patients with motor, visual, auditory, cognitive or the other health disabilities. These systems are primarily designed for alone living people. Flats are equipped by various electrical appliances with sensors, actuators and special wireless biomedical monitors. These devices work in a network that is connected to a remote center for collecting and processing data. Remote center makes diagnosis of the current situation and initiates action to correct the interference crisis situations.[1]

Information about physical activity of monitored person gains monitoring of the position of the person in the flat. There are identified cyclic changes with a period of approximately 24 hours. This is a circadian rhythm of activity (Circadian rhythm of activity, CAR). In the case we correlate this information with information about the type of room and activities envisaged in this room, we can define the personal circadian rhythm (CR). Circadian rhythm is influenced by social and biological rhythms at the same time. The correct analysis of the circadian rhythm becomes a source of important information about the behavior of the monitored person. We can detect a lot of useful information about the habits of persons under observation, such as sleep

N.T. Nguyen et al. (Eds.): *Adv. Methods for Comput. Collective Intelligence*, SCI 457, pp. 289–298.
DOI: 10.1007/978-3-642-34300-1_28

duration, frequency of toilet visits, the rhythm of food intake and other. It is important to monitor circadian rhythms in the remote home care systems.

2 Movement Monitoring System

Technical solutions to determine the position of persons in the apartment can be divided into two major subgroups according to the criterion of whether is placed on electronically monitored active distinctive element.

In the first subset the user is identified by electronically active element. User wears on his body active or passive transmission device - electronically recognizable identification device. These include methods for sensing the position of persons using RFID techniques and methods of using wireless communication devices with the application of algorithms for position determination.

The second sub-category includes techniques based on passive sensing. Person is identified without installing electronic monitoring devices on his body. This subgroup includes methods based on the PIR sensing, weight, image processing, capacitive mats and passive electronic gates.

The examples of implemented projects are HIS (CNRS Grenoble, France)[2] - PROSAFE (CNRS Toulouse, France)[3] and Monami project[4].

We designed and tested solution with PIR sensors in the own solution the first [5] even if PIR sensor has its limitations.

Therefore there are used two other technologies in our solution - Location Engine [6] and optoelectronic sensors. The scheme is at figure 1.

Location Engine technology determines the exact position of a moving transmitter. It uses a network of reference points - stationary transmitters with a defined position and one moving transmitter with unknown position. This technology is part of the commercial solutions based on ZigBee.

There was also designed and tested own solutions of optoelectronic door barriers placed in door frames, which detects the direction of movement. There are used two separate IR light beams to can we recognize the direction of movement through doorframe.

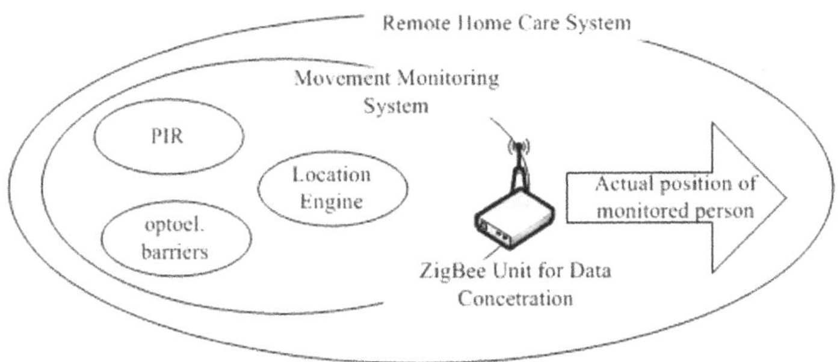

Fig. 1. Diagram of the subsystem monitoring position of the person in the apartment

3 The Circadian Rhythm Evaluation

The CR evaluation process is most important part of solution. Very significant research activities in this field are from TIMC of University in Grenoble, France [6] and from Seoul National University [7].

The criteria for significant circadian cycles of states are defined mostly by thresholds. Solvers use various mathematical and statistical procedures for determination of threshold. The prediction in CR evaluation could be very useful. There is described possibility of prediction in following work.

The fuzzy expert system is used in this work for classification and prediction of circadian rhythms changes.

The alarm processing is not handled at this work. A prerequisite is that the alarm processing system will be done away remote home care system. Alarm will be forwarded to the dispatch center and worked there either automatically or by a person on duty.

3.1 Fuzzy Expert Model Definition

There were defined four critical situations – four output values in this work:

- Faintness
- Hypo activity
- Hyper activity
- Unusual nocturnal activity

These values respond to the critical situations which are observed during human movement monitoring. Each defined critical situation has abbreviation, which is used in the fuzzy model definition, described below.

Faintness (FNT) is defined as the critical state of the monitored person when there is no moving activity caused by sudden or sustained decline in physical activity. There are detected events as the fall and subsequent loss of consciousness, fall and subsequent inability to move, loss of consciousness.

Hypo activity (HPO) is a condition where is detected a significant reduction in physical activity of monitored person. The reasons may be acute health problems and long-term changes in circadian rhythms caused by the development of chronic diseases.

Hyper activity (HPE) is a condition where the person is being noticeably increased his physical activity. These changes may be due to various acute events, injury, confusion, and the like. Hyperactivity is a diagnostic feature of elderly chronic disease.

Unusual nocturnal activity (UNA) is any suspicious activity of monitored persons in the night hours. An unusual nocturnal activity will be detected as increased activity in the night hours, and circadian rhythm chronology disorders (typical paths), but also as other unusual occurrence in flat areas during the night.

4 Fuzzy Expert System Structure

The fuzzy expert system uses seven input variables and one fuzzy output variable with four values as defined above. The fuzzy model is based on Mamdani rule base. There were defined 12 rules. The rules are described in the table below.

Each row in the table is one rule. For each rule is defined set of values of input variables. In the case the variable is not used in the rule, it is indicated by star symbol in the row.

Table 1. Fuzzy model rules

			Input Variable Value					
No.	IOM	DOI	POC	SOP	COP	TOA	CPC	OUTPUT
1	NON	LST	NON	LST	*	*	*	FNT
2	LOW	LONG	LESS	LONG	*	DAY	*	HPO
3	*	LONG	LESS	LONG	*	*	NST	HPO
4	HIGH	SRT	LOT	*	*	DAY	*	HPE
5	HIGH	SRT	*	LONG	*	DAY	*	HPE
6	*	*	LOT	SRT	*	DAY	NST	HPE
7	HIGH	SRT	LOT	*	*	MOR	*	UNA
8	HIGH	SRT	*	LONG	*	MOR	*	UNA
9	*	*	LOT	SRT	*	MOR	*	UNA
10	HIGH	SRT	LOT	*	*	EVE	*	UNA
11	HIGH	SRT	*	LONG	*	EVE	*	UNA
12	*	*	LOT	SRT	*	EVE	*	UNA

There are described input variables and their membership functions. The procedure of variable calculation is described too. Each input variable has own abbreviation, which is further used in fuzzy model.

4.1 The Intensity of Movement

Variable Intensity of Movement (IOM) is an expression of the number of movements recorded per minute. It is calculated as the sum of the recorded motion tracking system in the flat position for an elapsed minute. It is monitored by tracking a person in the apartment. The movements are not recorded by the position monitoring system in the apartment more than once per second. It is because of optimization of number of calculated samples. The maximum value of the intensity of movement observed is 60 persons per minute of recorded movements. Input variable intensity of movement (IOM) has fuzzy linguistic values none - low - high.

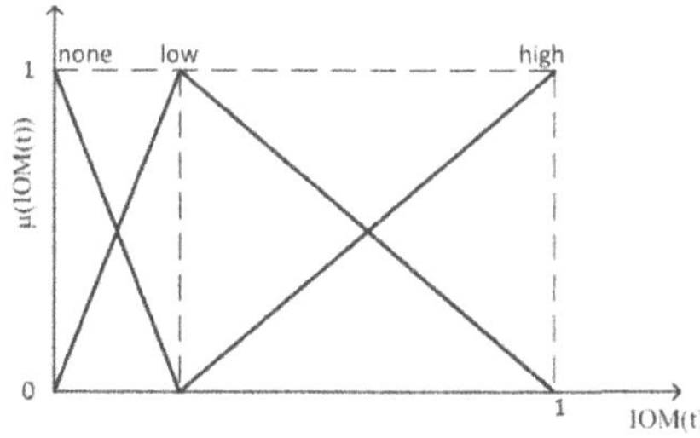

Fig. 2. Membership functions of variable intensity of movement

The normalization of IOM is calculated as Eq. 1. It is clear that is it is influenced by the time of the day.

$$IOM_{norm}(t) = \begin{cases} \dfrac{IOM(t)}{IOM_{\max,day}} & for\ t \in \langle t_{night}, t_{day} \rangle \\ \dfrac{IOM(t)}{IOM_{\max,night}} & for\ t \in \langle t_{night}, t_{day} \rangle \end{cases} \tag{1}$$

4.2 The Duration of Inactivity

The Duration of Inactivity (DOI) of monitored person in an apartment is the time between two detected movements of monitored person. It is the number of inactive seconds since last recorded physical activity in the apartment. In the event that the monitored person is outside the apartment, time of inactivity in this area is not measured. The movement monitoring system (referenced in chapter 2) can recognize the position of monitored person outside the flat by itself.

The input variable duration of inactivity has linguistic values short - long - lasting. The normalization process equation is the same as Eq.1. only with change of variable names and maximum values.

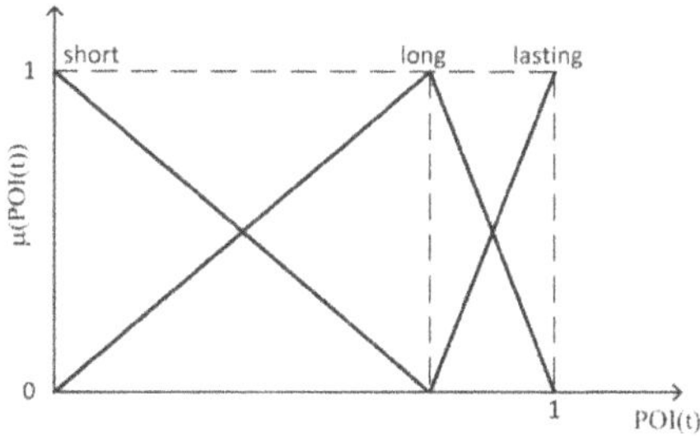

Fig. 3. Membership function of variable duration of inactivity

4.3 The Position Changes

This variable (POC) provides information about the number of changes of monitored person position the flat in different rooms during defined time. The variable doesn't include the information about the chronology of movement between the areas.

The calculation of the variable the Position Changes is always performed in one minute. The maximum value can be 60 position changes in one minute, because the monitoring system doesn't provide more information than one times per second. The input linguistic variable the position changes have values none - less - a lot.

To calculation of the normalized values of POC is done by the similar equation as mentioned in the chapter 4.1. The maximum values of POC for the day and night time are observed during measurement. It is possible to do analysis of the changes of maximal values too (This possibility is not a part of fuzzy model now).

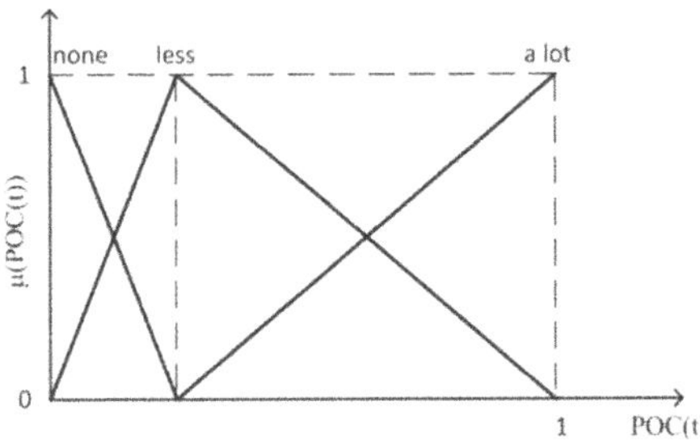

Fig. 4. Membership function of variable the position changes

4.4 The Stay in One Position

The time that monitored person spends in the room (one place in the flat) continuously, is a variable called Stay in One Position (SOP). It is expressed in minutes.

For each monitored area of the apartment is experimentally observed maximum time that the person spends in this area during the daytime and night time. These maximum values are then used for variable scaling and for it normalization. The example of calculation is equation 2. It is calculation for one part of flat – kitchen. The calculation for other parts of flat is similar.

$$SOP_{norm}(t) = \begin{cases} \dfrac{SOP(t)}{SOP_{\max,kitchen,day}} & for\ t \in \left\langle t_{night}, t_{day} \right\rangle \\ \dfrac{SOP(t)}{SOP_{\max,kitchen,night}} & for\ t \in \left\langle t_{night}, t_{day} \right\rangle \end{cases} \tag{2}$$

The input linguistic variable, stay in one position (SOP), takes the language of values short – long - lasting. Membership function of fuzzy set of variable values, stay in one position is shown below.

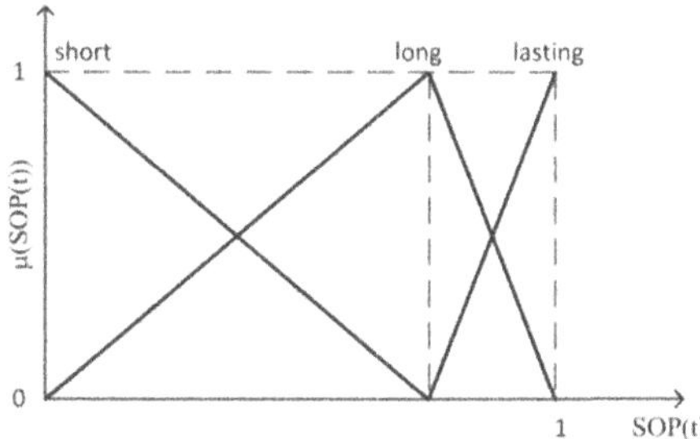

Fig. 5. Membership function variables stay in one position

4.5 The Character of Position

There was necessary to define dangerous areas of the apartment to can we establish the methodology for the evaluation of circadian rhythms and we can calculate the variable Character of Position (COP). Although each apartment has a different arrangement of rooms, we suggest that older people living alone will live in similar apartments.

There were defined two levels for apartment's hazardous area:

- Dangerous area
- Potentially dangerous area

Dangerous part of the apartment can be evaluated from two perspectives. The first is the probability of injury of monitored person, which depends on equipment and purpose of monitored area. The second aspect is to analyze the chronology of typical paths inside the apartment. Based on the analysis finished in thesis [8] there were designed fuzzy input variable the Character of Position (COP). COP has two values non-critical and critical. If there is a person watched in the critical area of the apartment, the variable takes the value of the critical nature of the position. The membership function for this variable has simple triangular type.

4.6 The Time of Activity

Variable Time of Activity (TOA) reflects in the fuzzy model influence on the time of day. The time of activity has a definite influence on the power of decision-making criteria. The evaluation of the situation is significantly different in day and night hours. The normalization of TOA is written at equation 3, where t is the time of the day.

$$TOA_{norm}(t) = \frac{t}{24} \qquad (3)$$

The input linguistic variable time of activity (TOA) takes linguistic values morning - day - evening. The values of linguistic variables are set intentionally due to the standardization process values TOA. Value Morning corresponds to the dawn activity period from midnight the night until awakening and daily activities. The value Evening expresses the twilight between the end of daily activities and midnight. In case when the person closes their daily activities after midnight, is to merge variables morning and evening to the value morning.

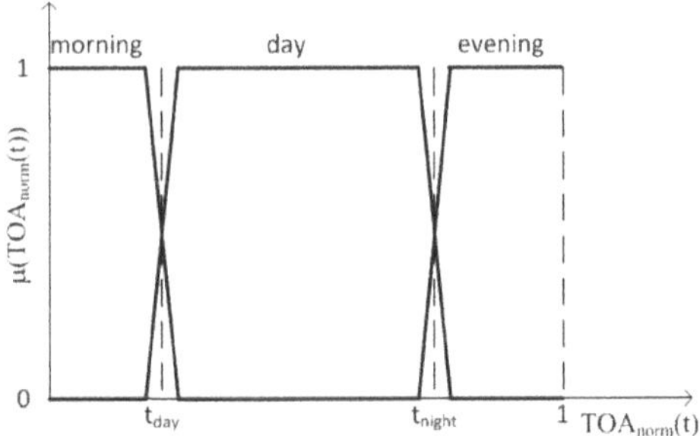

Fig. 6. Membership function of variable the time of activity

Membership function is defined by the times and seasons, which are not exact, whether it is day or night activity. Determining the times cannot be completely unambiguous. These times are changing every day. The calculation can be made based on the analysis of long-term record circadian rhythm.

4.7 Chronology of Position Changes

The variable chronology of position changes (CPC) indicates whether the monitored person moves normally in the apartment or not. The standard chronology of the movement of persons at home is considered as a condition where the person being found in an apartment in a flat area, which usually occurs at that time. Based on the analysis of long-term record is then possible to determine whether it is standard or nonstandard. This variable does not need scaling, because already in the calculation of the value of standardization. The detailed definition of this variable is in [8].

5 Tests of Designed Fuzzy Expert System

There was used MATLAB development environment for testing purposes. There were used fuzzy toolbox functions for working with designed fuzzy model. There were created a custom function that is ready for evaluation of circadian rhythm variations in real time.

The first tests were made for diversification abilities of fuzzy model. There were defined boundary test sets values of input variables which have known output variable value.

For further testing, the test set of values formed the input variables. There were created sets of input variables values with respect to predefined decision rules base of fuzzy model. For each decision rule there were created 7 to 10 test sets of input variables. The aim of these tests was to verify the predictive ability of generated fuzzy model. The example of results is on the picture below. The prediction possibility is visible from the figure – it is because of the increasing or decreasing of output variable values. The threshold for decision in this type of models is 0,6. The prediction could be done by the analysis of increasing or decreasing of output variable values.

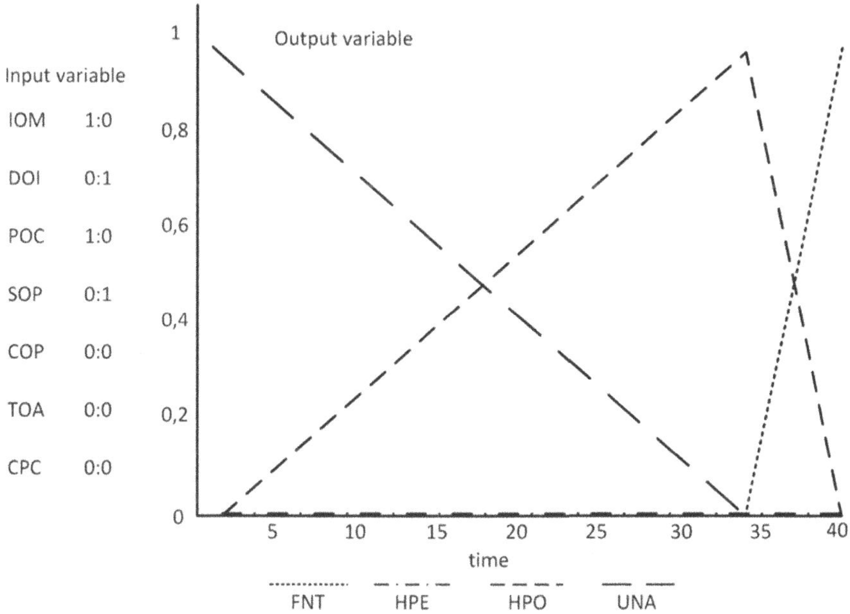

Fig. 7. Results of fuzzy expert system on model data

In the final of this work the test on real measured data were done. There were used data from our testing flat, where we measure 24 hour testing period. During this period were done typical deviations of circadian rhythm.

6 Conclusion

This work shows own technical solution and methodology of movement monitoring in the remote home care systems. It is based on a combination of known methods of movement monitoring of people in the flat position using PIR sensors, on improved existing measurement methods and new wireless technologies.

There were defined a fuzzy expert system with Mamdani rule base that can evaluate the circadian rhythm and its variations. There were defined own set of input linguistic variables, membership functions and the method of their preparation for the fuzzy model. Based on expert knowledge and experience base there has been defined decision rules fuzzy model and a core group of states evaluated and deviations circadian rhythm. The model was subjected to testing capabilities for diversification model development and testing of the output variable when changing the input variables. Tests have shown the usability of fuzzy model for analysis of circadian rhythm on both simulated data and on measured data in a controlled experiment testing. Tests also show a fuzzy model predictive ability, which is a significant benefit in the evaluation of circadian rhythms. Further test on this fuzzy expert system are in progress.

Acknowledgement. The work and the contributions were supported by the project SP2012/114 "Biomedical engineering systems VIIII" and TACR TA01010632 "SCADA system for control and measurement of process in real time". The paper has been elaborated in the framework of the IT4Innovations Centre of Excellence project, reg. no. CZ.1.05/1.1.00/02.0070 supported by Operational Programme 'Research and Development for Innovations' funded by Structural Funds of the European Union and state budget of the Czech Republic.

References

1. Chan, M., Campo, E., et al.: Smart homes - Current features and future perspectives Maturitas, vol. 64, pp. 90–97 (2009) ISSN: 0378-5122
2. Virone, G., Noury, N., Demongeot, J.: A System for Automatic Measurement of Circadian Activity Deviations in Telemedicine. IEEE TBME 49(12) (2002)
3. Chan, M., Campo, E., Lavaland, E., Esteve, D.: Validation of a remote monitoring system for the elderly: Application to mobility measurements. Technology and Health Care 10, 391–399 (2000)
4. Project MonAmi – Mainstreaming on Ambient Intelligence, `http://www.monami.info`
5. Cerny, M., Penhaker, M.: Wireless Body Sensor Network in Health Maintenance system. Journal Electronics and Electrical Engineering 11(9), ISSN 1392-1215
6. Poujaud, J., Noury, N.: Identification of inactivity behavior in Smart Homes. In: Proceedings 30th Annual International IEEE EMBS Conference, pp. 2075–2078. IEEE (2008) ISBN 978-1-4244-1815-2
7. Shin, J.H., Lee, B., Park, K.: Park Detection of Abnormal Living Patterns for Elderly Living Alone Using Support Vector Data Description. IEEE Trans. on Information Technology in Biomedicine 14(3), 438–448 (2011), doi:10.1109/TITB.2011.2113352
8. Cerny, M.: Circadian Rhythm Evaluation in Remote Home Care. Ph.D thesis, VSB – Technical University of Ostrava, Ostrava, Czech Republic (2011)

Part VI
Optimization and Control

Optimization of Container Handling Systems in Automated Maritime Terminal

Hamdi Dkhil[1,2], Adnan Yassine[1,3], and Habib Chabchoub[2]

[1] LMAH - Laboratory of Applied Mathematics, University of Le Havre, Le Havre, France
[2] MODEOR - University of Sfax, Tunisia
[3] Superior Institute of the Logistic Studies (ISEL)
hamdi.dkhil@etu-univ-lehavre.fr, adnan.yassine@univ-lehavre.fr, habib.chabchoub@fsegs.rnu.tn

Abstract. Container terminals play a crucial role in global logistic networks. Because of the ever-increasing quantity of cargo, terminal operators need solutions for different decisional problems. In the maritime terminal, at boat arrival or departure, we observe five main problems: the allocation of berths, the allocation of query cranes, the allocation of storage space, the optimization of stacking cranes work load and the scheduling and routing of vehicles. A good cooperation between the different installations in the terminal is important in order to minimize container handling time. In an automated container terminal using Automated Guided Vehicles (AGVs) Query Cranes (QCs) and Automated Stacking Cranes (ASCs) numerical solutions have become essential to optimize operators' decisions. Many recent researches have discussed the optimization of ACT equipment scheduling using different approaches. In this paper we propose three mathematical models and an exact resolution of QC-AGV-ASC planning, the problem of tasks in an automated container terminal. Our first objective is to minimize the makespan (the time when the last task is achieved). The second objective is to minimize the number of required vehicles.

1 Introduction

In an automated container terminal (ACT) the time of handling operations depends on the interactions between the different storage equipments. Different researches are established to improve the handling systems performance. We will interest next to the problems which consider our two objectives (minimizing the makespan and minimizing the AGV fleet size). These two objectives are treated in the AGV scheduling problem. The problem of AGV scheduling was treated in the general context of AGVS and in the particular context of ACT. AGVS is a materials handling system that uses automated vehicles which are programmed to achieve missions between different manufacturing and warehouse stations . It represents a very important innovation in international transport and logistics. ACT is one of the most famous examples of AGVS. Studies of AGVS optimization have different objectives: maximizing the throughput, maximizing the vehicle utilization, minimizing the inventory level, minimizing the transportation costs, and maximizing the space utilization. AGVS mathematical models have to respect some conditions to eliminate the traffic

N.T. Nguyen et al. (Eds.): *Adv. Methods for Comput. Collective Intelligence*, SCI 457, pp. 301–312.
DOI: 10.1007/978-3-642-34300-1_29

problems. Approaches used in AGVS optimization can be classified in two kinds: analytical approaches and simulation-based approaches. Analytical methods are mathematical techniques such as queuing theory, integer programming, heuristic algorithm, and Markov Chains. A number of analytical approaches to AGVS optimization have been proposed in the literature.

1.1 Problems of Minimizing AGV Fleet Size in AGVS and ACT

AGVs historically have not been produced in high volume. Then in AGVS determining the minimum number of vehicles required to achieve a set of tasks in efficient and economic way is crucial to improve the global system productivity. Muller[2] used rough estimates of total AGV travel times and transport frequency to resolve the AGV system case. The team of Maxwell and Muckstadt [3] discuss the deterministic case of the problem. They consider the random aspect of the problem: variation of arrival pattern of jobs and vehicles speed and they developed an integer programming formulation to minimize the number of required AGVs. In Rajota et al [4] other parameters are considered: load handling times, empty travel time… The team developed a mixed integer programming model. Sinriech and Tanchoco [5] have developed a multi-objective model which keep the total cost of AGV system down and increases the system utilization. The problem is treated by I FA Vis [6], in the ACT context he developed new planning concepts to minimize the AGV fleet size and he applied it to the container terminal case considering a deterministic model with defined time windows for each container load. He proposed two methods to solve the problem: an integer programming model and a formulation of the problem as a set of partitioning sub problem.

1.2 Minimizing Vehicle Fleet Size in Other Contexts

Two analog problems are discussed in literature. The first problem is the determination of the minimum number of operators required to accomplish a known schedule of tasks. This problem was treated by Phillips and Garcia-Diaz [7]. They use a bipartite network where the maximum flow indicates pairs of tasks assigned to the same operator. Then they propose the research of arcs of the maximum flow to obtain the list of tasks for each operator. Ford and Fulkurson [8] discus this problem and use a partial order of tasks: tasks i precedes task j if the start time of i is earlier than the start time of j and if the two tasks can be achieved by the same operator. They resolve the problem with the determination of minimum chain decomposition. The second analog problem is the tanker scheduling. Dantizig and Fulkerson [9] describe a deterministic model to solve the tanker scheduling problem with linear programming formulation and simplex algorithm. Ahuja et al [10] propose another approach to resolve the same problem: they introduce a minimum cost flow formulation of the problem and use a minimum cost flow algorithm to minimize the fleet size of the main problem.

1.3 Problems of Minimizing Makespan in AGVS and ACT

The problem of minimizing makespan is treated in the general AGVS context. In 1984 Ebeglu and Tanchoco [11] developed a dispatching rules method for AGVs scheduling. Tanchoco et al. [12] discussed real-time control strategies for multiple-load AGVs. Models and methods applied to AGVS seem to be generally applicable and need to be adjusted for more specific contexts. The researches of minimizing makespan in ACT are recent especially with the integrated aspect of QC-AGV-ASC problem (AGV or ALV). Chen et al. [13] treated the scheduling of AGVs. They developed a dispatching approach and simplified the QC task considering it available to AGV loading or unloading which cannot' assure the solution optimality for the multiples QCs case. Kim and Bae[14] developed a model with fixed pick up time for each container and they proposed heuristic solution for more general cases. Meersman [1] was perhaps the first researcher to consider the integrated QCs, AGV and ASC scheduling problem. He showed that this problem is NP-Hard and developed mathematical theorems for the problem of scheduling ASC-AGV-QC tasks. He studied static traffic layout (layout with one fixed path for each task) and dynamic traffic layout (layout with different possible paths for each task). Meersman used branch and bound and beam search algorithms to resolve the static traffic case using mathematical theorem results to establish valid inequalities. Bae et al [15] developed a dynamic berth scheduling method for minimizing the cost of the vehicles travel during the loading or unloading of ship. The approach take into account many constraints and real dynamic situations.

1.4 Multi-criteria AGVS Scheduling Models

With the increasing automation of manufacturing systems, the use of efficient and multi-criteria decision systems is very important to optimize productivity. AGV systems seem to be the most famous example. A good evaluation of the cost of AGVS must take into account different characteristics: vehicle dispatch, load and unload central controller, complex host interface, product tracking, multiple paths layout etc. In 1981 Dahlstrom and Maskin [16] and Muller [17] have addressed the economical aspects of AGVS; the two papers compared the cost of different material handling systems. Sinriech and Tanchoco [18] have developed a multi-objective model which keep the total cost of AGVS down and increases the utilization of the system. They assume that AGVS cost is a formulation of operating costs (maintenance, energy...) and design costs (vehicle supervisory controller, vehicles, batteries, chargers, communication links etc).

In the next parts we propose solutions for three terminal layouts and we use Meersman's results [1] to improve the mathematical modeling and the quality of our numerical solutions. We propose a model with two objectives: the optimization of task time for the QC-AGV-ASC problem and the minimization of the number of vehicles used. We use Meersman's mathematical results to perform our modeling and resolution and we propose new models for the scheduling problem using a partial container' order and resolving large problem instances. Different automated terminal layouts can be studied. Meersman presents two possible port architectures: a simple layout with static AGV traffic and a complex layout with multiple variable paths. Another case is studied by Vu D N and Kap H.K [19]; we can describe this case as a multiple fixed paths layout. In this paper we will look at the three layout cases and propose mathematical modeling adapted for each case. Then we will give the simulation results that we obtain after using Meersman's theorem of partial order (cf. part 4).

2 Terminal Architectures

In this part we use a notion of ASC Points and QC Points: ASC Points are the places where ASCs pick up containers from AGVs and QC Points are the places where QCs loads containers on the vehicles (these notions will be used in the next parts). For the two first models we consider also Point A as a final position in the path for every task.

We consider three terminal layout possibilities.

2.1 One-Path Layout

The model supposes static AGV traffic and does not take into account traffic security.

We consider that all AGVs have the same path for each task.

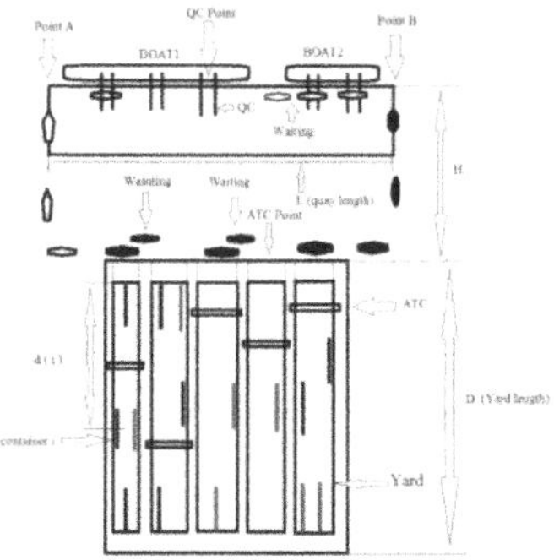

Fig. 1. One-Path layout

We can describe this case as a one-path layout. We consider the import case and the export case as symmetric and the scheduling problem is the same. In Fig.1 black vehicles represent the loaded AGVs and white vehicles the unloaded AGVs. Point A is the final point of every task. All AGVs have the same task path. We assume that the terminal's routes have two possible directions and that many AGVs can use the same path at the same time without risks. The AGVs start at QC point, then go to point B, then to the ASC point (where there is a possible waiting time) and finally they return to point A. Before starting its task, every AGV has to wait until the end of the last QC task.

With this model of terminal layout the optimization can minimize only the sum of waiting time at the QC and ASC points, the AGV paths are known and static. This layout is treated by Chang Ho Yang and all [16].

2.2 Multiple Fixed Paths Layout

Point A (show Fig.2) is the final point of every task. All AGVs have a known task path; they start at QC Point then choose the shortest path to the ASC Point, finally going to point A. The paths are not the same for all tasks but each path is initially known, they depend only on the ASC and QC positions. Before starting its tasks, every AGV has to wait sufficient time so as not to cause an accident with the predecessor AGV at QC.

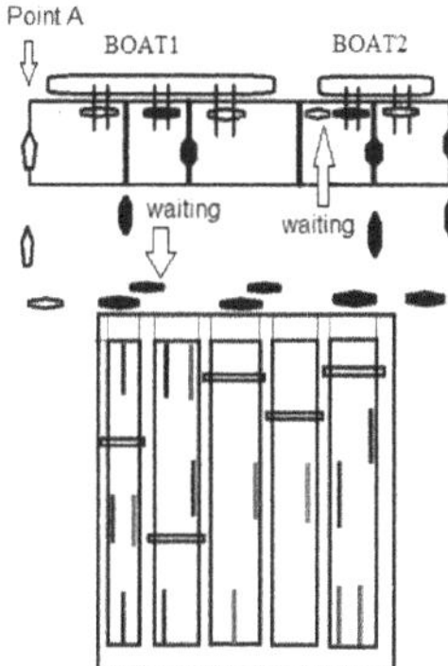

Fig. 2. Multiple fixed paths layout

With this model of terminal architecture we have to minimize only the sum of waiting times at the QC and ASC Points, because AGV routing are initially known.

For the two first cases (one-path and multiple fixed paths layouts), we optimize the AGV scheduling problem with the same linear model.

2.3 Multiple Variable Paths Layout

This third case is the most complex architectural model. The travel times are variable and unknown because for each task the AGV does not return to a common final point (Point A in FIG.2 and FIG.1) but moves directly to its next task. The travel time between the present task and the next is unknown and depends on the choice of the next task.

In the static and the semi-dynamic traffic model when we optimize container handling time, only the waiting time is important because in each case we can choose the first AGV (returned to Point A) for the next task .In the dynamic traffic case, the choice of AGV for some tasks is important because AGVs do not finish their tasks at the same point. Thus choosing the first free AGV for the next task is not a good idea: we have to release a double scheduling (TASKi, AGVj) if we resolve the problem with a branch and bound algorithm.

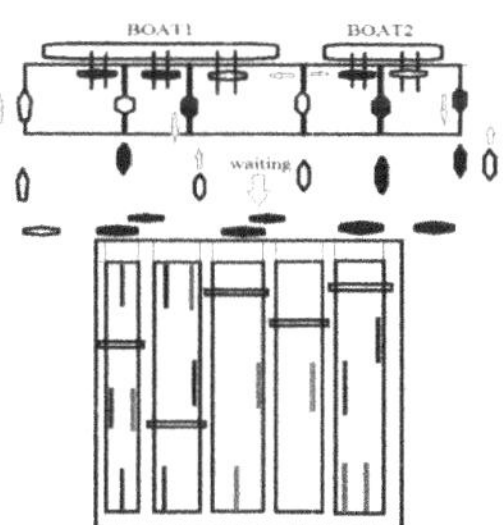

Fig. 3. Multiple variable paths layout

3 Data Generation

Data generation is based on terminal architecture and the handling speed of equipment.

In Fig.1, we demonstrate the distances which we use to generate data: L the quay length, D the yard length, H the distance between quay and storage zone.

The AGV and ASC transfer speed combined with the terminal dimensions give a clear idea about the data that we need for our modeling and simulations.

4 Important Theorem

Meersman [1] used a strategy of partial order to resolve large instances of the scheduling problem: the tasks of each ASC are totally ordered. He supposes a sufficient quantity of AGVs which can ensure an optimal schedule and he concluded an important theorem.

"Define the assignment order Π as the order in which the containers are assigned to the AGVs as they pass the common point. Moreover, define a suborder Πs as a subset of Π, such that if i is ordered before j in Πs, then i is ordered before j in Π, for all i, j € Πs. Theorem: For each ASC s € S, consider an optimal schedule. Let Πs denote the order in which ASC s handles its containers. Then there exists an optimal assignment order Π, such that Πs is a suborder of Π."

5 Mathematical Models

In parts 5.2 and 5.3, we consider that the number of AGVs is sufficient to complete an optimal schedule. We consider a total order in each set of ASC tasks and another total order in each set of QC tasks (we use models with buffer space of QC equal to 1). Then the set of all tasks has a partial order and for any task i the successor task and predecessor task in QC and the successor task and predecessor task in ASC are initially known. In the next part we consider that the matrixes ASCi,j and QCi,j are constant.

We define the following variables for all models.

V: the set of AGVs (Automated Guided Vehicles)
C: the set of ASCs (Automated Stacking Crane)
Q: the set of QCs (Query Crane)
M: the set of tasks
$AGV_{i,j}$: decision variable, if the container j is handled directly after the container i by the same AGV $AGV_{i,j}$ =1 else $AGV_{i,j}$ = 0
$QC_{i,j}$: If task i is succeeded directly by j in the same QC, $QC_{i,j}$ = 1 else $QC_{i,j}$ = 0, we consider this data known.
$ASC_{i,j}$: If task i is succeeded directly by j in the same ASC $ASC_{i,j}$= 1 else $ASC_{i,j}$ =0.We apply Meersman's theorem and we choose the order of tasks for every ASC (ASC order must respect QC order).

$T_{1,i}$: the travel time between the start point and the ASC point
$T_{2,i}$: the travel time between the ASC point and the final point
$T_{qc,i}$: the travel time between point A and QC Point
S_i: the ASC transfer time of task i (depend on ASC speed and d(i) (Fig.1))
$t_1(i)$: the start time of task i
$t_2(i)$: the completion time of task i
S_{qc}:the time that QC need to load container on AGV
S_{asc}: the time that ASC need to pick up container
S_s: safety waiting time (near QC Point).

t_0: start time.

5.1 Formulation of the Number of Used Vehicles

Consider |M| the number of tasks and |V| the number of AGVs, then:

$$|M| - \sum_{i\in M}\sum_{j\in M} AGV_{i,j} = |V|$$

Dem: $\sum_{i\in M}\sum_{j\in M} AGV_{i,j}$ is equal to the number of containers (or tasks) having direct predecessor in AGV then $|M| - \sum_{i\in M}\sum_{j\in M} AGV_{i,j}$ is equal to the number of containers or tasks not having a direct successor. A task with no direct successor is a first task for a fixed AGV then the number of those tasks is equal to the number of AGVs.

5.2 One-Path and Multiple Fixed Paths Mathematical Model

$$Min \max\{t_2(i)/i \in M\} \tag{1}$$

$$\sum_{j\in M} AGV_{i,j} \leq 1, \qquad \forall i \in M \tag{2}$$

$$\sum_{j\in M} AGV_{j,i} \leq 1, \qquad \forall i \in M \tag{3}$$

$$\sum_{i\in M}\sum_{j\in M} AGV_{i,j} = |M| - |V| \tag{4}$$

$$AGV_{i,i} = 0, \qquad \forall i \in M \tag{5}$$

$$t_1(i) \geq t_0, \quad \forall i \in M \tag{6}$$

$$t_1(i) + G(1 - \mathrm{AGV}_{j,i}) \geq t_2(j) + T_{qc,i}\ , \qquad \forall i,j \in M \tag{7}$$

$$\forall i,j \in M/ASC_{j,i} = 1:$$

$$t_2(i) = \max\left(t_1(i) + S_{qc} + T_{1,i} + S_{asc} + T_{2,i}\ , (t_2(j) - T_{2,j} + s_i)\right) \tag{8}$$

$$t_1(i) \geq QC_{i,j}(t_1(j) + S_{qc} + S_s), \qquad \forall i,j \in M \tag{9}$$

Constraint 1: minimize the completion time of the last tasks.
Constraints 2 and 3: limit the number of direct successor and direct predecessor, every container has one or zero direct successor and one or zero direct predecessor.
Constraint 4: if we use k AGVs , k containers will have exactly zero successor and k containers will have exactly zero predecessor because every AGV will have a first task and a last task(final task for it).Then for n containers, only n-k missions will be succeeded and only n-k missions will be preceded.
Constraint 5: No container can precede or succeed itself.
Constraint 6: No mission can start before t0.
Constraint 7: Relation between two successive tasks of an AGV. If container j is handled directly after containers i with the same AGV, then AGVi,j = 1 and we have:
t1(j) ≥ t2(i)+$T_{qc,j}$ or else the relation will be: t1(j) + G ≥ t2(i) +$T_{qc,j}$ and that is true because G is sufficiently large.
Constraint 8: This constraint is a result of 2 other constraints:

- t2 (i) ≥ t1 (i) +T1,i + Sqc + Sasc + T2,i : the final time of any mission is equal or greater than the start mission time plus the travel time plus the QC loading time plus the ASC loading time.
- V i,j Є M: t2 (j) - T2,j - Sasc ≥ QCi,j (t2(i) - T2,i + S(i)): t2 (j) - T2,j - Sasc is the start time of the ASC task j. t2(i) - T2,i + S(i): completion time of i.

Constraint 9: the difference between 2 successive QC tasks is greater than or equal to the loading time at QC plus a safety time.

5.3 Multiple Variable Paths Mathematical Model

$Y_{j,i}$ is the travel time between the final point of task j (ASC point) and the first point of task i (QC point). If we consider the first model we change only constraints (7) and (8) to obtain the dynamic traffic model.

$$\text{Min max}\ \{t_2(i)/i \in M\}$$

Constraints (2) to (6) and constraint (9) of the static traffic model

$$t_2(i) = \max\left(t_1(i) + S_{qc} + T_{1,i} + S_{asc}, (t_2(j) + s_i + s_{asc})\right), \forall i, j \in M/ASC_{j,i} = 1 \quad (10)$$

$$t_1(i) + G(1 - AGV_{j,i}) \geq t_2(j) + Y_{j,i}, \ \forall i, j \in M \quad (11)$$

5.4 Bi-objective Model

To resolve correctly the scheduling problem using the theorem of sub-orders, we need to use a sufficient number of AGVs for the optimal schedule. This number will depend on the travel distances, the AGV transfer speed and ASC transfer speed .In next model we can naturally use the theorem of the sub-orders because the minimal numbers of AGVs that we search has to satisfy the time optimality. In 2001 IFA's team developed a minimum flow algorithm to determine the number of AGVs required at a semi automated container terminal [6]. Our be-objective model is a good solution to resolve the scheduling problem in short run time and giving a small numbers off AGVs required. The value of k is important to resolve the problem; it depends on the number of tasks and on the equipment speed. and we have to choice a sufficiently great value. We replace (1) by (12) and (4) by (13) and (4) We obtain a new model which is more efficient and more intelligent. This model has two objectives: minimize the completion time of the last task and minimize the number of AGV necessary to complete the optimal scheduling. Constraints (2), (3), (4), (7) and (8) of the dynamic traffic model are used for this model.

$$Min\ (k \max\{t_2(i)/i \in M\} + (|M| - \sum_{i \in M}\sum_{j \in M} AGV_{i,j}) \quad (12)$$

$$\sum_{i \in M}\sum_{j \in M} AGV_{i,j} \leq |M| - 1 \quad (13)$$

6 Cplex Results

We choose cplex optimizer to test the performance of the models. The application of the sub-orders theorem combined with the use of constraint (4) give a possibility to resolve instances of hundreds of containers but with a use of a number of AGVs more than 10 per cent of the containers number. Using the third model we can resolve the scheduling problem with a small AGV set because the model has two objectives: minimize containers handling and transfer time and minimize the number of AGVs used. We resolve problem instances of 10 to 500 containers with a GAP of 0.15 to 0 percent. One of our most important results was the resolution of the bi-objective problems (minimizing stacking time and AGV resources) of 500 containers, 3 QC and 8 ASC. The GAP is not stable, the AGV and ATC speed and the paths routing time for some instances can increasing the GAP value. With the first presented model, using sufficient AGVs numbers (between 10 and 15 percent of the tasks numbers) we resolve small and big problem instances with optimal solution. The third model (two-objective model) is more efficacies for the instance with a limited numbers of vehicles. Results depend on the layout model: For the one-path layout problem instances less than 150 containers are generally easily resolved and the two objectives are reached with double optimality (Table 2). For the multiple variable paths layout problem instances the double optimality is harder and the run time is larger compared with the static case.

Table 1. LMAH-Model (with Cplex resolution) resolution compared to Meersman model (with branch and bound resolution)

	LMAH-Model	Meersman
Objectif(s)	2 objectives: minimizing "makespan" and minimizing AGV fleet size.	1 objectif: minimizing "makespane"
equipments	QC-AGV-ASC	QC-AGV-ASC
performance	A gap of 0 % for instances up to 500 containers 4 QCs and 12 ASCs. For these instances the runtime is between 0s and 60s	A gap of 0% to 8% for instances up to 170 containers up to 170 containers 27 ASCs and 24 AGVs. For these instances the runtime is between 0s and 658s
conditions	Consider a sufficient number of AGVs (optimal number) Consider the QC task as an AGV loading (time-lags)	Consider the QC task as an AGV loading (time-lags)

Table 2. Results of bi-objective modeling in the static traffic case Table 1. Results of bi-objective modeling with the one-path layout.

Instance	k	Makespan gap	Fleet size	Total gap	Run time
150/3/6	1	0%	18*	0%	4 s
250/4/12	1	0%	21*	0%	6 s
500/4/8	1	0%	23	0.11%	60 s

(*): optimal value Instance: number of containers / number of QCs / number of ASCs.

7 Conclusion

A new generation of terminal using automated container handling equipment needs solutions to optimize task scheduling and operating costs. Many storage strategies, statistical studies, mathematical models and algorithms are proposed by researchers. To resolve the planning of QC-AGV-ASC, we present an effective model for every kind of traffic layout. We propose an efficient bi-objective model, which is important to determine the optimal storage time and the minimal number of AGVs required. The bi-objective model can resolve large instances (up to 500 containers) with double optimality (giving the optimal makespan and the minimum number of required AGVs) in reasonable run time (less than 60 s). Our bi-objective model is perhaps the first model optimizing in on time the makespan and the AGV fleet size in a container terminal. Our models consider 3 types of handling equipment (AGV, QC and ASC) which is an efficient approach. For future works we will discuss metaheuristic programming resolutions and dispatching rules of a more advanced multi-criteria model.

References

[1] Meersmans, P.J.M.: Optimization of container handling systems. Ph.D. Thesis, Tinbergen Instutue 271 Erasmus University, Rotterdam (2002)

[2] Muler, T.: Automated Guided Vehicles. IFS (Publications) Ltd., Springer, UK, Berlin (1983)

[3] Maxwell, W.L., Muckstadt, J.A.: Design of automatic guided vehicle systems. IIE Transactions 14(2), 114–124 (1982)

[4] Rajotia, S., Shanker, K., Batra, J.L.: Determination of optimal AGV fleet size for an FMS. International Journal of Production Research 36(5), 1177–1198 (1988)

[5] Sinriech, D., Tanchoco, J.M.A.: An economic model determining AGV fleet size. International Journal of Production Research 30(6), 1255–1268 (1992)

[6] Vis, I.F.A., de Koster, R., Roodbergen, K.J., Peeters, L.W.P.: Determination of the number of automated guided vehicles required at a semi-automated container terminal 52, 409-417 (2001)

[7] Phillips, D.T., Garsia-Diaz, A.: Fundamentals of network Analysis. Prentice Hall, Inc., Engle-wood Cliffs (1981)

[8] Ford, L.R., Fulkerson, D.R.: Flows in Networks. Princeton University Press, Princeton (1962)

[9] Dantizing, G.B., Fulkerson, D.R.: Minimizing time number of tankers to meet a fixed schedule. Naval Research Logistics Quaterly 1, 217–222 (1945)
[10] Ahuja, R.K., Magnati, T.L., Orlin, J.B.: Network Flows, Theory, Algorithms, and Apliquations. Prentice Hall, New Jersy (1993)
[11] Egbelu, P.J., Tanchoco, J.M.A.: Characterization of automated guided vehicle dispatching rules. International Journal of Production Research 22, 359–374 (1984)
[12] Bilge, U., Tanchoco, J.M.A.: AGV Systems with Multi-Load Carriers: Basic Issues and Potential Benefits. Journal of Manufacturing Systems 16 (1997)
[13] Chen, Y., Leong, T.-Y., Ng, J.W.C., Demir, E.K., Nelson, B.L., Simchi-Levi, D.: Dispatching automated guided vehicles in a mega container terminal. The National University of Singapore/Dept. of IE & MS, Northwestern University (1997)
[14] Kim, K.H., Bae, J.: A dispatching method for automated guided vehicles to minimize delays of containership operations. International Journal of Management Science 5, 1–25 (1999)
[15] Bae, M.-K., Park, Y.-M., Kim, K.H.: A dynamic berth scheduling method. Paper Presented at the International Conference on Intelligent Manufacturing and Logistics Systems (IML 2007), Kitakyushu, Japan, February 26–28 (2007)
[16] Dahlstrom, K.J., Maskin, T.: Where to use AGV systems, manual fork lifts, traditional fixed roller conveyor systems respectively. In: Proceedings of the 1st International AGV Conference, 173–182 (1981)
[17] Muller, T.: Comparaison of operating costs between different transportation systems. In: Proceedings of the 1st International AGV Conference, pp. 145–155
[18] Sinirech, D., Tanchoco, J.M.A.: AN economic model for determining AGV fleet size. International Journal of Production Research 29(9), 1725–1268 (1992)
[19] Ngyuyen, V.D., Kim, K.H.: A dispatching method for automated lifting vehicles in automated port container terminals. Computers and industrial engineering 56, 1002–1020 (2009)

A Study of Bandwidth Guaranteed Routing Algorithms for Traffic Engineering

Phuong-Thanh Cao Thai[1] and Hung Tran Cong[2]

[1] Saigon University, Vietnam
ctpthanh@sgu.edu.vn
[2] Post & Telecommunications Institute of Technology, Vietnam
conghung@ptithcm.edu.vn

Abstract. Due to the fast growths of computer networks, traffic engineering (TE) which tries to satisfy both quality of services and resource utilization requirements is an important research area. Among TE mechanisms, routing algorithm – a strategy to select paths for traffic – plays a crucial role and there have been many TE routing proposals. This paper gives a thorough study of these algorithms by presenting their key ideas and mathematical descriptions, and then by various experiments so to analyse their performances with different metrics. The paper also discusses the trend of routing algorithms and future research directions.

Keywords: bandwidth guaranteed routing, traffic engineering.

1 Introduction

Besides traditional network services, next generation network applications such as voice over IP, video on demand and web games require certain quality of services (e.g. minimum available bandwidth guaranteed) described as customer service-level agreements (SLAs). Network providers try not only to satisfy those SLAs but also to optimize the network resources for profits. As a result, traffic engineering which is defined as techniques to manage traffic flows through networks with the joint goals of service performance and resource optimization has attracted much attention. One of the most important TE implementation is reactive routing that dynamically selects routes for data flows based on states of the network in order to balance traffic loads as well as conform to quality of services (QoS) requirements. Although there are several types of QoS criteria such as bandwidth, delay, and loss ratio, most academic works consider bandwidth as a primary constraint because others can be efficiently converted into bandwidth demand [1]. As a consequence, this paper focuses on reactive bandwidth guaranteed TE routing algorithms.

Specifically, key ideas of those algorithms are reviewed and, for the first time, their mathematical formulations are presented in the unified manner. Moreover, the performances are compared and analysed by various experiments with two metrics: accepted ratio and calculation time. The rest of the paper is organized as

N.T. Nguyen et al. (Eds.): *Adv. Methods for Comput. Collective Intelligence*, SCI 457, pp. 313–322.
DOI: 10.1007/978-3-642-34300-1_30

follow. In section 2, after definitions and notations, bandwidth guaranteed routing algorithms are classified into three categories: single pair aware, minimum interference and machine learning. Section 3 presents simulated experiments and analysis. Finally, section 4 presents conclusions and the future research directions.

2 Routing Algorithms

2.1 Problem Definition and Notations

A network topology with n node and m links is considered. Each link has its own capacity and residual bandwidth at a given time. Traffic demands, which require certain bandwidths from ingress to egress nodes, are handled by the routing algorithm so as to maximize the number of accepted demands. Table 1 lists the mathematical notations.

Table 1. Notations used in formulations

Symbol	Description
$G(N,L)$	A direct graph presents the network topology $N(\|N\|=n)$ is a set of nodes $L(\|L\|=m)$ is a set of links
$c(l)$	Capacity bandwidth of link l
$r(l)$	Residual bandwidth of link l
$d(s,d,b)$	A traffic demand from ingress node s to egress node d with required bandwidth b
D	A set of all ingress-egress pairs
p_{sd}	A routing path from s to d
P_{sd}	A set of all paths from s to d

The goal of routing algorithms is:

$$\begin{cases} \text{Maximize number of satisfied demands} \\ \qquad\qquad \text{subject to} \\ \text{Find } p_{sd} \text{ for } d(s,d,b)/\forall l \in p_{sd} : r(l) \geq b \end{cases} \tag{1}$$

Since ingress-egress pairs have commodity integral flows (i.e. common links as well as sequences of links), the routing problem summarized in (1) is NP-hard [2]. Moreover, the algorithms can be generalized as table 2.

2.2 Single Pair Aware Algorithms

The simplest solution is Minimum Hop Algorithm (MHA) where all weights are statically equal to 1. Dijkstra or Bellman-Ford algorithm is applied to find least hop counts paths. It means shortest paths are always selected and their links are quickly congested whereas others underutilize. To improve this property,

Table 2. General TE routing algorithm

Input	A network graph $G(N, L)$ with sets of links and residual bandwidths A traffic demand $d(s, d, b)$
Output	A satisfied bandwidth path from s to d, p_{sd}, toward the optimal goal in (1) Or no route satisfying the request
General algorithm	1. Calculate all link weights $w(l)$ 2. Remove links that have residual bandwidth less than b 3. Find the least cost path p_{sd} based on weights of remaining links

Widest Shortest Path algorithm (WSP) [3] selects paths having maximum bottleneck link bandwidth between the shortest equal-length ones. Additionally, Bandwidth Constrained Routing Algorithm (BCRA) [4] combines three parameters (link capacity, residual bandwidth, and path length) to calculate link weights.

$$w(l) = cost(l) * load(l) + 1$$
$$cost(l) = \frac{10^8}{C(l)}$$
$$load(l) = \frac{c(l) - r(l)}{c(l)}$$

Heavy weight values mean low capacities and/or heavy loads so those links are likely avoided, whereas hop count is reflected by the addition of 1. Experiments confirm that such combination of properties improves the routing performance.

Despite of using different network parameters, the above algorithms are classified as single-pair-aware because they greedily find good routes for the being demanded ingress-egress pair but not consider other ones. It might noticeably affect future requests. As a result, minimum interference solutions are proposed.

2.3 Minimum Interference Algorithms

The first Minimum Interference Routing Algorithm (MIRA) uses the maxflow-mincut characteristic [1]. When a routing request arrives, a maxflow residual graph is computed to determine a mincut set for each other ingress-egress pair. According to the maxflow-mincut theory [5], a bandwidth decrease of a link belonging to the mincut set will lead to the same amount reduction of the corresponding maxflow value. It means if such links are selected to route the current request, they will interfere with future demands of other pairs (i.e. narrow maxflow of them). MIRA defines those links as critical and assign more weights to them.

$$w_{sd}(l) = \sum_{(s',d') \in D \setminus (s,d)} \alpha_{s'd'} \quad \text{if } l \text{ is critical of } (s', d')$$

$\alpha_{s'd'}$ reflects the importance of the pair (s', d')

MIRA computes link weights from all ingress-egress pairs except the current demanded pair. Such current pair is excluded because the algorithm aims to prevent interference with the other ones. When all pairs are treated equally ($\alpha = 1$), a weight is the frequency of link's criticality, and the more critical a link is, the less it is chosen. Evaluations prove that MIRA outperform MHA in term of accepted ratio. Inspired by MIRA, different minimum interference algorithms are proposed.

Authors of NewMIRA [6] comment that MIRA only takes links of mincut sets into account, whereas all links that put up maxflows might affect future demands. Therefore, NewMIRA calculates a link's criticality by its load contribution to maxflow and the residual bandwidth.

$$w_{sd}(l) = \sum_{(s',d') \in D \setminus (s,d)} \frac{f_l^{s'd'}}{\theta^{s'd'}.r(l)}$$

$\theta^{s'd'}$ is the maxflow of the pair (s', d')

$f_l^{s'd'}$ is the subflow of $\theta^{s'd'}$ through link l

High interfering links are ones that largely contribute to maxflows and/or have small remaining bandwidths. Similar to MIRA, the NewMIRA also excludes the current pair (s,d) from weight calculation.

Dynamic Online Routing Algorithm (DORA) [7] does not use maxflows for interference but use the numbers of time links appearing in disjointed routing paths. Specifically, a link criticality of one (s,d) is decreased if that link is a part of any path from s to d, and increased if it belongs to paths of the other pairs.

$$criticality_s d(l) = \sum_{(i,j) \in D} \sum_{p_{ij} \in dP_{ij}} v_l$$

$$v_l = \begin{cases} 0 & \text{if } l \notin p_{ij} \\ -1 & \text{if } l \in p_{ij} \text{ and } ij \equiv sd \\ 1 & \text{if } l \in p_{ij} \text{ and } ij \not\equiv sd \end{cases}$$

dP_{ij} is the disjointed path set of the pair (i, j)

Realizing that the criticalities are computed solely by the network topology (i.e. by identification of disjointed paths); those values are prior-determined and only recalculated once the topology changes. This calculation is called the offline phase to differ from the online reactive routing phase. When a routing request arrives (i.e. the online phase), weights are formed from the corresponding criticalities and the current bandwidths. Because the interference is pre-determined, DORA selects routes more quickly than the above algorithms, especially when there are many ingress-egress pairs.

$w_{sd}(l) = (1 - \alpha).N_{C_{sd}(l)} + (\alpha).N_{RRB}$

$N_{C_{sd}(l)} \in [0, 100]$ is the normalization of the $criticality_{sd}(l)$

$N_{RRB} \in [0, 100]$ is the normalization of the reciprocal of $r(l)$

$\alpha \in [0, 1]$ is the proportion parameter

Similar to DORA, Bandwidth Guarantee with Low Complexity algorithm (BGLC) [8] has two phases. The critical values are directly proportional to the frequencies of links in all possible paths rather than the disjointed paths of just other pairs. In addition, the online phase also involves the residual bandwidths.

$$w(l) = criticality(l).\frac{1}{r(l)}$$

$$criticality(l) = \sum_{(i,j)\in D} \sum_{p_{ij}\in P_{ij}} \frac{v_l}{|P_{ij}|}$$

$$v_(l) = \begin{cases} 0 & \text{if } l \notin p_{ij} \\ 1 & \text{if } l \in p_{ij} \end{cases}$$

$|P_{ij}|$ is the number of all paths from i to j

Besides the minimum interference ideas, additional routing algorithms are recently proposed in the extent of machine learning applications.

2.4 Machine Learning Algorithms

Random race – a machine learning technique – is applied in Random Race based Algorithm for TE (RRATE) [9] to improve route computation time. The routing race approach is summarized as follow:

- The offline phase selects k shortest paths for each ingress-egress pair (s,d) as racing candidates and initialize a race reward value x_{i_sd} for the path i of (s,d).
- The online phase includes two stages: learning and post-learning. These stages are conducted separately for each ingress-egress pair.
- In the learning stage, when a demand $d(s,d,b)$ arrives, costs of the k selected paths are computed based on the number of critical links and the maximum residual bandwidths. Specifically, critical links are determined by the MIRA's maxflow-mincut definitions (i.e. critical links belong to mincut sets).

$$cost(p_{i\_sd}) = k_1.C_{i\_sd} + k_2/R_{i\_sd}$$

C_{i_sd} is the number of critical links in the path p_{i_sd}

$$C_{i\_sd} = \sum_{(s,d)\in D} \sum_{l\in p_{i\_sd}} v_l$$

$$v_(l) = \begin{cases} 1 & \text{if } l \text{ is critical of } (s,d) \\ 0 & \text{if } l \text{ is not critical of } (s,d) \end{cases}$$

R_{i_sd} is the maximum remaining bandwidth in p_{i_sd} if $d(s,d,b)$ is routed through p_{i_sd}

$$R_{i\_sd} = max_{l\in p_{i\_sd}}(r(l) - b)$$

k_1, k_2 are the moderation parameters

- High cost values mean there are more critical links and/or small remaining bandwidths. Therefore, routes are chosen in the increasing order of costs. For example, the smallest cost path is first checked for bandwidth requirement. If all links satisfy the demand then traffic is routed through that path; otherwise, the second smallest cost path is considered and so on. Additionally, whenever a path is selected, its corresponding $x_{i_s d}$ is accumulated by 1. The racing of those reward values continues until one of the paths reaches a pre-defined threshold N. Then, k paths of the (s,d) pair are sorted by the decreasing order of their respective rewards. Further requests of (s,d) are handled by the post-learning stage.
- In the post-learning stage, there is no computation but a route is selected within the sorted paths. Particularly, paths are verified against bandwidth requirement in the order of racing positions (i.e. the reward values). If the first route does not satisfy the demand, next ones are inspected. The process repeats until a satisfied route is found or all k paths are checked.
- Both two phases are reset if the network topology changes. However, normal networks do not change frequently so the post-learning stage of RRATE reduces the routing decision time.

Using the same random race technique, Paths Optimal Ordering Algorithm (POOA) [10] modifies RRATE in several aspects.

- The offline phase not only selects k shortest paths for each ingress-egress pair but also computes critical values. A link criticality relates to its subflows constituting the maxflows. Given that a link l belongs to one or more paths from s to d, the criticality is determined as:

$$criticality(l) = \frac{\sum_{(s,d)\in D} f_l^{sd}}{\sum_{(s,d)\in D} \theta^{sd}}$$

f_l^{sd} is the subflow of link l throw the maxflow θ^{sd}

- The learning state (online phase) calculates path costs by the criticalities and the residual bandwidths

$$cost(p_{i\_sd}) = \sum_{l\in p_{i\_sd}} \frac{criticality(l)}{r(l)}$$

Similar to RRATE, paths with low costs are considered first and the winning route is rewarded. However, the POOA racing threshold is not those reward values but is the whole position orders of k paths. In particular, after each demand is routed, the learning stage sorts k paths by their accumulated rewards. If that order is changed comparing to the order of the last demand, then the race is reset. In other words, the race ends on the condition that the path order continuously remains k times. (The POOA threshold is fixed to k instead of another N value of RRATE.)

– The post-learning state is the same as RRATE. It is noticed that the POOA cost computation is faster than RRATE's because the criticalities are pre-computed. Nevertheless, the later race process may be longer than the former due to the racing mechanism. Experiments will further compare and analyse them.

3 Experimets and Analysis

3.1 Simulation Environment

All the above algorithms are implemented in the network simulator NS2 [11] for experiments. Two different network topologies are simulated: one (figure 1(a)) is adapted from many previous TE routing works such as [1], [7], [10], and the other (figure 1(b)) inherits the real CESNET MPLS topology [12]. Both networks' links are bidirectional and have two types of capacity. The higher (the thicker links in the figure) is 4800, 10000 and the lower (the thinner ones) is 1200, 1000 bandwidth units respectively.

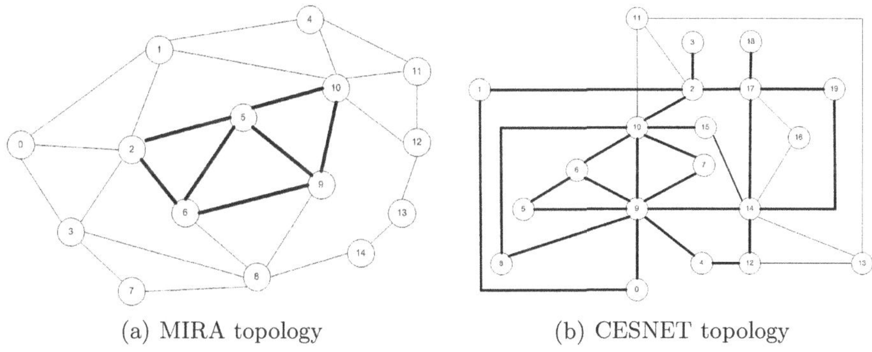

(a) MIRA topology (b) CESNET topology

Fig. 1. Network topologies

For each topology, three routing scenarios are evaluated. The first scenarios constantly demands 2000 static paths that stay in the network forever after being setup. The second one sets 2000 requests dynamically. Those requests arrive randomly according to the Poisson distribution of mean $\lambda = 80$ demands per time unit; whereas their holding (routing) time are distributed by the Exponential mean $\mu = 30$ time units. The third scenario is mixed between 200 static and 1800 dynamic requests. In this case, the distribution values are $\lambda = 40$ and $\mu = 10$. Furthermore, four ingress-egress pairs of (0, 12), (4, 8), (3, 1), and (4, 14); and eight pairs of (0, 18), (1, 11), (3, 16), (4, 7), (5, 13), (6, 19) , (15, 0) and (19, 8) are set for MIRA and CESNET topologies respectively. The former network has random bandwidth demands between 10, 20, 30, and 40 units. Meanwhile, the later arbitrarily needs 40, 80, 120, or 160 units for a request.

Two metrics are used to evaluate the algorithms in the aspect of the optimal goal described in (1). Firstly, percentage of accepted requests is compared. Obviously, the higher the accepted percent is, the better an algorithm performs.

The second metric is the average of computing time which is counted when a request arrives until it is accepted or rejected. This metric indicates the complexity of the reactive online routing phase, and should be minimized.

3.2 Evaluation Results

To obtain confident results, experiments are repeated several times with either different requests or algorithms' parameters. Comments in this paper describe the overall observation although there are few exceptions. Table 3 shows evaluating values with following parameters.

- For DORA, the bandwidth proportion $\alpha = 0.5$
- For RRATE, the moderation parameters $k_1 = k_2 = 0.5$, the number of preselected path $k = 25$, and the racing threshold $N = 10$
- For POOA, $k = N = 20$

Among single pair aware algorithms, the traditional ones (MHA and WSP) accept least number of demands in most experiments, whereas BCRA which considers more network's properties achieves better performances. There is an exception where WSP gains second highest percentage in the example results of mixed request experiment (table 3(c)). It might cause by equal-length paths in

Table 3. Comparison of accepted percent (in %) - computing time (in miliseconds) subject to number of requests (NoR)

(a) Results of static requests on MIRA network

NoR	MHA	WSP	BCRA	MIRA	NewMIRA	DORA	BGLC	RRATE	POOA
100	100-0.13	100-0.18	100-0.27	100-2.04	100-2.07	100-0.29	100-0.36	100-2.06	100-0.46
500	64.00-0.13	64.00-0.16	66.80-0.23	67.00-1.59	67.00-1.53	63.40-0.27	67.00-0.31	67.00-0.75	62.60-0.45
1000	36.20-0.12	37.20-0.14	41.40-0.22	41.50-1.24	41.50-1.14	39.70-0.27	41.50-0.33	38.20-0.44	39.20-0.43
1500	24.13-0.12	24.80-0.14	29.00-0.20	31.53-1.10	28.60-0.95	31.07-0.26	27.87-0.30	25.47-0.34	26.13-0.41
2000	18.10-0.12	18.60-0.13	21.75-0.19	23.65-1.02	21.45-0.86	25.35-0.25	20.90-0.28	19.10-0.29	19.60-0.40

(b) Results of dynamic requests on CESNET network

NoR	MHA	WSP	BCRA	MIRA	NewMIRA	DORA	BGLC	RRATE	POOA
100	100-0.16	100-0.21	100-0.35	100-4.74	100-2.52	100-0.37	100-0.31	100-5.25	100-0.59
500	86.00-0.17	83.80-0.21	84.80-0.33	84.60-4.55	89.60-2.61	85.80-0.37	86.20-0.30	94.00-3.06	88.20-0.45
1000	79.50-0.17	77.20-0.21	81.10-0.33	80.20-4.47	85.90-2.62	80.10-0.37	78.60-0.29	90.80-1.73	85.10-0.42
1500	79.27-0.17	76.73-0.21	82.53-0.33	81.07-4.67	87.13-2.85	80.33-0.36	79.93-0.30	90.40-1.21	86.20-0.40
2000	80.40-0.17	78.35-0.21	82.70-0.33	83.15-4.80	87.05-2.88	81.15-0.36	81.30-0.29	88.10-0.94	86.30-0.38

(c) Results of mixed requests on CESNET network

NoR	MHA	WSP	BCRA	MIRA	NewMIRA	DORA	BGLC	RRATE	POOA
100	100-0.59	100-0.68	100-0.35	100-5.30	100-2.49	100-0.50	100-0.30	100-5.37	100-0.59
500	77.00-0.25	78.00-0.30	76.80-0.33	76.80-4.67	76.40-2.39	78.40-0.39	76.40-0.29	78.00-3.21	76.00-0.47
1000	70.10-0.21	71.30-0.26	71.70-0.32	70.50-4.45	69.90-2.27	71.60-0.37	70.50-0.28	72.40-2.13	69.80-0.42
1500	66.60-0.20	69.87-0.24	68.20-0.32	67.13-4.39	65.93-2.22	68.20-0.37	68.00-0.28	70.47-1.73	66.13-0.42
2000	65.95-0.19	70.20-0.23	66.65-0.33	66.50-4.37	64.15-2.20	68.05-0.37	69.60-0.28	71.20-1.46	64.90-0.40

the network topology. Nevertheless, such result is not the overall trend. On the other hand, this category has the fastest computing time due to no additional calculation except the shortest path algorithm.

In addition, interference minimum algorithms attain considerable improvements of accepted percent over shortest path algorithms. Specifically, DORA accepts 7% higher number of requests than MHA in the static test (table 3(a)). Table 3(b) also depicts a 10% improvement of NewMIRA over WSP. However, there is no leading algorithm within the interference minimum category. For example, DORA obtains the highest value (25.35%) in the static experiment but the lowest one (81.15%) in the dynamic test. Meanwhile, in this test, NewMIRA outperforms the others with 87.05% comparing to the second highest of MIRA only at 83.15%. In the term of route selecting time, there is a significant difference between one-phase and two-phase algorithms. Particularly, the computing time of MIRA is even 10 ten times greater than DORA and BGLC's because MIRA recalculates maxflows for each routing demand whereas the others predetermine link's criticalities and need much less time to form link weights.

Finally, the machine learning algorithms (RRATE and POOA) also have good performances. Especially, RRATE achieve the best accepted percent in many test sets. Values of the dynamic and mixed requests experiments (table 3(b) and 3(c)) are the examples. Moreover, the average computing times of RRATE and POOA are compatible to the others. After the learning stage (e.g. after the first 100 requests in table 3(b)), the time significantly reduces. On the other hand, performances of these algorithms are greatly affected by the heuristic racing parameters (k and N values of RRATE; k value of POOA). For example, in the experiment of mixed requests on CESNET topology, when k is changed to 20 and N to 5, the average computing time of RRATE at 2000 requests decreases three times from 1.46 to 0.50 ms, whereas the corresponding percentage even increases from 71.2 to 71.5 %. However, it is almost impossible to choose the best values for all cases because the algorithms' performances depend on not only the network topology but also the routing requests.

4 Conclusion

This paper presents a study of traffic engineering routing algorithms. First, the routing problem is defined in the aspect of quality of services and traffic engineering objectives. Then, well known algorithms are described within three different categories: single pair aware, interference minimum, and machine learning. Finally, after thorough experiments, following conclusions are obtained:

- Traditional shortest path algorithms are not good enough for traffic engineering requirements. The application of network properties such as links' capacities, residual bandwidths can improve the routing effectiveness.
- The interference idea which computes links' criticalities based on the effects of current selections on future routing demands clearly enhances the number of accepted requests. Nevertheless, network topology and routing requests vary algorithms' performances so there is no best solution.

– Recently, machine learning techniques have been introduced for further improvements of TE routing algorithms. Because the requests themselves greatly impact the routing decisions, using history demands and routing data is a promising direction for traffic engineering problem.

References

1. Kar, K., Kodialam, M., Lakshman, T.: Minimum interference routing of bandwidth guaranteed tunnels with MPLS traffic engineering applications. IEEE Journal on Selected Areas in Communications 18(12), 2566–2579 (2000)
2. Even, S., Itai, A., Shamir, A.: On the complexity of time table and multi-commodity flow problems. In: Proceeding SFCS 1975 Proceedings of the 16th Annual Symposium on Foundations of Computer Science, pp. 184–193. IEEE (1975)
3. Guerin, R.A., Orda, A., Williams, D.: QoS routing mechanisms and OSPF extensions. In: Global Telecommunications Conference, GLOBECOM 1997, vol. 3, pp. 1903–1908. IEEE (1997)
4. Kotti, A., Hamza, R., Bouleimen, K.: Bandwidth constrained routing algorithm for MPLS traffic engineering. In: International Conference on Networking and Services (ICNS 2007), Athens, Greece, p. 20 (2007)
5. Ahuja, R.K., Magnanti, T.L., Orlin, J.B.: Network flows: Theory, algorithms, and applications. Prentice-Hall, Englewood Cliffs (1993)
6. Wang, B., Su, X., Chen, C.: A new bandwidth guaranteed routing algorithm for MPLS traffic engineering. In: Conference Proceedings of 2002 IEEE International Conference on Communications, ICC 2002 (Cat. No.02CH37333), pp. 1001–1005 (2002)
7. Boutaba, R., Szeto, W., Iraqi, Y.: DORA: efficient routing for MPLS traffic engineering. Journal of Network and Systems Management 10, 309–325 (2002), doi:10.1023/A:1019810526535
8. Alidadi, A., Mahdavi, M., Hashmi, M.R.: A new low-complexity QoS routing algorithm for MPLS traffic engineering. In: 2009 IEEE 9th Malaysia International Conference on Communications (MICC), Kuala Lumpur, Malaysia, pp. 205–210 (2009)
9. Oommen, B.J., Misra, S., Granmo, O.: Routing Bandwidth-Guaranteed paths in MPLS traffic engineering: A multiple race track learning approach. IEEE Transactions on Computers 56(7), 959–976 (2007)
10. Lin, N., Lv, W.-F., Yang, T.: Paths optimal ordering algorithm based on MPLS traffic engineering. In: 2009 Ninth International Conference on Hybrid Intelligent Systems, Shenyang, China, pp. 492–495 (2009)
11. Issariyakul, T., Hossain, E.: Introduction to network simulator NS2. Springer, New York (2011)
12. Novák, V., Adamec, P., Šmrha, P., Verich, J.: CESNET2 IP/MPLS backbone network design and deployment in 2008 (2008)

Application of a Clustering Based Location-Routing Model to a Real Agri-food Supply Chain Redesign

Fethi Boudahri[1], Wassila Aggoune-Mtalaa[2,⋆], Mohammed Bennekrouf[1], and Zaki Sari[1]

[1] MELT Laboratory, University of Tlemcen, Algeria
[2] Public Research Centre Henri Tudor, L-1855 Luxembourg, G.D Luxembourg
wassila.mtalaa@tudor.lu

Abstract. The supply planning of agricultural products and in particular perishable products is a critical issue in the supply chain management field due to high safety and quality risks associated with the delays in the products delivery. This work is concerned with the planning of a real agri-food supply chain for poultry products. More precisely the problem is to redesign the existing supply chain and to optimize the distribution planning. To this aim a clustering-based location-routing model is applied in a sequential manner. Furthermore, environmental costs of road transportation in terms of CO_2 emissions are taken into account in the computations. The proposed integrated approach permits to minimise the total costs of the agri-food supply chain not only in terms of economy but also in terms of ecology.

Keywords: clustering based location-routing, agri-food supply chains, optimisation, CO_2 emissions.

1 Introduction

Agri-food supply chains (ASC) is the term used to define logistics networks which encompasse production and distribution activities of agricultural or horticultural products from the farm to the consumers, see [2]. In particular the management of perishable products is a relevant issue in the ASC management domain, since the sellers cannot wait for the best favourable market conditions unless the quality and safety of their products deteriorate, see [9]. These products must therefore be rapidly shipped from the sellers to the customers. Moreover the demand of consumers on healthy products is ever increasing and regulations of the authority are more and more stringent. In addition to these factors, the variability in the demand and price of these products with limited shelf life adds complexity to the supply planning of these already critical supply networks. This explains the growing attention paid to the ASC in the literature, see [3].

⋆ Corresponding author.

N.T. Nguyen et al. (Eds.): *Adv. Methods for Comput. Collective Intelligence*, SCI 457, pp. 323–331.
DOI: 10.1007/978-3-642-34300-1_31

Still integrated approaches for ASC design and distribution planning are limited. Furthermore, most of the models encountered in the litterature need to be enriched by the consideration of uncertainty as well as shelf life features that directly impact the quality and freshness of the products. Last, examples of real applications of successful approaches are rare.

In this work, a real agri-food supply chain of poultry products for the city of Tlemcen in Algeria is studied. More precisely, the aim of this work is coordination of decisions for location, allocation and transportation of products to achieve an efficient and green logistic network design and distribution planning. The approach chosen is a location-routing formulation which consists in addressing a vehicle routing problem in which the optimal number and location of the slaughterhouses are to be determined simultaneously with the distribution routes to the customers in order to minimize the total costs, see [8]. Last the problem is formulated in a way that helps minimizing environmental costs of the supply chain in terms of greenhouse gases emissions.

This work is organized as follows. First, the problem is described. Then, the mathematical model is presented. Computational results are presented in section 3. The last section concludes the paper and provides some directions for future research.

2 Problem Presentation

In this section, the problem to redesign a real agri-food supply chain for poultry products in the city of Tlemcen is presented. The integrated approach to solve it is then described.

2.1 Problem Description

The network considered in this work is composed of customers and slaughterhouses from the city of Tlemcen and surronding areas. The aim is to redesign the supply chain in order to make savings and improve the distribution of chicken and turkey from the slaughterhouses to the retailers. As noted by [7] in their review on location analysis applied to agriculture, there is a need to consider the uncertainty in the strategic planning models applied to the agricultural industry, in particular to plan and forecast the demand from customers. In this work, to address uncertainty arising from the demand on poultry products from customers, the choice was made to capture the real demands on these products by studying them. To this purpose, more than three years have been required to observe the real flows of chicken and turkey distributed from the slaughterhouses to the retailers. The conclusion was that the distribution planning lacked efficiency leading sometimes to unsatisfied demands and increased prices at retailers.

In addition to that, an integrated model was chosen to better plan the activities of the supply chain and generate savings. As the problem is complex by nature, the proposed solution approach has been divided into three main steps. Indeed a clustering based location-routing model is addressed in a sequential manner to optimise the products flows. The first problem is to group customers which can be retailers into clusters provided that the demands on poultry products for each customer which belongs to a cluster is satisfied. The study performed on real data allows to suppose that these demands are known in advance. Then, when the capacity for the delivery of each cluster is calculated, a capacitated plant location problem is solved in order to decide which slaughterhouse has to be open, closed or reopen for the considered planning horizon. Last, a classical routing problem is tackled in order to define which routes are the cost-effective ones.

Limited capacities at potential slaughterhouses and for the refrigerated delivery trucks are additional assumptions made for this study. Let us recall that the demand on poultry products is supposed to be know in advance.

Last the problem is formulated in a way that helps minimizing environmental costs of the supply chain. More precisely, the overall CO_2 emissions caused by transport are quantified and their costs are added to the classical transportation costs.

The sets and indexes used in the model are as follows:

i $\in$ I, index of customers (which can represent butchers' shops),
j $\in$ J, index of customers' clusters,
k $\in$ K, index of potential location for slaughterhouses,
p $\in$ P, index of classes of products
v $\in$ V, index of vehicles classified according to their Authorized Gross Weight.

2.2 First Step Capacitated Centered Clustering Problem

In a first stage, the problem of grouping the retailers into clusters is formulated as a capacitated centered clustering problem. The originality and efficiency of this approach come from the fact that it limits dissimilarity among the formed groups since these clusters are centered at the "average" of their points' coordinates, see [4]. The parameters, decision variable and formulation of the first problem are presented in what follows.

V_p= average volume for a product of class p in m^3. The volume of each product is compared to a standard size in order to optimize the allocation of the customers with regard to the capacity volume of the refrigerated trucks used for the delivery.
$Q_{p,j}$= capacity of the refrigerated trucks allocated to cluster j for products of type p, in m^3
(x_a,y_a)= geometric position of site a,
n_j= number of customers in cluster j,
$dem_{p,i}$= demand on class p products at customer i.

The decision variable for the model is Y_{ij} which value equals 1 if customer i is assigned to cluster j, 0 otherwise.

The capacitated centered clustering problem, consists in minimizing:

$$\sum_{i\in I}\sum_{j\in J}\left[(x_i-x_j)^2+(y_i-y_j)^2\right]Y_{ij} \tag{1}$$

Such that:

$$\sum_{j\in J} Y_{ij}=1 \qquad \forall i\in I \tag{2}$$

$$\sum_{i\in I} Y_{ij}=n_j \qquad \forall j\in J \tag{3}$$

$$\sum_{i\in I} x_i Y_{ij}\le n_j x_j \qquad \forall j\in J \tag{4}$$

$$\sum_{i\in I} y_i Y_{ij}\le n_j y_j \qquad \forall j\in J \tag{5}$$

$$\sum_{i\in I}(dem_{p,i}*V_p)\,Y_{ij}\le Q_{p,j} \qquad \forall j\in J,\forall p\in P \tag{6}$$

$$(x_i,y_i)\in\mathbb{R},\quad (x_j,y_j)\in\mathbb{R} \qquad \forall i\in I,\forall j\in J \tag{7}$$

$$n_j\in\mathbb{N} \qquad \forall j\in J \tag{8}$$

$$Y_{ij}\in\{0,1\} \qquad \forall i\in I,\forall j\in J \tag{9}$$

The objective function (1) represents the total euclidian (square) distance between each customer and the centre of the cluster to which it belongs. One should note that the geometric position of the centre is unknown a priori since it depends on the customers allocated to the cluster.

Constraints (2) specify that each customer is allocated to exactly one cluster. Inequalities (3) set the number of customers at each cluster. Inequalities (4) and (5) define the geometric position of each cluster's centre.

Constraints (6) require that the volume of poultry products for each cluster does not exceed the capacity of the trucks associated to that cluster. Last, constraints (7) to (9) define the space boundaries for the set of parameters and variables of the problem.

It should be stressed that after the customers have been grouped into clusters, begins a second step calculation during which they have to be allocated to the slaughterhouses to be open.

2.3 Second Step Location-Allocation Formulation

The parameters, variables and formulation of the problem are as follows:

F_k= fixed cost of setting up and operating slaughterhouse k,
$dem_{p,j}$= demand on class p products at cluster j.

O_p= operational time required for the treatment of a product of class p,
$Q_{p,k}$= capacity of treatment at slaughterhouse k for class p products,
$N_{p,jk}$= Number of class p products shipped from site k to site j.
w_p= average weight of a product of class p in ton. As in the previous phase, the weight of each product is compared to this standard weight.
d_{jk}= euclidian distance from site k to site j in km,
$c_{jk}^{v,km}$= transportation costs per kilometer from site k to site j by vehicle v. These transportation costs involve costs for operating vehicle v, infrastructures costs, fuel consumption when v is empty and tolls.
$c_{jk}^{v,tkm}$= transportation costs per ton.kilometer from site k to site j by vehicle v. These costs are for fuel consumption per ton and environmental costs. These latter are calculated in two steps. First the emission factor per vehicle per ton.kilometer are assessed with the quantification method developed jointly by ADEME and Entreprises pour l'Environnement, see [1] and [6]. Then carbon dioxide emissions due to transportation are priced with the European Trading Scheme of carbon allowances on the European Energy Exchange, see [5].

The decision variables for the model are the following:

X_k=1, if a slaughterhouse is set up at site k, 0, otherwise,
Z_{jk}=1, if cluster j is allocated to slaughterhouse k, 0 otherwise.

The capacitated plant location problem consists in minimizing quantity:

$$\sum_{k\in K} F_k X_k + \sum_{p\in P}\sum_{v\in V}\sum_{j\in J}\sum_{k\in K}\left(w_p N_{p,jk} c_{jk}^{v,tkm} + c_{jk}^{v,km}\right) d_{jk} Z_{jk} \tag{10}$$

Such that:

$$\sum_{k\in K} Z_{jk} = 1 \ \ \forall j \in J \tag{11}$$

$$\sum_{k\in K} N_{p,jk} * Z_{jk} = dem_{p,j} \ \ \forall p \in P, \forall j \in J \tag{12}$$

$$\sum_{j\in J} N_{p,jk} * O_p \leq Q_{p,k} * X_k \ \ \forall p \in P, \forall k \in K \tag{13}$$

$$N_{p,jk} \geq 0 \ \ \forall j \in J, \forall k \in K, \forall p \in P \tag{14}$$

$$X_k \in \{0,1\} \ \ \forall k \in K \tag{15}$$

$$Z_{jk} \in \{0,1\} \ \ \forall j \in J, \forall k \in K \tag{16}$$

The objective of the problem (10) is a cost function composed of fixed and variable costs. The fixed costs are linked to the opening of slaughterhouses and include investment costs for the land, the land tax and the slaughter units. The variable costs include the economical as well as the ecological transportation costs. Equalities (11) specify that a cluster j can be served by only one slaughterhouse. The set of constraints (12) stipulates that the demand on class p product at cluster j must be fully met. Inequalities (13) are capacity constraints at slaughterhouse k. Inequalities (14) guarantee non negativity of the products flows. Last, constraints (15) and (16) impose binary conditions.

2.4 Third Step Traveling Salesman Problem

Last a classical TSP is applied to each cluster: Given the number of customers and the costs of traveling from any customer to another, the aim is to find the cheapest round-trip route that visits each customer exactly once and then returns to the starting point. Let us recall that the transportation costs include economical as well as environmental costs as described in the previous section. The results of the application of this model are presented in what follows.

3 Experimental Results

In this section the model is applied to the problem. LINGO 12 has been used to solve the three programs and to obtain exact solutions by using Branch and Bound with default parameters of the solver.

Results of the first step calculations are summarized in table 1. The problem was to group 113 customers into clusters. For each of the resulting customers' clusters, the number of assigned customers as well as their reference number

Table 1. Capacitated centered clustering results

Cluster N^0	n_j	Cluster's centre position	Assigned customers number
1	1	(8819.92,5632.02)	5
2	1	(8854.72,5646.22)	7
3	1	(8856.22,5617.00)	6
4	0	/	/
5	8	(7758.02,6679.99)	3/4/11/17/18/102/103/104
6	12	(9007.11,6033.94)	10/12/13/14/15/16/20/88/89/90/91/105
7	8	(7540.73,6046.07)	1/53/54/56/57/58/59/112
8	0	/	/
9	7	(5112.69,5767.09)	46/64/66/67/68/69/70
10	8	(7568.44,5825.43)	2/9/19/49/51/52/55/60
11	10	(10043.23,9490.09)	79/80/81/82/83/84/92/93/94/95
12	8	(7478.42,5177.42)	23/26/30/32/33/34/50/111
13	5	(7146.53,5353.10)	21/22/27/47/65
14	10	(7532.64,4847.30)	24/25/28/29/31/35/36/37/38/48
15	0	/	/
16	12	(10685.80,9767.27)	71/72/73/74/75/76/77/78/85/86/87/110
17	10	7726.06,9343.55)	96/97/98/99/100/101/106/107/108/109
18	10	(6974.03,4724.97)	39/40/41/42/43/44/45/61/62/63
19	1	(7478.42,5177.42)	113

Table 2. Location-allocation results

Slaughterhouse number	Location decision	Allocated cluster number	Corresponding capacity for Product 1	Product 2
1	Open	2	680	20
		3	680	20
		11	420	16
		13	375	19
		14	395	21
		16	395	15
		19	454	5
2	Closed	none	0	0
3	Open	12	380	19
		18	340	28
4	Closed	none	0	0
5	Open	1	680	20
		9	405	16
6	Closed	none	0	0
7	Open	4	680	20
		5	370	24
		6	430	10
		7	400	19
		10	380	22
		17	295	14

are given. The coordinates of the cluster centre also appear in the table. Out of the nineteen clusters generated, sixteen have been assigned customers. The biggest clusters are cluster number six and number sixteen with twelve retailers assigned to both. Results of the first problem, as well as the fixed investment costs, capacities at the slaughterhouses, and distances between the potential slaughterhouses and the centre of each cluster are the data used for the second problem. The two types of products considered correspond to chicken and turkey meat. The average turkey volume is supposed to be five time the chicken's one.

Results of the second step computations are presented in table 2. Four slaughterhouses have been open out of the seven existing locations. The three remaining slaughterhouses had to be closed.

Some results of the third step calculations appear in figure 1. Four clusters are represented: the ones referenced as clusters 7, 10, 11 and 18. For each of them, the cost-effective routes for the trucks are described begining from the starting point. For instance, for cluster number 7, the truck is supposed to deliver first customer 1, then customer 57, followed by customer 59 and 54, and so on. The route finishes with customer 112 and last customer 1.

These results have been presented to the actors of the studied agri-food supply chain. The implementation of the proposed solution have led to significant success and savings.

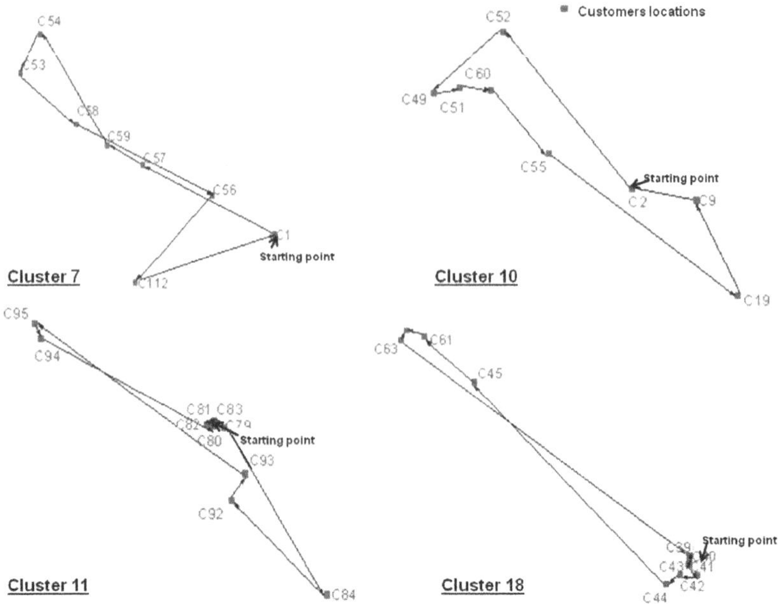

Fig. 1. Travelling Salesman Problem results

4 Conclusion

In this paper, a clustering-based location-routing approach has been applied with success to redesign a real agri-food supply chain for poultry products in the city of Tlemcen in Algeria. To this aim, a three step mathematical model has been built and solved in a sequential manner. Once the customers have been grouped into clusters, the slaugterhouses to set up, to close or to reopen have been identified, and the clusters of retailers have been allocated to them. Then, a classical TSP has been applied to each cluster in order to calculate the best routing to deliver the customers. Last, environmental costs of the supply chain have been included in the commputations.

A perspective to this work would be to involve another echelon to this real supply chain with the farmers locations. This would lead to greater size instances of this problem which is NP-Hard. That would require the development of some heuristics and metaheuristics to solve it.

References

1. ADEME: Bilan Carbone "Entreprises et Collectivités". Guide des facteurs d'émissions_version 6.1 (2010), `http://www.ademe.fr`
2. Aramyan, C., Ondersteijn, O., Van Kooten, O., Lansink, A.O.: Performance indicators in agri-food production chains. In: Quantifying the Agri-Food Supply Chain, pp. 49–66. Springer, Netherlands (2006)

3. Ahumada, O., Villalobos, J.R.: Application of planning models in the agri-food supply chain: A review. European Journal of Operational Reseearch 195, 1–20 (2009)
4. Chaves, A.A., Lorena, L.A.N.: Hybrid evolutionary algorithm for the Capacitated Centered Clustering Problem. Expert Systems with Applications 38, 5013–5018 (2011)
5. European Energy Exchange, `http://www.eex.com/en/`
6. Entreprises Pour l'Environnement, `http://www.epe-asso.org/`
7. Lucas, M.T., Chhajed, D.: Applications of location analysis in agriculture: A survey. Journal of the Operational Research Society 55, 561–578 (2004)
8. Nagy, G., Salhi, S.: Location-routing: Issues, models and methods. European Journal of Operational Research 177, 649–672 (2007)
9. Rong, A., Akkerman, R., Grunow, M.: An optimization approach for managing fresh food quality throughout the supply chain. International Journal of Production Economics 131, 421–429 (2011)

Crisis Management Model and Recommended System for Construction and Real Estate

Artūras Kaklauskas[1,*], Edmundas Kazimieras Zavadskas[1], Paulius Kazokaitis[1], Juozas Bivainis[1], Birute Galiniene[2], Maurizio d'Amato[3], Jurga Naimaviciene[1], Vita Urbanaviciene[1], Arturas Vitas[2], Justas Cerkauskas[1]

[1] Vilnius Gediminas Technical University, Sauletekio av. 11, Vilnius, LT-10223, Lithuania
[2] Vilnius University, 3 Universiteto St, LT-01513 Vilnius, Lithuania
[3] Technical University of Bari, via Amendola, 126/B - 70126 Bari, Italy
arturas.kaklauskas@st.vgtu.lt

Abstract. Integrated analysis and rational decision-making at the micro-, meso- and macro-levels are needed to mitigate the effects of recession on the construction and real estate sector. Crisis management involves numerous aspects that should be considered in addition to making economic, political and legal/regulatory decisions. These must include social, culture, ethical, psychological, educational, environ- mental, provisional, technological, technical, organizational and managerial aspects. This article presents a model and system for such considerations and discusses certain composite parts of it.

Keywords: construction, real estate, crisis management, quantitative and qualitative methods, global development trends, alternatives, Lithuania, Model, System, forecasting.

1 Introduction

Various econometrics (e.g., Keynesian models, time-series analysis using multiple regression, Box-Jenkins analysis, Time-varying Parameter Model, duration statistical model, multivariate Logit model, competing-risks hazard models with time-varying covariates, dummy variable approach) and operations research (statistical analysis (discriminant analysis [1], Logit and Probit regression models [2]), artificial neural network models (fuzzy clustering and self-organizing neural networks [3], the back-propagation neural networks model [4]), multiple criteria decision making [5]-[7], artificial intelligence (the support vector machine [8], k-nearest neighbor algorithm [4], and decision tree [9], etc.) methods and models in the construction and real estate sector as well as in separate segments are being applied for crisis management worldwide today. Technical approaches of operations research (decision support systems [5], [10], [11], expert systems [12], mathematical programming [13], multicriteria decision methods [5]) are used for crisis management in different construction and real estate fields.

---

* Corresponding author.

N.T. Nguyen et al. (Eds.): *Adv. Methods for Comput. Collective Intelligence*, SCI 457, pp. 333–343.
DOI: 10.1007/978-3-642-34300-1_32

It may be noted that above researchers from various countries engaged in the analysis of crisis in the construction and real estate sectors but they did not consider the research object that is being analyzed by the authors of this present research. The latter may be described as a life cycle of the construction and real estate industry, the stakeholders involved and as the micro, meso and macro environments that have some particular impact on a life cycle in making an integral whole. A complex analysis of the formulated research object was performed with the help of Construction and Real Estate Crisis Management Model and Recommended System which were especially developed for this purpose.

This paper is structured as follows: After this introduction, Section 2 describes Construction and Real Estate Crisis management Model. A sketch of the Recommended Construction and Real Estate Crisis Management System appears in Section 3. Finally Section 4 provides some concluding remarks.

2 Construction and Real Estate (CARE) Crisis Management Model

The traditional analysis of a crisis in construction and real estate is based on economic, legal/regulatory, institutional and political aspects. Social, cultural, ethical, psychological and educational aspects of crisis management receive less attention. To perform an integrated analysis of the life cycle of a crisis in the construction and real estate sectors, the cycle must be analyzed in an integrated manner based on a system of criteria (see Figure 1).

The aim of this research was to produce a construction and real estate (hereafter – CARE) crisis management model for Lithuania by undertaking a complex analysis of the micro, meso and macro environmental factors affecting it and to present recommendations on increasing its competitive ability. The research was performed by studying the expertise of advanced industrial economies and by adapting such to Lithuania while taking into consideration its specific history, development level, needs and traditions. A simulation was undertaken to provide insight into the development of an effective environment for the CARE Crisis management Model by choosing rational micro, meso and macro factors.

The word, model, implies “a system of game rules” by which CARE crisis management could be used to its best advantage in Lithuania’s development.

This research includes the following six stages.

Stage I. A comparative description is written on CARE crisis management in developed countries and in Lithuania which includes: a system of criteria that characterizes crisis management efficiency which is determined by using relevant literature and expert methods; a description based on this system of criteria in conceptual (textual, graphical, numerical, etc.) and quantitative forms on the present state of crisis management in developed countries and in Lithuania.

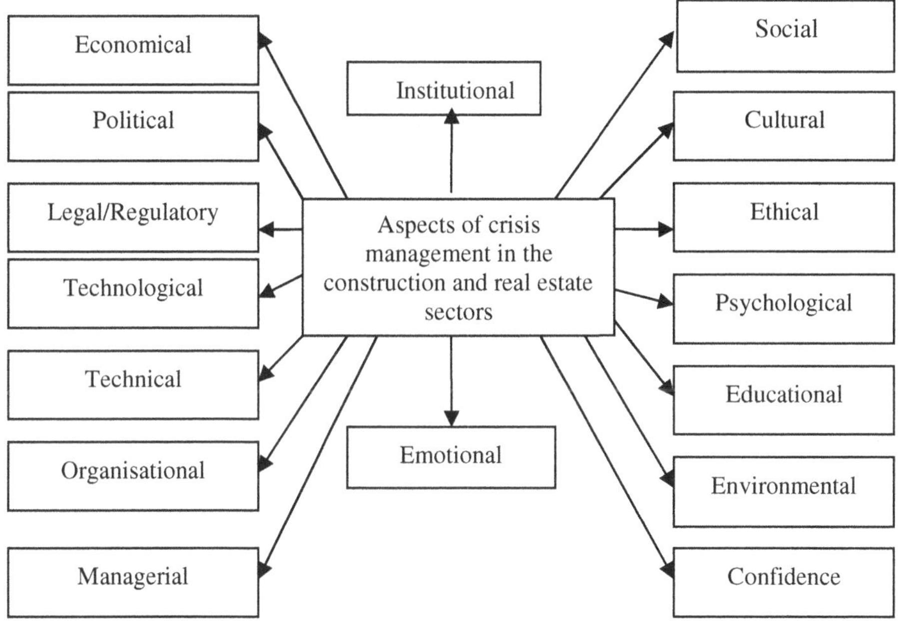

Fig. 1. Aspects of crisis management in the construction and real estate sectors

Stage II. A comparison and contrast of CARE crisis management in developed countries and in Lithuania are performed which include: an identification of global development trends (general regularities) in crisis management; an identification of crisis management differences between developed countries and Lithuania; a determination of the pluses and minuses of these differences for Lithuania; establishment of the best crisis management practice for Lithuania based on actual conditions; an estimation of the deviation between the knowledge stakeholders have about the best practice worldwide and their practice-in-use.

Stage III. Some general recommendations are developed on how to improve efficiency levels for CARE stakeholders and construction (real estate) firms.

Stage IV. Certain recommendations for CARE stakeholders and construction (real estate) firms are submitted. Each of the general recommendations proposed in Stage III contain several specific alternatives.

Stage V. A multiple criteria analysis is performed on the components of CARE crisis management, and a selection is made of the most efficient version of the life cycle of crisis management. After this, the gained compatible and rational components of one type of crisis management are joined into a full, crisis management process.

Stage VI. Transformational learning is performed, and mentality and actual behavior in practice are redesigned.

3 Construction and Real Estate Crisis Management Recommended System

Based on the analysis of existing information, expert and decision support systems and in order to determine most efficient versions of Recommended Construction and Real Estate Crisis Management (RCRECM) System consisting of a database, database management system, model-base, model-base management system and user interface was developed (Figure 2).

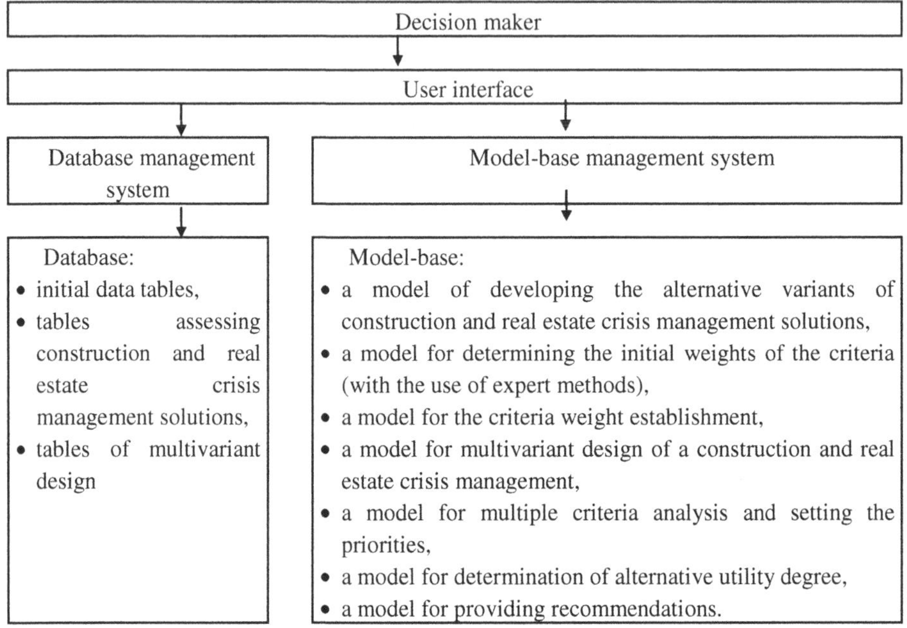

Fig. 2. The components of RCRECM system

In order to thoroughly analyze the alternatives available and obtain an efficient compromise solution it is often necessary to define them on the basis of economic, political and legal/regulatory decisions, social, culture, ethical, psychological, educational, environmental, provision, technological, technical, organizational and managerial information (Figure 1). The presentation of information needed for decision making in RCRECM system may be in conceptual (digital (numerical), textual, graphical (diagrams, graphs, drawing, etc), photographical, sound, visual (video)) and quantitative forms.

RCRECM system has a relational database structure when the information is stored in the form of tables. These tables contain quantitative and conceptual information. Logically linked parts of the table make a relational model.

The tables of construction and real estate crisis management variant assessment contain the variants available and their quantitative and conceptual description. Quantitative description of the alternatives deals with the systems and subsystems of criteria fully defining the variants as well as the units of measurement and values and initial weights. Conceptual description defines the alternatives available in a commonly used language giving the reasons and providing grounds for choosing a particular criterion, calculation its value, weight and the like. In this way, RCRECM system enables the decision maker to get various conceptual and quantitative information on construction and real estate crisis management from a database and a model-base allowing him to analyze the above factors and make an efficient solution.

The stakeholders of a construction and real estate sector have their specific needs and financial situation. Therefore, every time when using RCRECM system they may make corrections of the database according to the aims to be achieved and the financial situation available.

The above tables are used as a basis for working out the matrices of decision making. These matrices, along with the use of a model-base and models, make it possible to perform multivariant design and multiple criteria evaluation of construction and real estate crisis management alternatives resulting in the selection of most beneficial variants. In order to design and realise an effective construction and real estate crisis management project the alternatives available should be analysed. Computer-aided multivariant design requires the availability of the tables containing the data on the interconnection of the solutions as well as their compatibility, possible combination and multivariant design.

Based on the above tables of multivariant construction and real estate crisis management design possible variants are being developed. When using a method of multivariant design suggested by the author until 10 000 000 construction and real estate crisis management alternatives may be obtained. These versions are checked for their capacity to meet various requirements. Those which can not satisfy these requirements raised are excluded from further consideration. In designing a number of variants of construction and real estate crisis management the problem of weight compatibility of the criteria arises. In this case, when a complex evaluation of the alternatives is carried out the value of a criterion weight is dependent on the overall criteria being assessed as well as on their values and initial weights.

Since the efficiency of a construction and real estate crisis management variant is often determined taking into account economic, legal/regulatory, institutional, political social, cultural, ethical, psychological and educational and other factors a model-base of a decision support system should include models enabling a decision maker to do a comprehensive analysis of the variants available and make a proper choice. The following models of model-base are aimed to perform this function: a model of developing the alternative variants of construction and real estate crisis management solutions; a model for determining the initial weights of the criteria (with the use of expert methods); a model for the criteria weight establishment; a model for multivariant design of a construction and real estate crisis management; a model for multiple criteria analysis and setting the priorities; a model for determination of alternative utility degree; a model for providing recommendations.

Multiple criteria decision-making methods that the authors herein have developed for the Recommended Construction and Real Estate Crisis Management System are as follows [14]: a method of complex determination of the weight of the criteria taking into account their quantitative and qualitative characteristics.; a method of multiple criteria complex proportional evaluation of the alternatives; a method of defining the utility and market value of an alternative; a method of multiple criteria multivariant design of a alternative life cycle.

Based on the above models, a RCRECM system can make until 10 000 000 construction and real estate crisis management alternative versions, performing their multiple criteria analysis, determining utility degree and selecting most efficient variant.

To demonstrate the application of the Recommended Construction and Real Estate Crisis Management System a few specific examples from Lithuania have been solved (see Figure 3).

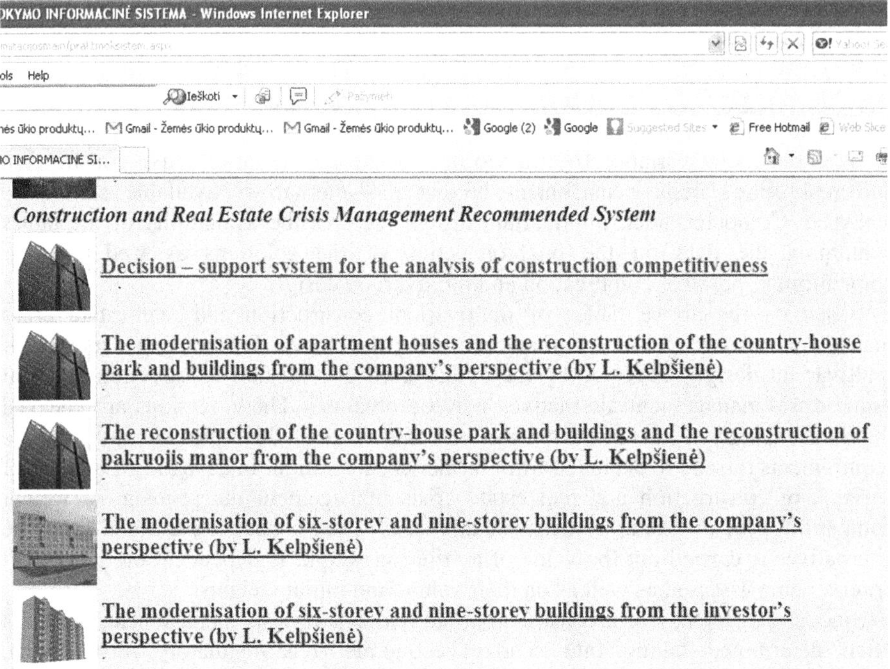

Fig. 3. A few specific examples solved by using the Recommended Construction and Real Estate Crisis Management System

The more alternative versions are investigated before making a final decision, the greater is the possibility to achieve a more rational end result. Basing oneself on possessed information and the RCRECM system it is possible to perform multiple criteria analysis of construction and real estate crisis management alternatives components and select the most efficient versions. After this, the received compatible and rational components of a construction and real estate crisis management alternatives are joined up into projects. Having performed multiple criteria analysis of projects made up in such a way, one can select the most efficient ones. Strong and weak sides of investigated projects are also given an analysis. Facts of why and by what degree one version is better than the other are also established. All this is done basing oneself on conceptual and quantitative information.

As example, two models (a model for multiple criteria analysis and setting priorities and a model for determining project utility degree) are described further. A model for multiple criteria analysis and setting priorities performs the multiple criteria analysis of options and sets project priorities based on the weighted criteria and tacit and explicit knowledge. The model for determining the alternative utility degree sets the utility degree for each analysed option. The Complex Proportional Assessment method (COPRAS) [14] is used for these purposes.

The results of the comparative analysis of projects are presented as a grouped decision-making matrix where the columns contain n alternative projects, while all the pertinent quantitative and conceptual information is found in Table 1. Any alternative that has a poorer criterion value than that required is rejected. In order to perform a complete study of a project, a complex evaluation is needed of its aspects (see Figure 1). Quantitative and conceptual descriptions provide this information. The diversity of aspects being assessed should include a variety of data presented as needed for decision-making. Therefore the necessary conceptual information may be presented in numerical, textual, graphical (schemes, graphs, diagrams), virtual and augmented realities or equation formats and as audio or videotapes. An analysis should include all - the criteria used for conceptual descriptions, their definitions and the reasons for the choice of a criteria system and their values and weights. Conceptual information about the possible ways of performing a multivariant design is needed to make a more complete and accurate evaluation. Quantitative information is based on criteria systems and subsystems, units of measure, values and initial weights of the project alternatives. Conceptual information is a more flexible and less accurate means for expressing estimates than numbers are.

Table 1. Sorted decision-making matrix for a multiple criteria analysis of alternatives

Quantitative project information									
Analysed criteria	*	Weight	Units	Analysed projects					
				a_1	a_2		a_j		a_n
Quantitative criteria	z_1	q_1	m_1	x_{11}	x_{12}		x_{1j}		x_{1n}
	z_2	q_2	m_2	x_{21}	x_{22}		x_{2j}		x_{2n}
	...	...	...	...	...		...		...
	z_i	q_i	m_i	x_{i1}	x_{i2}		x_{ij}		x_{in}
	...	...	...	...	...		...		...
	z_t	q_t	m_t	x_{t1}	x_{t2}		x_{tj}		x_{tn}
Qualitative criteria	z_{t+1}	q_{t+1}	m_{t+1}	$x_{t+1\,1}$	$x_{t+1\,2}$		$x_{t+1\,j}$		$x_{t+1\,n}$
	z_{t+2}	q_{t+2}	m_{t+2}	$x_{t+2\,1}$	$x_{t+2\,2}$		$x_{t+2\,j}$		$x_{t+2\,n}$
	...	...	...	...	...		...		...
	z_i	q_i	m_i	x_{i1}	x_{i2}		x_{ij}		x_{in}
	...	...	...	...	...		...		...
	z_m	q_m	m_m	x_{m1}	x_{m2}		x_{mj}		x_{mn}
Conceptual information pertinent to projects (i.e. text, graphics, video, virtual and augmented reality)									

Sign z_i (±) shows, respectively, the better/poorer value of the criterion of requirements for better stakeholders satisfaction.

This method assumes direct and proportional dependence of significance and priority by versions investigated on a system of criteria adequately describing the alternatives and on the values and weights of the criteria. A system of criteria is determined, and experts calculate the values and initial weights of criteria. All this information can be adjusted by interested parties considering their goal pursuits and existing capabilities. Hence the results of the assessment of alternatives fully reflect the initial project data that was jointly submitted by experts and interested parties. The determination of the significance and priority of alternatives is carried out in 5 stages.

Stage 1: The weighted, normalised decision-making matrix D is formed. The purpose of this stage is to receive dimensionless weighted values from the comparative indexes. When the dimensionless values of the indexes are known, all criteria, originally having different dimensions, can be compared. The following formula is used for this purpose:

$$d_{ij} = \frac{x_{ij} \cdot q_i}{\sum_{j=1}^{n} x_{ij}}, \quad i=\overline{1,m}; \quad j=\overline{1,n}. \tag{1}$$

where x_{ij} - the value of the i-th criterion in the j-th alternative of a solution; m - the number of criteria; n - the number of the alternatives compared; q_i - significance of i-th criterion.

The sum of dimensionless weighted index values d_{ij} of each criterion x_i is always equal to the significance q_i of this criterion:

$$q_i = \sum_{j=1}^{n} d_{ij}, \quad i=\overline{1,m}; j=\overline{1,n}. \tag{2}$$

In other words, the value of significance q_i of the investigated criterion is proportionally distributed among all alternative versions a_j according to their values x_{ij}.

Stage 2. The sums of weighted normalized indexes describing the j-th version are calculated. The versions are described by minimizing indexes S_{-j} and maximizing indexes S_{+j}. The lower value of minimizing indexes is better (price of the plot and building, etc.). The greater value of maximizing indexes is better (comfortability and aesthetics of the building, etc.). The sums are calculated according to the formula:

$$S_{+j} = \sum_{i=1}^{m} d_{+ij}; \quad S_{-j} = \sum_{i=1}^{m} d_{-ij}, \quad i=\overline{1,m}; j=\overline{1,n}. \tag{3}$$

In this case, the values S_{+j} (the greater is this value (project 'pluses'), the more satisfied are the interested parties) and S_{-j} (the lower is this value (project 'minuses'), the better is goal attainment by the interested parties) express the degree of goals attained by the interested parties in each alternative project. In any case the sums of 'pluses' S_{+j} and 'minuses' S_{-j} of all alternative projects are always respectively equal to all sums of significances of maximizing and minimizing criteria:

$$S_{+}=\sum_{j=1}^{n} S_{+j} = \sum_{i=1}^{m}\sum_{j=1}^{n} d_{+ij},$$
$$S_{-}=\sum_{j=1}^{n} S_{-j} = \sum_{i=1}^{m}\sum_{j=1}^{n} d_{-ij}, \quad i=\overline{1,m};\ j=\overline{1,n}. \tag{4}$$

Stage 3: The significance (efficiency) of the compared versions is determined by describing the characteristics of positive alternatives ("pluses") and negative alternatives ("minuses"). The relative significance Q_j of each alternative a_j is found according to the formula:

$$Q_j=S_{+j}+\frac{S_{-\min}\cdot\sum_{j=1}^{n} S_{-j}}{S_{-j}\cdot\sum_{j=1}^{n}\frac{S_{-\min}}{S_{-j}}}, \quad j=\overline{1,n}. \tag{5}$$

Stage 4: The priorities of the alternatives are determined. The greater is the Q_j the higher is the efficiency (priority) of the project alternative.

The analysis of the method presented makes it possible to state that it may be easily applied to evaluating projects and selecting the most efficient of them while being fully aware of the physical meaning of the process. Moreover, it allowed to formulate a reduced criterion Q_j which is directly proportional to the relative effect of the compared criteria values x_{ij} and significances q_i on the end result.

Significance Q_j of project a_j indicates the degree of satisfaction of demands and goals pursued by the interested parties — the greater is the Q_j the higher is the efficiency of the project (see Table 2).

Table 2. Results of the multiple criteria analysis of alternatives

Quantitative project information									
Analysed criteria	*	Weight	Units	Analysed projects					
				a_1	a_2		a_j		a_n
X_1	z_1	q_1	m_1	d_{11}	d_{12}		d_{1j}		d_{1n}
X_2	z_2	q_2	m_2	d_{21}	d_{22}		d_{2j}		d_{2n}
X_3	z_3	q_3	m_3	d_{31}	d_{32}		d_{3j}		d_{3n}
...	...	...	...	...	...		...		...
X_i	z_i	q_i	m_i	d_{i1}	d_{i2}		d_{ij}		d_{in}
...	...	...	...	...	...		...		...
X_m	z_m	q_m	m_m	d_{m1}	d_{m2}		d_{mj}		d_{mn}
Sum of maximising normalised rated indicators (project advantages)				S_{+1}	S_{+2}		S_{+j}		S_{+n}
Sum of minimising normalised rated indicators (project disadvantages)				S_{-1}	S_{-2}		S_{-j}		S_{-n}
Significance of the project alternative				Q_1	Q_2		Q_j		Q_n
Priority of the project alternative				P_1	P_2		P_j		P_n
Project's utility degree				N_1	N_2		N_j		N_n

Sign z_i (±) shows, respectively, the better/poorer value of the criterion of requirements for better stakeholders satisfaction.

Stage 5: The degree of project utility directly associates with its relevant quantitative and conceptual information. If one project is characterised as the best comfort, aesthetics and price indexes, while another is shown with better maintenance and facilities management characteristics, both will have obtained the same significance values as a result of the multiple criteria evaluation; this means their utility degree is also the same. With an increase (decrease) in the significance of the project analysed, the project's degree of utility also increases (decreases). The degree of project utility is determined by comparing the project analysed with the most efficient project. In this case, all the utility degree values related to the project analysed will range from 0% to 100%. This will facilitate a visual assessment of the project's efficiency.

The formula used for calculating alternative aj utility degree *Nj* is the following:

$$N_j = \left(Q_j : Q_{max}\right) \cdot 100\%. \quad (6)$$

here Q_j and Q_{max} are the significances of the property obtained from the equation 5.

4 Conclusions

Research object that is being analyzed by the authors of this present research may be described as a life cycle of the construction and real estate industry, the stakeholders involved and as the micro, meso and macro environments that have some particular impact on a life cycle in making an integral whole. A complex analysis of the formulated research object was performed with the help of Construction and Real Estate Crisis Management Model and Recommended System which were especially developed for this purpose.

References

1. Haslem, J.A., Scheraga, C.A., Bedingfield, J.P.: An analysis of the foreign and domestic balance sheet strategies of the U.S. banks and their association to profitability performance. Management International Review (1992)
2. Canbas, S., Cabuk, A., Kilic, S.B.: Prediction of commercial bank failure via multivariate statistical analysis of financial structures: the Turkish case. European Journal of Operational Research 166, 528–546 (2005)
3. Alam, P., Booth, D., Lee, K., Thordarson, T.: The use of fuzzy clustering algorithm and self-organizing neural networks for identifying potentially failing banks: an experimental study. Expert Systems with Applications 18(3), 185–199 (2000)
4. Tam, K.Y.: Neural network models and the prediction of bank bankruptcy. Omega: The International Journal of Management Science 19(5), 429–445 (1991)
5. Pasiouras, F., Gaganis, C., Zopounidis, C.: Multicriteria classification models for the identification of targets and acquirers in the Asian banking sector. European Journal of Operational Research 204(2), 328–335 (2010)
6. Niemira, M.P., Saaty, T.L.: An Analytic Network Process model for financial-crisis forecasting. International Journal of Forecasting 20(4), 573–587 (2004)

7. García, F., Guijarro, F., Moya, I.: Ranking Spanish savings banks: A multicriteria approach. Mathematical and Computer Modelling 52(7-8), 1058–1065 (2010)
8. Boyacioglu, M.A., Kara, Y., Baykan, O.K.: Predicting bank financial failures using neural networks, support vector machines and multivariate statistical methods: a comparative analysis in the sample of savings deposit insurance fund (SDIF) transferred banks in Turkey. Expert Systems with Applications 36(2), 3355–3366 (2009)
9. Frydman, H., Altman, E.I., Kao, D.: Introducing recursive partitioning for financial classification: the case of financial distress. Journal of Finance 40(1), 269–291 (1985)
10. Chen, X., Wang, X., Wu, D.D.: Credit risk measurement and early warning of SMEs: An empirical study of listed SMEs in China. Decision Support Systems 49(3), 301–310 (2010)
11. Gao, S., Xu, D.: Conceptual modeling and development of an intelligent agent-assisted decision support system for anti-money laundering. Expert Systems with Applications 36(2/1), 1493–1504 (2009)
12. Lin, S.L.: A new two-stage hybrid approach of credit risk in banking industry. Expert Systems with Applications 36(4), 8333–8341 (2009)
13. Siriopoulos, C., Tziogkidis, P.: How do Greek banking institutions react after significant events?—A DEA approach. Omega: The International Journal of Management Science 38(5), 294–308 (2010)
14. Kaklauskas, A.: Multiple criteria decision support system for building life cycle. Habilitation Work. Technika, Vilnius (1999)

Clinical Activity and Schedule Management with a Fuzzy Social Preference System

Pawel Wozniak, Tomasz Jaworski, Pawel Fiderek, Jacek Kucharski, and Andrzej Romanowski

Institute of Applied Computer Science, Lodz University of Technology
androm@kis.p.lodz.pl
http://ubicomp.pl

Abstract. This work covers the design of an inference system for hospital use by suggesting an automated approach to managing hospital schedules and surgical teams. The authors present a solution for advising head nurses on proper surgical team compositions taking the complex nature of social relationships into account. This paper illustrates an interdisciplinary attempt at solving a persistent problem within a clearly defined and unique environment. Knowledge from the fields of computational intelligence, fuzzy logic, ubiquitous computing, interaction design and operations research is utilised to aid medical professionals. An innovative use of fuzzy logic methods in a real-life application is proposed. The solution is evaluated through simulations.

1 Problem Background and Motivation

This paper focuses on assembling the best possible surgical team for a given procedure, especially in a case where surgeons of different specialities cooperate. The authors suggest using a fuzzy inference system for that purpose. The need for the system is a byproduct of cooperation with one of Poland's largest hospitals in Lodz that focused on improving cross-ward cooperation during emergency procedures dedicated to foetal disorders and birth defect prevention. The hospital specialises in obstetrics and pediatrics. The original goal of the solution dedicated to the improvement of these procedures involved designing a medical data sharing system with remote imaging data presentation, personnel selection, tracking and personalized data delivery, as depicted in [1]. During the user studies conducted for the overall system design, it was identified that the final stage of the emergency procedure, *i.e.* the cardio surgery of the infant, could be performed with greater efficiency if several of the surgery related issues were automated (operating theater schedules, personnel arrangement, *etc.*).

1.1 State of the Art

Until now, information within the institutions involved in the collaboration was exchanged using landline telephony and personal consults. The main problem of

N.T. Nguyen et al. (Eds.): *Adv. Methods for Comput. Collective Intelligence*, SCI 457, pp. 345–354.
DOI: 10.1007/978-3-642-34300-1_33

this arrangement was that the patient, a pregnant female, was diagnosed at one department, then the infant was delivered in a different clinic of the hospital. If the new born child was to be operated, the procedure was conducted at the pediatric surgery ward in yet another building. The child was transported via an underground tunnel. Therefore, from the moment of ultrasound or echocardiography diagnosis, the information about the anticipated surgery could be automatically propagated and properly processed using ICT. The reasons for a dedicated IT solution instead of standard systems are discussed in [1] along with detailed information on the technical design of the system and its use of context-awareness.

Optimising hospital resources and perfecting clinical schedules is a well analysed problem and several approaches have been reported so far. However, since hospitals are generally considered a safety-critical environment, introducing novel interactive clinical systems is usually subject to evaluation limitations [2]. Consequently, theory and controlled studies are of great significance in this matter — theoretical models have been developed.

Harper [3] focuses on limiting costs and maximising the use of infrastructure by using operations research to construct a complete model for hospital operations. This activity is aimed at helping in long-term decision making and not supporting day-to-day operations. Several other works claim to help elective surgery scheduling and operating theatre management. Devi *et al.* [4] aimed at predicting surgery times using a variety of factors and Khanna *et al.* [5] have designed an automated system for elective procedures.

As this work concerns social relationships which are by nature ambiguous, fuzzy logic appears to be the proper choice of modeling method. Fuzzy techniques plan an important role in the design of ambient itelligence and context-aware systems [6], [7], [8]. Some examples of successful application of fuzzy methods in context-aware systems can be found [9], also with the use of type-2 fuzzy sets for modeling of more complex imprecision of information [10]. Moreover, some recently published works (*e.g.* [11], [12] and [13]), prove that the fuzzy set theory and fuzzy logic is a very promising tool in the field of team building. The combination of these two approaches seems to establish a new possible field of development of intelligent context-aware systems. It is worth mentioning that fuzzy techniques, applied broadly in various engineering areas, have also been adopted by medical community [14].

2 System Description

The general solution facilitates notifying the institutions, assembling medical teams and exchanging opinions between physicians through context-aware technology [15], [1]. The principal requirement of the design of the system was to minimise the influence on the regular hospital procedures and possible alterations of physician habits, hence the proposed solution is an example of minimalist design.

Fig. 1 presents the architecture of the surgical team manager module. This tool helps the ward's administrative personnel (usually the head nurse) to decide on

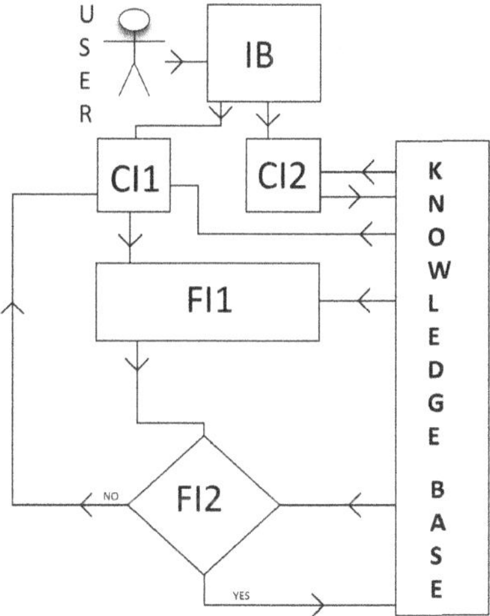

Fig. 1. An overview diagram of the surgical team manager

which surgeons, anaesthesiologists and assistants to notify so that the procedure is carried out in the most efficient manner. The system takes several aspects of the hospital environment into account, namely:

- the type of the surgery to be performed
- the number of specialised staff members required to perform the procedure
- the urgency of the surgery (includes planned and emergency procedures)
- the complexity of the procedure

Furthermore, the solution records and references activities within the unit and uses this information in the decision process. The data (denoted as *Knowledge base* in Fig. 1) includes: operating theatre occupancy, personnel workload, professional experience of the physicians, surgical specialities and interpersonal relations between staff members.

2.1 Description of the Decision Process

The system consists of several decision blocks which utilise both crisp rules and fuzzy inference. The overall composition can be seen in Fig. 1. In each step, the potential staff pool is reduced with a complete surgical team at the end of the process. The individual blocks are described below.

IB is an input block representing the information that needs to be explicitly provided by the head nurse. This includes the type of the surgery, the severity of the case and any special circumstances.

CI1 is the first crisp decision step where the entire staff pool is analysed for proper skills and availability. The *CI2* block aims at finding a free operating theatre suitable for the procedure and booking the room.

In *FI1*, the system limits the potential staff pool by comparing the complexity of the procedure with the experience of the staff thus assuring proper supervision by a consultant. *FI2* is responsible for making inferences based on the personal relation between the staff. The method used is described further in this chapter.

When a satisfactory surgical team is assembled, the suggestion is communicated to the user and the personnel can be notified by means of SMS messages or fixed-line telephony. A feedback loop is used in the design since the head nurse may want to alter the squad of the teams or the reasoning system may produce unsatisfactory results in the first instantiation (*e.g.* a fully competent team is assembled, yet there is a known rivalry between the chief surgeon and the chosen assistant). A given team member is eliminated form the process (by means of choosing the one whose schedule is most busy or experience most limited) and the reasoning procedure is repeated.

2.2 Modelling the Staff's Personal Preferences with Fuzzy Logic

Taking the surgical staff's interpersonal relations into account when selecting the proper surgical team is the key factor in the system. In this section, an approach to solving this issue by means of fuzzy logic is introduced. The idea of fuzzy logic, as defined by Zadeh [16] involves reasoning that is approximate rather then exact and precise.

In case of the system under consideration, degrees of truth are used to describe social relations between the potential members of the surgical team. For instance, let us assume that a social relation matrix is available for almost all potential team members. Such a matrix would contain a set of values that one could interpret as a personal preference relation between persons A and B, defined as $LIKE(A, B)$ with the following properties:

- if person A does not know person B then

$$LIKE(A, B) = 0 \tag{1}$$

- personal preferences are not always reciprocated and may be asymmetric

$$\exists(A, B \in Personnel) : LIKE(A, B) \neq LIKE(B, A) \tag{2}$$

A sample social preference relation matrix for an anonymised personnel group is presented in Table 1. This example assumes that the relation is defined on range from -50 (profound hostility) to 50 (declared friendship). To utilize the fuzzy logic approach, a the fuzzy variable $liking_F$ is defined with reference to the fuzzy sets (labels), depicted in Fig. 2.

At this point, every crisp social preference relation is expressed in terms of $liking_F$ (a fuzzy description). The following form is used:

$$liking^F(l) = \{\mu_{friend}(l); \mu_{acquaintance}(l); \mu_{neutral}(l); \mu_{reluctance}(l); \mu_{hostility}(l)\} \tag{3}$$

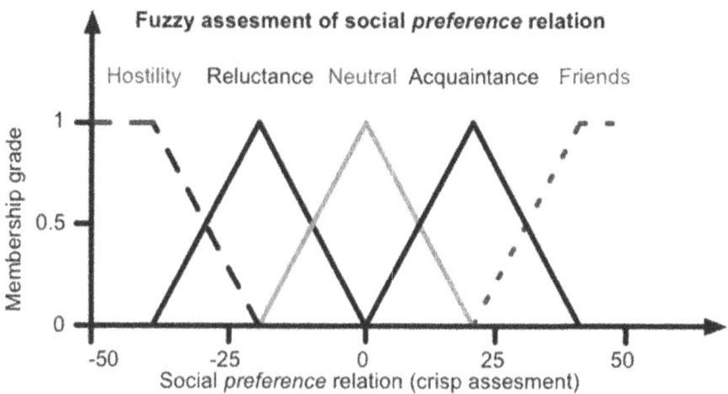

Fig. 2. The fuzzy variable $liking_F$ and its labels (linguistic variables)

Table 1. The social preference matrix for potential surgical team members

		Adam	Bronek	Celina	Darek	Edward	Filip	Gienia	Hanna	Idzi
Surgeons	Adam		20	10	-5	10	20	10	22	-5
	Bronek	35		20	25	20	40	30	-17	22
	Celina	-20	25		-20	20	35	25	-10	10
Anesth.	Darek	25	35	-15			12	5	-10	25
	Edward	-10	40	30			0	0	0	5
Assistant Staff	Filip	35	10	35	-10	0		10	15	30
	Gienia	15	30	0	0	0	0		10	-25
	Hanna	20	-20	5	30	20	10	0		
	Idzi	25	15	10	5	5	35	-20		

where μ_N is a characteristic function (*i.e.* the membership function) of the fuzzy set N and l is the $LIKE$ relation from Table 1.. With such a definition, every relation $LIKE(\circ)$ has its individual fuzzy description $liking^F$ which consists of 5 values.

For instance, let us consider two surgeons (Celina and Adam) with defined social preference relations $LIKE(Celina, Adam) = -20$ and $LIKE(Adam, Celina) = 10$. On the basis of Eq. 3 and Fig. 2, one can infer that Adam considers Celina a colleague($\mu_{neutral} = 0.5$; $\mu_{acquiantance} = 0.5$), but Celina is rather reluctant to him ($\mu_{neutral} = 0.5; \mu_{reluctant} = 0.5$).

When every personal preference relation in Table 1 can be expressed with fuzzy logic, a measure of the surgical team's *cooperative capability* can be derived. The measure the authors used is:

$$G^F(T) = \bigvee_{(A,B) \in P(T)} mutual^F_{liking^F}(A, B) \tag{4}$$

where T denotes the surgical team, (A, B) are pairs of team members, $P(T)$ is the set of all team member pairs of T. The fuzzy operator $\wedge$ was implemented in a classical way — by using max. Similarly, the operator $\vee$ (used below) utilises min. In Eq. 4 the cooparative capability of the team is expressed as a fuzzy sum of the relation $liking^F$. The new operator $mutual_R^F$ defined in Eq. 5 can be interpreted as a mutualisation of the relation R between persons A and B. The mutualisation is accomplished by means of selecting the most pessimistic version of the personal preference relation $liking^F$. In:

$$mutual^F_{liking^F}(A, B) = liking^F(min(LIKE(A, B), LIKE(B, A))) \tag{5}$$

min is not fuzzy operator, but an implementation of the worst case of the $liking$ relation. In the case of Celina and Adam (see Table 1), $mutual^F$ will be equal to:

$$mutual^F_{liking^F}(Celina, Adam) = liking^F(-20) = \left\{ \begin{array}{l} \mu_{friend} = 0 \\ \mu_{colleague} = 0; \\ \mu_{neutral} = 0.5 \\ \mu_{reluctance} = 0.5 \\ \mu_{hostility} = 0 \end{array} \right\} \tag{6}$$

The approach can be further extended by introducing weights to the relations between particular team members. A sample solution is presented in Fig. 3. It can be clearly seen that the personal relations between the lead surgeon and the practitioner is key while less importance is assigned to the personal preferences of the assisting staff. The presented weight are based on user studies performed by the authors. Eq. 4 can be extended with the use of relation weights by using:

$$G^F(T) = \left\{ \begin{array}{c} \mu_L \\ L \in LABELS(liking^F) \end{array} : \frac{\sum_{(A,B) \in P(T)} W(A, B) \mu_{mutual^F_{liking^F}(A,B),L}}{\sum_{(A,B) \in P(T)} W(A, B)} \right\} \tag{7}$$

where W(A,B) is the weight of the personal preference relation between A and B, assuming that $W(A, B) = W(B, A)$. Eq. 7 is a weighted average of the importance of the personal preference relation with weight defined in Fig. 3.

Finally, in order to obtain a scalar measure of the cooperative capability of the team the fuzzy quantity $G^F(T)$ must be defuzzyfied. The authors suggest using the Center of Gravity method for performing this task. The value obtained

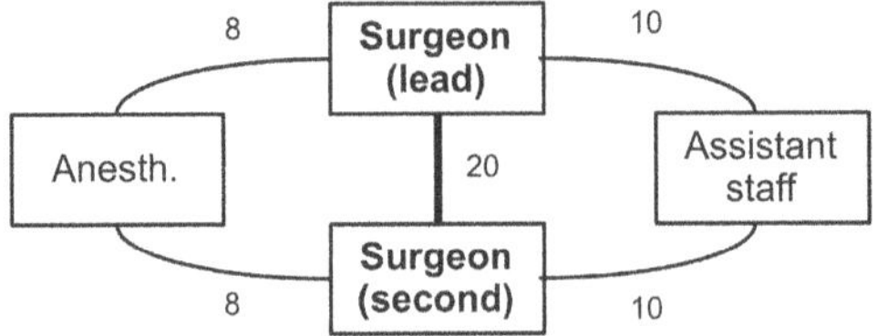

Fig. 3. Sample weight distribution of personal preference relations in a surgical team

will belong to the same interval as the values of the *liking* relation in Table 1. The lower the measure of the operational capability $G(T)$ of the team, the lower the probability of assembling such a team in practice should be.

3 Evaluation and Discussion

3.1 Simulations

Several simulations were performed in order to confirm the feasibility of the proposed solution. An experiment was conducted using a sample surgical team consisting of four members (two surgeons, an anaesthesiologist and an assisting staff member). The staff pool allocated of this task consisted of nine professionals (3 surgeons, 2 anaesthesiologist and 4 assisting staff). The data used in the simulation is the set presented in Table 1. The data was anonymised by changing the names of the staff members. The total number of possible surgical teams is compute using the following method:

The number of possible variations of each personnel type in the staff pool is calculated with:

$$V_{personnel}^{positions} = \frac{\mid personnel \mid!}{(\mid personnel \mid -positions)!} \tag{8}$$

where *positions* is the number of positions of a given time in a surgical team and *personnel* is the set of staff members that can perform a given task. Having obtained the count for all task types, the total number of teams is obtained using:

$$team\_count(T) = \prod_{p \in Positions(T)} V_{Personnel(p)}^{p} \tag{9}$$

where T denotes the surgical team, $Positions(T)$ is the set of positions that need to be filled and $Personnel(p)$ is the set of staff members that can perform task p.

3.2 Results

The simulation analysed 144 possible teams which were graded according the the suggested algorithm. The results are presented in Table 2. Considering the relationship matrix in Table 1 and the weights presented in Fig. 3, it can be observed that the choice of the most important team members, *i.e.* the surgeons is largely affected by their relationships with the other team members. Even though, in the first four cases, Bronek is the favoured member due to his good relations with Celina and Adam, in case 5 ($grade = 20.73867$) a team containing Celina and Adam was selected. The reason for that is their good relations with other team members, especially Edward and Filip.

Case 6 ($grade = -1.71573$) presents a borderline occurrence where both surgeons are antagonistic towards each other and their relations with the other team members are also far from optimal.

Table 2. Results of the system performance simulation. The first column contains the objecive measure of the team's usefulness (from −50 to 50). The number of teams with a given score is shown along with the names of the team members. A given column represent assigning a particular role to the staff member.

Grade	No. of teams	Sample team composition			
		Lead surgeon	Assisting surgeon	Anesthesiologist	Assisting staff
23.41	2	Bronek	Celina	Edward	Filip
		Celina	Bronek	Edward	Filip
22.42	2	Adam	Bronek	Darek	Filip
		Bronek	Adam	Darek	Filip
22.32	2	Bronek	Celina	Edward	Gienia
		Celina	Bronek	Edward	Gienia
21.71	2	Adam	Bronek	Darek	Idzi
		Bronek	Adam	Darek	Idzi
20.74	2	Adam	Celina	Edward	Filip
		Celina	Adam	Edward	Filip
-1.72	2	Adam	Celina	Darek	Gienia
		Celina	Adam	Darek	Gienia
-1.69	2	Adam	Celina	Darek	Idzi
		Celina	Adam	Darek	Idzi
0.52	2	Adam	Bronek	Edward	Hanna
		Bronek	Adam	Edward	Hanna
1.9	2	Adam	Bronek	Darek	Hanna
		Bronek	Adam	Darek	Hanna
3.67	2	Adam	Celina	Darek	Hanna
		Celina	Adam	Darek	Hanna

There are sixteen unique values in Table 2 and two teams have the highest possible score — *grade* = 23.41420. The choice between the two can be made using other factors (*e.g.* by comparing workloads) or the choice may be random.

3.3 Directions for Future Work

The authors believe that this obstacle can be overcome in the near future as electronic medical records (EMRs) have begun to pervade hospital environments worldwide [17]. With proper data mining techniques and assuming that access to a long history of past cases is available, most of the required information can be acquired computationally. Moreover, most efficient physician/assistant combinations can be identified so that the past success rate of surgical teams is considered in the selection process. This way, the subjective assessment of interpersonal relations that needs to be obtained via a survey or an interview can be substituted with an objective measure of how particular staff members cooperate to produce clinical results. Such a solution would also make the system resistant to a case in which it would rarely assign highly skilled professionals who work with each other efficiently despite personal disagreements to a case.

4 Conclusions

Understanding the peculiarities of an institution as large as the hospital in question is a complex task. The problem of choosing proper personnel for surgical procedures emerged while designing a holistic solution aimed increasing the operational efficiency of remote wards. The design presented in this paper illustrates that performance can be improved in some cases without redesigning the entire infrastructure. It can be concluded that social preferences play a crucial role in assembling surgical teams and may constitute an obstacle on road toward seamless functioning in a hospital. Owing to that, the authors suggest modeling the social relationships in a clinical environment as an intrinsic part of the decision-making process.

In this paper, the authors have illustrated how applied fuzzy logic can be used to increase the efficiency of day-to-day hospital resource management. Fuzzy modeling is used to help the administrators make interpersonal relations a part of their supervising effort. Simulation results prove the feasibility of the system design, but several new questions emerge. The evaluation of the system is largely limited due to practical constraints. Several implementation issues, as discussed in the paper, will have to be solved by more research.

References

1. Romanowski, A., Wozniak, P.: Medical imaging data sharing for emergency cases. In: Romanowski, A., Sankowski, D. (eds.) Computer Science in Novel Applications, pp. 281–307. Lodz Univeristy of Technology Press (2012)
2. Magrabi, F.: Using cognitive models to evaluate safety-critical interfaces in healthcare. In: CHI 2008 Extended Abstracts on Human Factors in Computing Systems, CHI EA 2008, pp. 3567–3572. ACM, New York (2008)
3. Harper, P.: A framework for operational modelling of hospital resources. Health Care Management Science 5, 165–173 (2002)
4. Devi, S.P., Rao, K.S., Sangeetha, S.S.: Prediction of surgery times and scheduling of operation theaters in optholmology department. J. Med. Syst. 36, 415–430 (2012)
5. Khanna, S., Cleaver, T., Sattar, A., Hansen, D., Stantic, B.: Multiagent Based Scheduling of Elective Surgery. In: Desai, N., Liu, A., Winikoff, M. (eds.) PRIMA 2010. LNCS, vol. 7057, pp. 74–89. Springer, Heidelberg (2012)
6. Acampora, G., Loia, V.: A proposal of ubiquitous fuzzy computing for ambient intelligence. Inf. Sci. 178, 631–646 (2008)
7. Ye, J., Dobson, S., McKeever, S.: Review: Situation identification techniques in pervasive computing: A review. Pervasive Mob. Comput. 8, 36–66 (2012)
8. Anagnostopoulos, C.B., Ntarladimas, Y., Hadjiefthymiades, S.: Situational computing: An innovative architecture with imprecise reasoning. J. Syst. Softw. 80, 1993–2014 (2007)
9. Zhou, S., Chu, C.H., Yu, Z., Kim, J.: A context-aware reminder system for elders based on fuzzy linguistic approach. Expert Syst. Appl. 39, 9411–9419 (2012)
10. Doctor, F., Hagras, H., Callaghan, V.: A type-2 fuzzy embedded agent to realise ambient intelligence in ubiquitous computing environments. Inf. Sci. 171, 309–334 (2005)

11. Topaloglu, S., Selim, H.: Nurse scheduling using fuzzy modeling approach. Fuzzy Sets Syst. 161, 1543–1563 (2010)
12. Wi, H., Oh, S., Mun, J., Jung, M.: A team formation model based on knowledge and collaboration. Expert Syst. Appl. 36, 9121–9134 (2009)
13. Strnad, D., Guid, N.: A fuzzy-genetic decision support system for project team formation. Appl. Soft Comput. 10, 1178–1187 (2010)
14. Sadegh-Zadeh, K.: The fuzzy revolution: Goodbye to the aristotelian weltanschauung. Artif. Intell. Med. 21, 1–25 (2001)
15. Bardram, J.E.: Applications of context-aware computing in hospital work: examples and design principles. In: Proceedings of the 2004 ACM Symposium on Applied Computing, SAC 2004, pp. 1574–1579. ACM, New York (2004)
16. Zadeh, L.A.: Fuzzy Sets, Fuzzy Logic, and Fuzzy Systems: Selected Papers by Lotfi A. Zadeh. World Scientific Publishing Co., Inc., River Edge (1996)
17. Hochheiser, H., Shneiderman, B.: Electronic medical records: usability challenges and opportunities. Interactions 18, 48–49 (2011)

Fuzzy $\overline{x}$ and s Charts: Left & Right Dominance Approach

Thanh-Lam Nguyen[1], Ming-Hung Shu[2], Ying-Fang Huang[2], and Bi-Min Hsu[3]

[1] Graduate Institute of Mechanical and Precision Engineering, National Kaohsiung University of Applied Sciences, Kaohsiung 80778, Taiwan
green4rest.vn@gmail.com
[2] Department of Industrial Engineering and Management, National Kaohsiung University of Applied Sciences, Kaohsiung 80778, Taiwan
[3] Department of Industrial Engineering and Management, Cheng Shiu University, Kaohsiung 83347, Taiwan

Abstract. Traditionally, variable control charts are constructed based on precise data collected from well-defined quality characteristics of manufacturing products. However, in the real world, there are many occasions that the stated quality characteristic of products, such as surface roughness of optical lens, contains certain degree of imprecise information which is called fuzzy data. As a result, the traditional method for constructing the variable control charts exist a limitation of dealing with fuzzy data. Therefore, in this paper, $\overline{x}$ and s control charts for fuzzy data are proposed. The fuzzy control limits are obtained on the basis of employing a well-known principle called resolution identity in the fuzzy theory. And in order to evaluate of the control charts, a fuzzy ranking method, named left and right dominance, is presented to classify the underlying manufacturing process condition. Finally, a practical example is provided to demonstrate the applicability of the proposed methodologies.

Keywords: Fuzzy numbers, Fuzzy control chart, Left and Right Dominance.

1 Introduction

A control chart, one of the major tools in Statistical process control (SPC), is widely used to monitor and examine manufacturing processes as well as ultimately control product quality. Its power lies in the ability to detect process shifts and identify abnormal conditions in the on-line manufacturing. It reduces the variability of the process to achieve high stability and improvement in productivity and capability which results in not only the decreased manufacturing costs due to the low percentage of nonconformities and defects but also the increase in customer satisfaction [1-2].

Typically, a control chart consists of a center line (CL) which is the estimated process target value, and two control lines- the upper control limit (UCL) and lower control limit (LCL) which are the boundaries of the normal variability used to test if the majority of the observations are in control. The UCL and LCL are usually set at a distance of ± three-sigma away from the CL. From the control chart, if the entire statistic points of collected sample data fall within the limits, the process is classified as in statistical control; and if any of the points out-lies the limits, the process is said to be affected by assignable causes to be investigated for a good solution [1].

N.T. Nguyen et al. (Eds.): *Adv. Methods for Comput. Collective Intelligence*, SCI 457, pp. 355–366.
DOI: 10.1007/978-3-642-34300-1_34

The data collected should represent the various levels of the quality characteristics associated with the product. The characteristics might be measurable on numerical scales; and each of them is called a variable. When dealing with a variable, it is usually necessary to monitor both its mean value and variability. The process mean value is normally controlled under the control chart for means, called $\bar{x}$ -chart; whereas its variability can be monitored with control chart for standard deviation, called s-chart.

The traditional control charts are constructed for precise data. However, in the real world, the data sometimes cannot be recorded or collected precisely which is known as fuzzy data. Therefore, the traditional control charts are to be expanded to deal with the fuzzy data [2]. In manufacturing process with fuzzy data, a few researchers, Shu & Wu [2], Gülbay & Kahraman [3], Gülbay et al. [4], Kanagawa et al. [5], and Wang & Raz [6-7], have suggested constructing the fuzzy control charts based on the fuzzy set theory. Shu & Wu pointed out the limitations on the approaches proposed by the other researchers before suggesting their approach called fuzzy dominance to verify if the fuzzy mean and variance of each subgroup fuzzy data lie within their respective fuzzy control limits to evaluate the process condition in the fuzzy environment [2]. In this paper, another approach for the evaluation is proposed based on the ranking fuzzy numbers with the left and right dominance offered by Chen & Lu [8]. And an empirical study on the turning process in producing optical lens in Taiwan is conducted as an illustration for this approach.

2 Control Charts for Real-Valued Data

We now briefly review the development of the equations for constructing the control limits on the $\bar{x}$ and s control charts. Let x be the key quality characteristic following a normal distribution $N(\mu, \sigma^2)$, where parameters μ and σ are usually unknown in practice, meaning that they must be estimated by using the sample data, at least 20 to 25 samples taken from a process thought to be it control. Assume that we have collected the sample data $A = \left\{ A_i \middle| A_i = (x_{i1}, x_{i2}, \ldots, x_{in}), i = \overline{1,m} \right\}$ which consist of m subgroups and each subgroup contains n observations of x.

2.1 $\bar{x}$ Control Chart

We first review the $\bar{x}$ control chart. Let $\bar{x}_i$ be the average of the ith sample.

$$\bar{x}_i = \frac{1}{n} \sum_{j=1}^{n} x_{ij} \text{. for } i = \overline{1,m} \tag{1}$$

Then the process average, the best estimator of μ , is the grand average

$$\bar{\bar{x}} = \frac{1}{m} \sum_{i=1}^{m} \bar{x}_i . \tag{2}$$

The sample mean $\bar{x}$ and variance s^2 are respectively unbiased estimators of the population mean μ and variance σ^2 . Or, with the expected value operator E, we have

$$E(\overline{x}) = \mu. \text{ and } E(s^2) = \sigma^2. \tag{3}$$

However, the sample standard deviation s is not an unbiased estimator of the population standard deviation σ. It can be shown that

$$E(s) = (\frac{2}{n-1})^{1/2} \frac{\Gamma(n/2)}{\Gamma[(n-1)/2]} \sigma = c_4 \sigma. \tag{4}$$

The values of c_4 for sample sizes $2 \le n \le 25$ are available in the textbook [1].

Let s_i is the standard deviation of the ith sample.

$$s_i = \sqrt{\frac{\sum_{i=1}^{m} (\overline{x}_i - \overline{x})^2}{m-1}}. \tag{5}$$

The average of the m standard deviations is:

$$\overline{s} = \frac{1}{m} \sum_{i=1}^{m} s_i. \tag{6}$$

Consequently, from Eq. (4) and (6), an unbiased estimator $\hat{\sigma}$ of σ, is given by:

$$\hat{\sigma} = \overline{s} / c_4. \tag{7}$$

The $\overline{x}$-chart with the usual three-sigma control limits are given by

$$UCL_{\overline{x}} = \overline{x} + 3\overline{s} / c_4 \sqrt{n} \qquad CL_{\overline{x}} = \overline{x} \qquad LCL_{\overline{x}} = \overline{x} - 3\overline{s} / c_4 \sqrt{n} \tag{8}$$

2.2 *s* Control Chart

The parameters of the s chart are determined by [1]:

$$UCL_s = \overline{s} + 3\overline{s}\sqrt{1-c_4^2} / c_4 \qquad CL_s = \overline{s} \qquad LCL_s = \overline{s} - 3\overline{s}\sqrt{1-c_4^2} / c_4 \tag{9}$$

3 Fuzzy Numbers

Definition 1 (Fuzzy set) [9]**.** The fuzzy subset $\tilde{a}$ of $\Re$ is defined by a function $\xi_{\tilde{a}} : \Re \to [0,1]$. $\xi_{\tilde{a}}$ is called a membership function.

Definition 2 (α-cut set) [9]**.** The α-cut set of $\tilde{a}$, denoted by $\tilde{a}_{\alpha c}$, is defined by $\tilde{a}_{\alpha c} = \{x \in \Re : \xi_{\tilde{a}}(x) \ge \alpha\}$ for all $\alpha \in (0,1]$. The 0-cut set $\tilde{a}_{0c}$ is defined as the closure of the set $\{x \in \Re : \xi_{\tilde{a}} > 0\}$.

Definition 3 (α-level set) [9]**.** The α-level set of $\tilde{a}$, denoted by $\tilde{a}_\alpha$, is defined by $\tilde{a}_\alpha = \{x \in \Re : \xi_{\tilde{a}}(x) = \alpha\}$ for all $\alpha \in [0,1]$. Consequently, the α-level set $\tilde{a}_\alpha$ is a usual set which is written in the form of $\tilde{a}_\alpha = \left[\tilde{a}_\alpha^L, \tilde{a}_\alpha^U\right]$.

Definition 4 (Convex set) [2]. A fuzzy set $\tilde{a}$ is said to be convex if the following relation holds true:

$$\xi_{\tilde{a}}(\lambda x_1 + (1-\lambda)x_2) \geq \min(\xi_{\tilde{a}}(x_1), \xi_{\tilde{a}}(x_2)) \text{, for } \lambda \in [0,1] \text{ and } x_1, x_2 \in \Re .$$

Definition 5 (Normalized set) [9]. A fuzzy subset $\tilde{a}$ is said to be normalized if there exists $x \in \Re$ such that $\xi_{\tilde{a}}(x) = 1$.

Definition 6 (Fuzzy number) [2]. A fuzzy number $\tilde{a}$ is a convex normalized fuzzy subset of the real line $\Re$ with an upper semi-continuous membership function of bounded support. Its membership function $\xi_{\tilde{a}}(x)$ is said to be upper semi-continuous if $\tilde{a}_\alpha = \{x \in \Re : \xi_{\tilde{a}}(x) \geq \alpha\}$ is a closed subset of $\Re$ for each $\alpha \in (0,1]$.

Definition 7 (Triangular fuzzy number) [9]. A Fuzzy number $\tilde{a}$ is called a triangular fuzzy number, donated by $\tilde{a} = (a_1, a_2, a_3)$ if its membership function $\xi_{\tilde{a}}(x)$ has the following form:

$$\xi_{\tilde{a}}(x) = \begin{cases} (x-a_1)/(a_2-a_1) & if \quad a_1 \leq x \leq a_2, \\ (a_3-x)/(a_3-a_2) & if \quad a_2 \leq x \leq a_3, \\ 0 & otherwise, \end{cases}$$

The triangular fuzzy number $\tilde{a}$ can be expressed as "around a_2" or "being approximately equal to a_2". a_2 is called the core value of $\tilde{a}$, a_1 and a_3 are called the left and right spread value of $\tilde{a}$, respectively.

Remark 1. Let $\tilde{a}$ be a fuzzy number; then, the following statements hold true:

(i) $\tilde{a}_\alpha^L \leq \tilde{a}_\alpha^U$ for all $\alpha \in [0,1]$

(ii) $\tilde{a}_\alpha^L$ is increasing with respect to α on $[0,1]$; that is $\tilde{a}_\alpha^L \leq \tilde{a}_\beta^L$ for $\alpha < \beta$.

(iii) $\tilde{a}_\alpha^U$ is increasing with respect to α on $[0,1]$; that is $\tilde{a}_\alpha^U \geq \tilde{a}_\beta^U$ for $\alpha < \beta$.

If $\tilde{a}$ is a fuzzy number, $l(\alpha) = \tilde{a}_\alpha^L$ and $u(\alpha) = \tilde{a}_\alpha^U$ are continuous functions on $[0,1]$, then $\tilde{a}$ is a fuzzy real number.

Remark 2 (Resolution identity). Let $\tilde{a}$ be a fuzzy subset of $\Re$ with membership function $\xi_{\tilde{a}}$. Then, the membership function can be expressed as the following form:

$$\xi_{\tilde{a}}(x) = \sup_{\alpha \in [0,1]} \alpha . 1_{\tilde{a}_\alpha}(x)$$

where $1_{\tilde{a}_\alpha}(x)$ is the indicator function of set $\tilde{a}_\alpha$.

4 Fuzzy Control Charts for Fuzzy Data

Let $\tilde{x}_{i1}, \ldots, \tilde{x}_{in}$ be fuzzy observations (fuzzy data) $\left(i = \overline{1,m}\right)$, which are assumed to be fuzzy real numbers. Obviously, upper and lower control limits are needed to be constructed by using these fuzzy data. For any given $\alpha \in [0,1]$, we can obtain the corresponding real-valued data $(\tilde{x}_{ij})_\alpha^L$ and $(\tilde{x}_{ij})_\alpha^U$ for $i = \overline{1,m}$ and $j = \overline{1,n}$ based on the Eqs.(1)-(7), which results in the followings:

$$\overline{x}_{i;\alpha}^K = \frac{1}{n}\sum_{j=1}^{n}(\tilde{x}_{ij})_\alpha^K \qquad \overline{\overline{x}}_\alpha^K = \frac{1}{m}\sum_{i=1}^{m}\overline{x}_{i;\alpha}^K$$

$$\overline{s}_{i;\alpha}^K = \sqrt{\frac{\sum_{j=1}^{n}\left((\tilde{x}_{ij})_\alpha^K - \overline{x}_{i;\alpha}^K\right)^2}{n-1}} \qquad \overline{s}_\alpha^K = \sqrt{\frac{\sum_{i=1}^{m}\left(\overline{x}_{i;\alpha}^K - \overline{\overline{x}}_\alpha^K\right)^2}{m-1}} \tag{10}$$

where K is either U or L.

4.1 Fuzzy $\overline{x}$ Control Chart

Substituted with the fuzzy results in Eq.(10), the parameters for the fuzzy $\overline{x}$ control chart in Eq.(8) are obtained by:

$$u_{\overline{x};\alpha}^U \equiv (ucl_{\overline{x}})_\alpha^U = \overline{\overline{x}}_\alpha^U + \lambda\overline{s}_\alpha^U \qquad c_{\overline{x};\alpha}^U \equiv (cl_{\overline{x}})_\alpha^U = \overline{\overline{x}}_\alpha^U \qquad l_{\overline{x};\alpha}^U \equiv (lcl_{\overline{x}})_\alpha^U = \overline{\overline{x}}_\alpha^U - \lambda\overline{s}_\alpha^U \text{ and}$$

$$u_{\overline{x};\alpha}^L \equiv (ucl_{\overline{x}})_\alpha^L = \overline{\overline{x}}_\alpha^L + \lambda\overline{s}_\alpha^L \qquad c_{\overline{x};\alpha}^L \equiv (cl_{\overline{x}})_\alpha^L = \overline{\overline{x}}_\alpha^L \qquad l_{\overline{x};\alpha}^L \equiv (lcl_{\overline{x}})_\alpha^L = \overline{\overline{x}}_\alpha^L - \lambda\overline{s}_\alpha^L \tag{11}$$

where $\lambda = 3 / c_4\sqrt{n}$

Construct the Fuzzy Upper Control Limit $\tilde{u}_{\overline{x}}$

By using the results of Eq. (11), let's consider a closed interval A_α which is defined as:

$$A_\alpha = \left[\min\left\{u_{\overline{x};\alpha}^L, u_{\overline{x};\alpha}^U\right\}, \max\left\{u_{\overline{x};\alpha}^L, u_{\overline{x};\alpha}^U\right\}\right] \equiv [l(\alpha), u(\alpha)]$$

where $l(\alpha) = \min\left\{u_{\overline{x};\alpha}^L, u_{\overline{x};\alpha}^U\right\}$ and $u(\alpha) = \max\left\{u_{\overline{x};\alpha}^L, u_{\overline{x};\alpha}^U\right\}$. (12)

According to the result of the resolution identity, presented in Remark 2, the membership function of the fuzzy upper control limit can be defined as

$$\xi_{\tilde{u}_{\overline{x}}}(c) = \sup_{\alpha\in[0,1]} \alpha \cdot 1_{A_\alpha}(c) \tag{13}$$

Since each $\tilde{x}_{ij}$ is a fuzzy real number, $(\tilde{x}_{ij})_\alpha^L$ and $(\tilde{x}_{ij})_\alpha^U$ are continuous with respect to α on $[0,1]$, saying that $\overline{\overline{x}}_\alpha^L$, $\overline{s}_\alpha^L$, $\overline{\overline{x}}_\alpha^U$, and $\overline{s}_\alpha^U$ are continuous with respect to α on $[0,1]$. Under these facts, the α-level set $(\tilde{u}_{\overline{x}})_\alpha$ of fuzzy upper control limit $\tilde{u}_{\overline{x}}$ can be simply written as

$$(\tilde{u}_{\overline{x}})_\alpha = \left\{c : \xi_{\tilde{u}_{\overline{x}}}(c) \geq \alpha\right\} = \left[\min_{\alpha\leq\beta\leq1} l(\beta), \max_{\alpha\leq\beta\leq1} u(\beta)\right] = \left[(\tilde{u}_{\overline{x}})_\alpha^L, (\tilde{u}_{\overline{x}})_\alpha^U\right] \tag{14}$$

where $l(\beta)$ and $u(\beta)$ are shown in Eq. (12).

Furthermore, from Eq. (14), the relationship between $(\tilde{u}_{\bar{x}})^L_\alpha$ and $\tilde{u}^L_{\bar{x};\alpha}$ is

$$(\tilde{u}_{\bar{x}})^L_\alpha = \min_{\alpha\le\beta\le1} l(\beta) = \min_{\alpha\le\beta\le1} \min\left\{ u^L_{\bar{x};\beta}, u^U_{\bar{x};\beta} \right\}. \tag{15}$$

Similarly, the relationship between $(\tilde{u}_{\bar{x}})^U_\alpha$ and $\tilde{u}^U_{\bar{x};\alpha}$ is

$$(\tilde{u}_{\bar{x}})^U_\alpha = \min_{\alpha\le\beta\le1} l(\beta) = \max_{\alpha\le\beta\le1} \max\left\{ u^L_{\bar{x};\beta}, u^U_{\bar{x};\beta} \right\}. \tag{16}$$

Construct the Fuzzy Lower Control Limit $\tilde{l}_{\bar{x}}$

Similarly, from Eq. (11), let's consider the closed interval B_α which is defined as:

$$B_\alpha = \left[\min\left\{ l^L_{\bar{x};\alpha}, l^U_{\bar{x};\alpha} \right\}, \max\left\{ l^L_{\bar{x};\alpha}, l^U_{\bar{x};\alpha} \right\} \right]$$

The membership function of the fuzzy lower control limit $\tilde{l}_{\bar{x}}$, can be defined as

$$\xi_{\tilde{l}_{\bar{x}}}(c) = \sup_{0\le\alpha\le1} \alpha \cdot 1_{B_\alpha}(c)$$

Therefore, the endpoints of the α-level closed interval of the fuzzy lower control limit $\tilde{l}_{\bar{x}}$ are determined by $(\tilde{l}_{\bar{x}})_\alpha = \left[(\tilde{l}_{\bar{x}})^L_\alpha, (\tilde{l}_{\bar{x}})^U_\alpha \right]$

where $$(\tilde{l}_{\bar{x}})^L_\alpha = \min_{\alpha\le\beta\le1} \min\left\{ l^L_{\bar{x};\beta}, l^U_{\bar{x};\beta} \right\} \text{ and } (\tilde{l}_{\bar{x}})^U_\alpha = \max_{\alpha\le\beta\le1} \max\left\{ l^L_{\bar{x};\beta}, l^U_{\bar{x};\beta} \right\} \tag{17}$$

4.2 Fuzzy *s* Control Chart

Substituted with the fuzzy results in Eq.(10), the parameters for the fuzzy s control chart in Eq.(9) are obtained by:

$$u^U_{s;\alpha} \equiv (ucl_s)^U_\alpha = \delta \overline{s}^U_\alpha \qquad c^U_{s;\alpha} \equiv (cl_s)^U_\alpha = \overline{s}^U_\alpha \qquad l^U_{s;\alpha} \equiv (lcl_s)^U_\alpha = \gamma \overline{s}^U_\alpha \qquad \text{and}$$

$$u^L_{s;\alpha} \equiv (ucl_s)^L_\alpha = \delta \overline{s}^L_\alpha \qquad c^L_{s;\alpha} \equiv (cl_s)^L_\alpha = \overline{s}^L_\alpha \qquad l^L_{s;\alpha} \equiv (lcl_s)^L_\alpha = \gamma \overline{s}^L_\alpha \tag{18}$$

where $\delta = 1 + 3\sqrt{1-c_4^2}/c_4$ and $\gamma = 1 - 3\sqrt{1-c_4^2}/c_4$.

The construction of the fuzzy control limits for fuzzy s control chart is done the same as that for fuzzy $\bar{x}$ control chart. The results are summarized as the followings:

+ The endpoints of the α-level closed interval $(\tilde{u}_s)_\alpha = \left[(\tilde{u}_s)^L_\alpha, (\tilde{u}_s)^U_\alpha \right]$ of fuzzy upper control limit $\tilde{u}_s$ are:

$$(\tilde{u}_s)^L_\alpha = \min_{\alpha\le\beta\le1} \min\left\{ u^L_{s;\beta}, u^U_{s;\beta} \right\} \text{ and } (\tilde{u}_s)^U_\alpha = \max_{\alpha\le\beta\le1} \max\left\{ u^L_{s;\beta}, u^U_{s;\beta} \right\}. \tag{19}$$

+ The endpoints of the α-level closed interval $(\tilde{l}_s)_\alpha = \left[(\tilde{l}_s)^L_\alpha, (\tilde{l}_s)^U_\alpha \right]$ of fuzzy lower control limit $\tilde{l}_s$ are:

$$(\tilde{l}_s)^L_\alpha = \min_{\alpha\le\beta\le1} \min\left\{ l^L_{s;\beta}, l^U_{s;\beta} \right\} \text{ and } (\tilde{l}_s)^U_\alpha = \max_{\alpha\le\beta\le1} \max\left\{ l^L_{s;\beta}, l^U_{s;\beta} \right\} \tag{20}$$

Remark 3. In order to realize whether the fuzzy average $\tilde{\bar{x}}_i$ and fuzzy mean standard deviation $\tilde{\bar{s}}_i$ are within the fuzzy control limits, we need to calculate their membership functions as the following:

Consider the closed interval C_α, which is defined as:

$$C_\alpha = \left[\min\left\{ \bar{x}_{i;\alpha}^L, \bar{x}_{i;\alpha}^U \right\}, \max\left\{ \bar{x}_{i;\alpha}^L, \bar{x}_{i;\alpha}^U \right\} \right]$$

The membership function of $\tilde{\bar{x}}_i$ is defined as:

$$\xi_{\tilde{\bar{x}}_i}(c) = \sup_{0 \le \alpha \le 1} \alpha \cdot 1_{C_\alpha}(c)$$

Therefore, the endpoints of the α-level closed interval $(\tilde{\bar{x}})_\alpha = \left[(\tilde{\bar{x}})_\alpha^L, (\tilde{\bar{x}})_\alpha^U \right]$ of fuzzy average $\tilde{\bar{x}}_i$ are:

$$(\tilde{\bar{x}})_\alpha^L = \min_{\alpha \le \beta \le 1} \min\left\{ \bar{x}_{i;\beta}^L, \bar{x}_{i;\beta}^U \right\} \text{ and } (\tilde{\bar{x}})_\alpha^U = \max_{\alpha \le \beta \le 1} \max\left\{ \bar{x}_{i;\beta}^L, \bar{x}_{i;\beta}^U \right\} \quad (21)$$

Similarly, the endpoints of the α-level closed interval $(\tilde{\bar{s}})_\alpha = \left[(\tilde{\bar{s}})_\alpha^L, (\tilde{\bar{s}})_\alpha^U \right]$ of fuzzy mean standard deviation $\tilde{\bar{s}}_i$ are:

$$(\tilde{\bar{s}})_\alpha^L = \min_{\alpha \le \beta \le 1} \min\left\{ \bar{s}_{i;\beta}^L, \bar{s}_{i;\beta}^U \right\} \text{ and } (\tilde{\bar{s}})_\alpha^U = \max_{\alpha \le \beta \le 1} \max\left\{ \bar{s}_{i;\beta}^L, \bar{s}_{i;\beta}^U \right\} \quad (22)$$

5 Left and Right Dominance Approach

In order to verify if the fuzzy mean and variance of each subgroup fuzzy data lie within their respective fuzzy control limits to evaluate the process condition in the fuzzy environment, ranking fuzzy numbers is usually used. There have been some different methods for the ranking suggested; and in this paper, the left and right dominance approach offered by Chen & Lu [8] is used.

For a fuzzy number $\tilde{a}$, the α-level sets $\tilde{a}_\alpha = \left\{ x \in \Re \middle| \xi_{\tilde{a}}(x) \ge \alpha \right\}, \alpha \in [0,1]$ are convex subsets of $\Re$. The lower and upper limits of the kth α-level set for the fuzzy number $\tilde{a}$ are respectively defined as

$$l_{i,k} = \inf_{x \in \Re} \left\{ x \middle| \xi_{\tilde{a}}(x) \ge \alpha_k \right\} \quad \text{and} \quad r_{i,k} = \sup_{x \in \Re} \left\{ x \middle| \xi_{\tilde{a}}(x) \ge \alpha_k \right\} \quad (23)$$

where $l_{i,k}$ and $r_{i,k}$ are respectively left and right spreads.

The left dominance $D^L(\tilde{a}_i, \tilde{a}_j)$ of $\tilde{a}_i$ over $\tilde{a}_j$ is defined as the average difference of the left spreads at some α-level sets and right dominance $D^R(\tilde{a}_i, \tilde{a}_j)$ of $\tilde{a}_i$ over $\tilde{a}_j$ is similarly defined as the average difference of the right spreads at some α-level sets. The formulas to calculate the dominances are as the following:

$$D^L\left(\tilde{a}_i,\tilde{a}_j\right)=\frac{1}{t+1}\sum_{k=0}^{t}\left(l_{i,k}-l_{j,k}\right) \quad \text{and} \quad D^R\left(\tilde{a}_i,\tilde{a}_j\right)=\frac{1}{t+1}\sum_{k=0}^{t}\left(r_{i,k}-r_{j,k}\right) \tag{24}$$

where t+1 is the number of α -level sets used to calculate the dominance.

At the index of optimism $\beta\in[0,1]$, the total dominance $D_\beta\left(\tilde{a}_i,\tilde{a}_j\right)$ of $\tilde{a}_i$ over $\tilde{a}_j$ with can be obtained by:

$$D_\beta\left(\tilde{a}_i,\tilde{a}_j\right)=\frac{1}{t+1}\left[\left(\beta\sum_{k=0}^{t}r_{i,k}+(1-\beta)\sum_{k=0}^{t}l_{i,k}\right)-\left(\beta\sum_{k=0}^{t}r_{j,k}+(1-\beta)\sum_{k=0}^{t}l_{j,k}\right)\right] \tag{25}$$

$D_\beta\left(\tilde{a}_i,\tilde{a}_j\right)$ indicates that it is in fact a comparison function and the right dominance is more important than the left one if the index of optimism β is greater than 0.5 and vice versa. Therefore, β is used to reflect a decision-maker's degree of optimism. In ranking fuzzy numbers, the following rules are used:

(i) If $D_\beta\left(\tilde{a}_i,\tilde{a}_j\right)>0$, then $\tilde{a}_i>\tilde{a}_j$;

(ii) If $D_\beta\left(\tilde{a}_i,\tilde{a}_j\right)=0$, then $\tilde{a}_i=\tilde{a}_j$; and,

(iii) If $D_\beta\left(\tilde{a}_i,\tilde{a}_j\right)<0$, then $\tilde{a}_i<\tilde{a}_j$.

Remark 4. For any *i, j* and β, the following statements hold true:

(i) $D_\beta\left(\tilde{a}_i,\tilde{a}_j\right)=-D_\beta\left(\tilde{a}_j,\tilde{a}_i\right)$

(ii) $D_\beta\left(\tilde{a}_i,\tilde{a}_j\right)=D_\beta\left(\tilde{a}_i,\tilde{a}_t\right)+D_\beta\left(\tilde{a}_t,\tilde{a}_j\right)=D_\beta\left(\tilde{a}_i,\tilde{a}_t\right)-D_\beta\left(\tilde{a}_j,\tilde{a}_t\right)$

(iii) If $D_\beta\left(\tilde{a}_i,\tilde{a}_j\right)>0$ and $D_\beta\left(\tilde{a}_j,\tilde{a}_t\right)>0$, then $D_\beta\left(\tilde{a}_i,\tilde{a}_t\right)>0$.

Remark 5. At a certain value of the index of optimism $\beta>0.5$, based on the fuzzy standard deviation, the manufacturing process can be categorized as the following:

(i) The process is in-control if $D_\beta\left(\tilde{u}_s,\tilde{\bar{s}}_i\right)>0$ and $D_\beta\left(\tilde{\bar{s}}_i,\tilde{l}_s\right)>0$

(ii) The process is out-of-control if $D_\beta\left(\tilde{u}_s,\tilde{\bar{s}}_i\right)<0$ or $D_\beta\left(\tilde{\bar{s}}_i,\tilde{l}_s\right)<0$

(iii) The process is rather out-of-control if $D_\beta\left(\tilde{u}_s,\tilde{\bar{s}}_i\right)=0$ or $D_\beta\left(\tilde{\bar{s}}_i,\tilde{l}_s\right)=0$

(iv) The process is rather in-control if one of the two following situations happens:

$$\left(D_\beta\left(\tilde{u}_s,\tilde{\bar{s}}_i\right)=0 \text{ and } D_\beta\left(\tilde{\bar{s}}_i,\tilde{l}_s\right)>0\right) \text{ or } \left(D_\beta\left(\tilde{u}_s,\tilde{\bar{s}}_i\right)>0 \text{ and } D_\beta\left(\tilde{\bar{s}}_i,\tilde{l}_s\right)=0\right)$$

Remark 6. At a certain value of the index of optimism $\beta>0.5$, based on the fuzzy average, the manufacturing process can be categorized as the following:

(i) The process is in-control if $D_\beta\left(\tilde{u}_{\bar{x}},\tilde{\bar{x}}_i\right)>0$ and $D_\beta\left(\tilde{\bar{x}}_i,\tilde{l}_{\bar{x}}\right)>0$

(ii) The process is out-of-control if $D_\beta\left(\tilde{u}_{\bar{x}}, \tilde{\bar{x}}_i\right) < 0$ or $D_\beta\left(\tilde{\bar{x}}_i, \tilde{l}_{\bar{x}}\right) < 0$

(iii) The process is rather out-of-control if $D_\beta\left(\tilde{u}_{\bar{x}}, \tilde{\bar{x}}_i\right) = 0$ or $D_\beta\left(\tilde{\bar{x}}_i, \tilde{l}_{\bar{x}}\right) = 0$

(iv) The process is rather in-control if one of the two following situations happens:

$$\left(D_\beta\left(\tilde{u}_{\bar{x}}, \tilde{\bar{x}}_i\right) = 0 \text{ and } D_\beta\left(\tilde{\bar{x}}_i, \tilde{l}_{\bar{x}}\right) > 0\right) \text{ or } \left(D_\beta\left(\tilde{u}_{\bar{x}}, \tilde{\bar{x}}_i\right) > 0 \text{ and } D_\beta\left(\tilde{\bar{x}}_i, \tilde{l}_{\bar{x}}\right) = 0\right)$$

6 Practical Example

In recent years, optical lenses have become key components in many industrial products, such as digital cameras. On a lens, it is usually desirable to have as little roughness on a lens surface as possible so that light is scattered as little as possible. Surface roughness is one of the most important factors in evaluating the quality of a lens. Moreover, roughness is usually considered a good measurement of the performance of a mechanical component, since irregularities on the surface may form nucleation sites for cracks or corrosion [10]. Profiles of roughness and waviness are shown in Fig. 1.

The value of surface roughness cannot be recorded or collected precisely due to human error or some unexpected situations; as a result, the measured value should be regarded as a fuzzy number. Roughness is measured by the size or height of the irregularities with respect to an average line as depicted in Fig. 2.

In this study, fuzzy surface roughness index is denoted by a triangular fuzzy number. Twenty-five samples, each of size ten, have been measured for their roughness height in the manufacturing process. Based on these data, the fuzzy control limits and the control charts are processed with computer software called Matlab. The fuzzy control limits for s-chart and $\bar{x}$ -chart are shown in Table 1 and Table 2, respectively; whereas the fuzzy s control chart and $\bar{x}$ -chart are shown in Fig. 3 and Fig. 4.

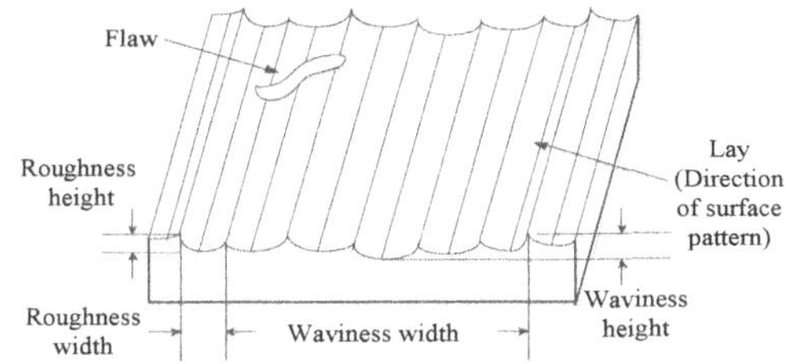

Fig. 1. Roughness and waviness profiles

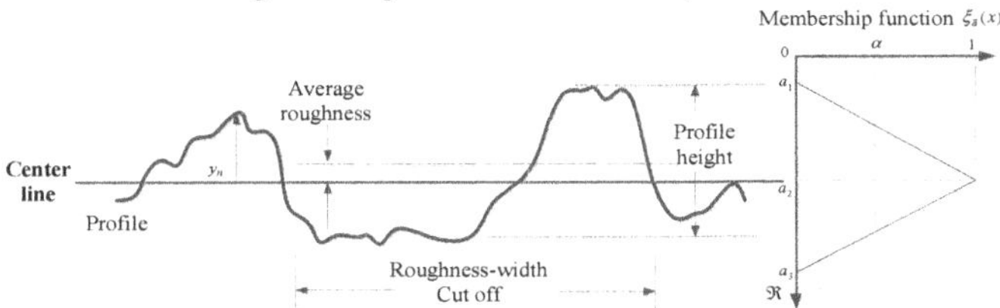

Fig. 2. Schematic of surface profile as produced by a stylus device

In this paper, there are only two values of the index of optimism investigated (0.6 and 0.8) which are all above 0.5 so that the manufacturing process can be categorized as stated in Remark 5 and 6. Based on the results shown in Table 3 and combined with the Remark 5 and 6, there are twenty-four out of twenty-five samples that are said to be in-control and there is only one sample (18th) to be said rather in-control. The calculation of left and right dominances well match the s-chart shown in Fig. 3 and $\overline{x}$ -chart shown in Fig. 4. Therefore, it can be said that the current manufacturing process is in-control. However, it is necessary to operate with caution to the line that produces the 18th sample.

Table 1. The fuzzy control limits for fuzzy s-chart

α-level	$(\tilde{l}_s)_\alpha^L$	$(\tilde{l}_s)_\alpha^U$	$(\tilde{u}_s)_\alpha^L$	$(\tilde{u}_s)_\alpha^U$
0.0	0.2007	0.3528	1.2127	2.1317
0.1	0.2102	0.3609	1.2700	2.18 2
0.2	0.2207	0.3695	1.3332	2.2321
0.3	0.2320	0.3785	1.4017	2.2825
0.4	0.2443	0.3877	1.4761	2.3410
0.5	0.2583	0.3983	1.5604	2.4042
0.6	0.2740	0.4074	1.6555	2.4601
0.7	0.2 14	0.4191	1.7606	2.5291
0.8	0.3103	0.4310	1.8750	2.6023
0.9	0.3308	0.4416	1.9985	2.6467
1.0	0.3528	0.4550	2.1317	2.6641

Table 2. The fuzzy control limitsfor fuzzy $\overline{x}$ -chart

α-level	$(\tilde{l}_{\overline{x}})_\alpha^L$	$(\tilde{l}_{\overline{x}})_\alpha^U$	$(\tilde{u}_{\overline{x}})_\alpha^L$	$(\tilde{u}_{\overline{x}})_\alpha^U$
0.0	0.7298	1.7145	2.9300	2.9300
0.1	0.8371	1.8938	3.0325	3.0 25
0.2	0.9531	2.0673	3.1397	3.1397
0.3	1.0597	2.2349	3.2515	3.2515
0.4	1.1633	2.3976	3.3673	3.3673
0.5	1.2568	2.5526	3.4869	3.4869
0.6	1.3534	2.7004	3.6101	3.6101
0.7	1.4480	2.8400	3.7368	3.7368
0.8	1.5399	2.9709	3.8669	3.8669
0.9	1.6287	3.0921	4.0002	4.0002
1.0	1.7145	3.1965	4.1369	4.1369

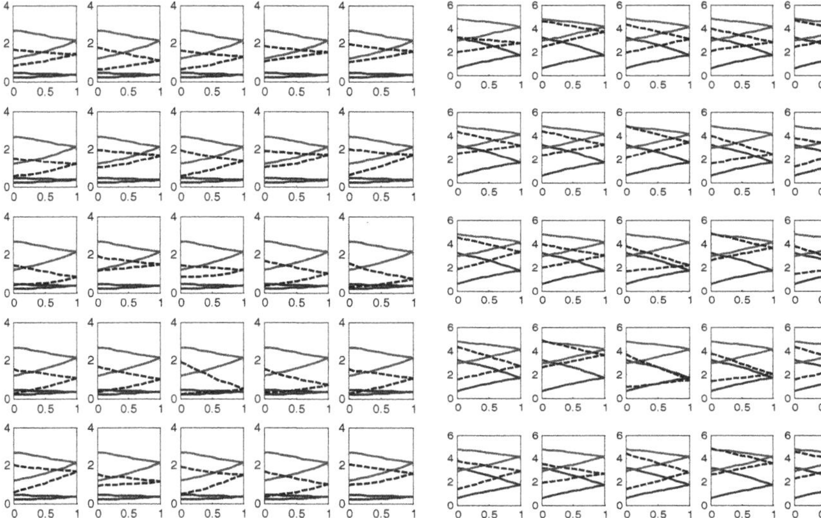

Fig. 3. The fuzzy *s*- chart for roughness height in manufacturing process of optical lens

Fig. 4. Fuzzy $\overline{x}$ -chart for roughness height in manufacturing process of optical lens

Table 3. Dominance at different β values in s-chart and $\overline{x}$ -chart

β	Sample Number	1	2	3	4	5	6	7	8	9	10	11	12
0.6	D1(*)	0.600	0.623	0.624	0.515	0.445	0.701	0.321	0.412	0.365	0.383	0.721	0.610
	D2(*)	0.812	0.722	0.712	0.859	0.889	0.705	0.842	0.725	0.853	0.689	0.454	0.756
	D3(*)	0.722	0.114	0.215	0.433	0.120	0.325	0.302	0.111	0.513	0.520	0.252	0.319
	D4(*)	0.453	0.721	0.564	0.458	0.526	0.664	0.623	0.576	0.281	0.454	0.402	0.423
0.8	D1(*)	0.640	0.680	0.693	0.595	0.535	0.799	0.427	0.526	0.485	0.509	0.854	0.749
	D2(*)	0.852	0.772	0.770	0.923	0.958	0.778	0.919	0.805	0.936	0.775	0.543	0.848
	D3(*)	0.751	0.120	0.229	0.468	0.131	0.356	0.334	0.123	0.574	0.586	0.285	0.363
	D4(*)	0.471	0.762	0.603	0.495	0.573	0.729	0.689	0.641	0.315	0.511	0.455	0.481

β		13	14	15	16	17	18	19	20	21	22	23	24	25
0.6	D1(*)	0.668	0.631	0.690	0.661	0.620	***0.589***	0.721	0.682	0.330	0.613	0.361	0.494	0.322
	D2(*)	0.635	0.527	0.501	0.352	0.302	***0.000***	0.308	0.380	0.708	0.820	0.712	0.660	0.761
	D3(*)	0.562	0.090	0.689	0.218	0.071	***0.581***	0.414	0.321	0.383	0.422	0.386	0.058	0.208
	D4(*)	0.317	0.725	0.157	0.465	0.722	***0.000***	0.121	0.262	0.322	0.358	0.260	0.784	0.622
0.8	D1(*)	0.813	0.781	0.845	0.821	0.785	***0.759***	0.896	0.861	0.514	0.801	0.552	0.690	0.522
	D2(*)	0.729	0.623	0.600	0.453	0.405	***0.000***	0.415	0.489	0.819	0.933	0.826	0.775	0.878
	D3(*)	0.643	0.104	0.796	0.253	0.083	***0.680***	0.486	0.379	0.453	0.501	0.460	0.070	0.250
	D4(*)	0.363	0.834	0.182	0.540	0.841	***0.000***	0.142	0.308	0.381	0.426	0.310	0.938	0.746

(*) ***Note:*** $D1 = D_\beta\left(\tilde{u}_s, \tilde{\bar{s}}_i\right)$ $D2 = D_\beta\left(\tilde{\bar{s}}_i, \tilde{l}_s\right)$ $D3 = D_\beta\left(\tilde{u}_{\bar{x}}, \tilde{\bar{x}}_i\right)$ $D4 = D_\beta\left(\tilde{\bar{x}}_i, \tilde{l}_{\bar{x}}\right)$.

7 Conclusions

In monitoring a manufacturing process, the conventional control charts are only applicable for real-valued data and categorizing the process in either in-control or out-of-control; however, due to some certain problems in measuring, the data are not precisely collected, which is said to be fuzzy data. Therefore, in the fuzzy environment, the traditional control charts turn out to be inappropriate. Hence, we propose the fuzzy control charts ($\overline{x}$ and s-chart) whose control limits are obtained based on the results of the resolution identity in the fuzzy set theory. To monitor the process based on fuzzy control charts, it is suggested to use the fuzzy left and right dominance approach which can categorize the process in four statuses: in-control, out-of-control, rather out-of control and rather in-control; so that the decision-makers could have proper actions to make the quality of manufactured products under control.

References

1. Montgomery, D.C.: Introduction to Statistical Quality Control, 6th edn. Wiley, Asia (2009)
2. Shu, M.H., Wu, H.C.: Fuzzy $\overline{X}$ and R Control charts: Fuzzy Dominance Approach. Comput. Ind. Eng. 61, 676--685 (2011)
3. Gülbay, M., Kahraman, C.: An Alternative Approach to Fuzzy Control Charts: Direct Fuzzy Approach. Inf. Sci. 177, 1463–1480 (2007)
4. Gülbay, M., Kahraman, C., Ruan, D.: α –cut Fuzzy Control Charts for Linguistic Data. Int. J. Intell. Syst. 19, 1173--1196 (2004)

5. Kanagawa, A., Tamaki, F., Ohta, H.: Control Charts for Process Average and Variability Based on Linguistic Data. Int. J. Prod. Res. 31(4), 913–922 (1993)
6. Wang, J.H., Raz, T.: Applying Fuzzy Set Theory in the Development of Quality Control Charts. In: Int. Ind. Eng. Conf. Proc., pp. 30–35 (1988)
7. Wang, J.H., Raz, T.: On the Construction of Control Charts Using Linguistic Variables. Int. J. Prod. Res. 28(3), 477–487 (1990)
8. Chen, L.H., Lu, H.W.: An Approximate Approach for Ranking Fuzzy Numbers Based on Left and Right Dominance. Comput. Math. Appl. 41, 1589–1602 (2001)
9. Lee, K.H.: First Course on Fuzzy Theory and Applications, 1st edn. Springer (2004)
10. Lou, M.S., Chen, J.C., Li, C.M.: Surface Roughness Prediction Technique for CNC End-Milling. J. Ind. Technol. 15, 1–6 (1998)

A High Energy Efficiency Approach Based on Fuzzy Clustering Topology for Long Lifetime in Wireless Sensor Networks

Quynh-Trang Lam[1], Mong-Fong Horng[2], Trong-The Nguyen[2], Jia-Nan Lin[3], and Jang-Pong Hsu[3]

[1] Department of Industrial Management, National Kaohsiung University of Applied Sciences, Kaohsiung, Taiwan
[2] Department of Electronics Engineering, National Kaohsiung University of Applied Sciences, Kaohsiung, Taiwan
[3] Advance Multimedia Internet, Inc., Taiwan

Abstract. Fuzzy logic has been successfully applied in various fields of daily life. Fuzzy logic is based on non-crisp set. The characteristic function of non-crisp set is permitted to have to range value between 0 and 1. In a cluster each node is definitely not only belong a cluster but also belong more than a cluster like as the non-crisp set. Therefore, classification cluster in wireless sensor network (WSN) is a complex problem. Fuzzy c-mean algorithm (FCM) is a highly suitable for classification cluster. The paper proposes for integration of Fuzzy Logic Controller and FCM to give a solution to improve the energy efficiency of WSN. Moreover, through the simulation results the lifetime of cluster is increased by more than 55%. The paper shows that the proposed approach has been confirmed that is the better choice of high energy efficiency for longer lifetime in cluster of WSN.

Keywords: Fuzzy logic, Fuzzy c-mean algorithm, wireless sensor network, slave nodes, master nodes, Fuzzy Logic Controller, network lifetime.

1 Introduction

Wireless sensor network consists of the limited number of clustering in the area. In each cluster, sensor nodes are distributed deployed to collect interesting information from the environment [8]. There are two kinds of transmitting information between source and destination nodes in WSN. The first method is indirect transmission. Nodes to nodes communicate with an indirectly-connected. In clustered networks, usually a representing node, called cluster head (CH), is assigned to each cluster [10]. The data collected by CH in different clusters connects with each other, and then usually multi-hop to the sink [3]. The method is more effective than direct method. Direct method is by a mean of a direct-connection between node A and node B. The data transfer indirect method like data transmission in peer–to–peer network. However, in an indirect method, CH has to cover all of the nodes in this cluster. CH has to spend energy consumption more than normal nodes. Thus, CH nodes can easily become a bottleneck [6],

N.T. Nguyen et al. (Eds.): *Adv. Methods for Comput. Collective Intelligence*, SCI 457, pp. 367–376.
DOI: 10.1007/978-3-642-34300-1_35

if the number of nodes is too large. CH can easily be exhausted and the connection can be postponed. The large number of CH or a few CH also affects the energy consumption and network lifetime in the cluster. So, the number of CH in WSN is very important. The technical issue is how many CH are enough in WSN.

In this paper, the authors propose a solution that uses Fuzzy Logic Controller (FLC) to control the network lifetime and use an approach of FCM clustering to determine the number of clusters in WSN. This paper also suggests a new approach to define the number nodes of the cluster. FCM has solved some complex problems such as the classification and targeted treatment of patients, identifying unknown functions of genes and recovering image. The FCM algorithm is one of the most widely-used fuzzy clustering algorithms [1], [2]. The FCM algorithm tries to divide a set of object data X= $\{x_1, x_2,\dots x_n\}$ into the collection of c fuzzy clusters with respect to some given criteria [9]. Assumption has an original set of data, using FCM algorithm returns the list of c clusters. Each cluster has a center (V) and a partition trix μ_{ij} which shows the belonging of elements in X to the ith cluster. The FCM includes 4 steps which are explained clearly in Section 3. In this paper, the FCM has been used to determine the number of clusters in WSN. The cluster centers play a role of the cluster head (CH) or master nodes and the others nodes play a role of slave nodes. FLC has been used to control the network lifetime (T). T depends on the number of master nodes, slave nodes and the distance of nodes. If the number of nodes is too small in a large area, the cluster is large and energy consumption is large. Therefore, the network lifetime becomes short. In the next of the paper, related work is explored in Section 2.

2 Related Works

2.1 Fuzzy Logic in Wireless Sensor Network

Father of Fuzzy logic is Lotfi A. Zadeh (February 4, 1921, in Azerbaijan) [4]. He received the 2009 Benjamin Franklin Medal in Electrical Engineering for his invention of fuzzy logic. Fuzzy logic was developed from fuzzy theory to perform an approximation rather than arguing exactly the classical predicate logic. Fuzzy logic can be considered as the application of fuzzy set theory to handle real-world value of the complex problem. In everyday life, there are many situations where the exact classical logic is not applicable. For example, in a high building, the balance of the heating, ventilating and air condition (HVAC) is a problem. The temperature with different people is a fuzzy set (non-crisp set) and it always changes. WSN is a solution to balance of HVAC and people in a high building. Sensors are located in each room. Sensors collect information about the environment available such as temperatures and humidifiers and number of people, status of windows, curtains and then send to the control system. Using the FLC, the dampers and volume dampers can be controlled automatically. If the temperatures are low and the numbers of people in the room are small, the volume damper of air condition system is opened at high temperature.

The most important in FLC is Fuzzy Rule. For an example, the table of Fuzzy rule for an example of air conditioners consisting of 6 rules is built below:

Table 1. A Fuzzy rule consisting of 6 rules for air conditioners

	P. Small	P. Crowded
T. High	Low temperature	Very low temperature
T. Average	Average temperature	Low temperature
T. Low	High temperature	Very high temperature

Fuzzy rules and Fuzzy membership functions are applied to combine the multiple objectives and to handle the vague terms. There are at least two rules in FLC [4]. There is normally an output controller consisting of a finite collection of premise. For an example, the Fuzzy rule base with two premises and an output is shown below:

IF (master_nodes is small) AND (area is large) THEN (lifetime is short) (1)

Where "small, medium, high" is the set of "master_nodes", "small, medium, large" is the set of area, and the fuzzy set "short" of "lifetime". "Master_nodes, area" also is called linguistic variables and "small, medium, high" is called linguistic values.

If there is more than a rule, for an example, rule R1 and rule R2 are defined on master_nodes×slave_nodes and slave_nodes×area, where X= {1, 2...30}, Y= {30, 31...130} and Z= {1, 2...10}. The strength of the rule is computed by Eq. (2) as

$$\mu R1°R2(x, z) = \max \{\min (\mu R1(x, y), \mu R2(y, z))\} \quad (2)$$

Where $R_1°R_2 = \{(x, z), \max (\min (\mu_{R_1}(x,y), \mu_{R_2}(y, z)))|x \in X, y \in Y, z \in Z\}$is a fuzzy set.

The Fuzzy rule base plays an important role in Inference Systems of FLC. The general structure inside of a FLC consists of fuzzification, an inference system (or rule base), and defuzzification.

- Fuzzification: This is the first part in the FLC. The function of fuzzification is to change input data of membership function from number to the linguistic value based on rules.
- Rule base: The rules are explained above.
- Defuzzification: The results of Fuzzy rule base are the fuzzy values; these fuzzy values are needed to change into a numeric value to control objects.

2.2 Fuzzy C-Mean Algorithm

Fuzzy c -means clustering is an unsupervised clustering algorithm [9], similar to the k-means algorithm with a goal as the same strategy to divide cluster. However, k-means is an algorithm based on crisp set and FCM is an algorithm based on non-crisp set. The main advantage of FCM is high performance. The crisp set only accepts the value of true or false, 0 or 1 [4]. Contrary to crisp set, the non-crisp set is allowed to have values between 0 and 1, which denotes the degree of membership of an element in a given set. The degree of membership is permitted to have any values between 0 and 1 [4]. Thus, Fuzzy logic accepts 50% true value or half true value. Because of this difference, FCM has a success in solving many complex problems such as identifying fingerprint, identifying faces, recovering images, separating clusters of color, classification of diseases. However, there are still disadvantages of FCM. The time of convergence is long with a big data set, noise sensitivity and sensitive element in the data set so that the real center may be placed far from the results of calculating.

The FCM is the fuzzy equivalent of the nearest mean hard clustering algorithm which Dunn (1973) was proved in [5]. FCM minimized the following objective function $J_m(\mu, V)$ with respect to membership functions μ_{ij} and cluster centroids v_i:

$$J_m(\mu,V) = \sum_{i=1}^{c} \sum_{j=1}^{n} \mu_{ij}^{m} d^2(x_j, v_i) \tag{3}$$

Where $d^2(x_j, v_i) = \|x_j - v_i\|^2 = (x_j - v_i)^T A (x_j - v_i)$ denotes the measure of the Euclidean distance between x_j^{th} object and the cluster center v_i. Squared distances satisfy that objective function is non-negative, $J > 0$, A is a p×p positive definite matrix, p is the dimension of the vectors x_j, x_j are data set of dimensional vector $X = \{X_j \mid 1 \leq j \leq n\} \subset R^s$, c is the number of clusters, n is the number of vectors, $m > 1$ is the weight exponent, v_i is the centroid of ith clusters, $v_i \in R^s$, $V = [v_1, v_2, \ldots v_c]$ presents cluster center, μ_{ij} is the value of the objective of membership of x_j to the centroid v_i, $\mu_{ij} \in [0,1]$.

Normally, the objective function $J_m(\mu, V)$ of FCM is to find out the membership function μ_{ij} to minimize $J_m(\mu, V)$. Hence, objective function $J_m(\mu, V)$ yield to minimum value as:

$$\mu_{ij} = \frac{1}{\sum_{k=1}^{c} \left(\frac{d_{ij}}{d_{kj}}\right)^{\frac{2}{m-1}}} \tag{4}$$

Where c is the number of clusters, $1 \leq I \leq c$, k is the number of vectors, $1 \leq k \leq n$, μ_{ij} is the values of the objective of membership of x_j to the centroid v_i, d_{ij} is the distance between x_j and the centroid v_i, d_{kj} is the distance between x_j and the centroid v_k.

$$\mu_{ij} = 0 \tag{5}$$

$$\sum_{i \in i_k} \mu_{ik} = 1 \tag{6}$$

$$v_i = \frac{\sum_{k=1}^{n} (\mu_{ik})^m x_k}{\sum_{k=1}^{n} (\mu_{ik})^m} \tag{7}$$

Where μ_{ik} is the value of the membership of x_k to the centroid v_k, $\mu_{ik} \in [0,1]$, x_k is data set of dimensional vector X, $m > 1$ is the weight exponent, n is the number of objects.

The FCM algorithm includes 4 steps:

- Initialize membership $\mu = [\mu_{ij}]$ matrix where μ satisfy that $\mu_{ij} \in [0, 1]$
- Find the fuzzy centroid v_i for $i = \{1, 2, 3\ldots c\}$
- Update the fuzzy membership μ_{ij}
- Repeat steps 2, 3 until $J_m(\mu, V)$ is no longer decreasing

The pseudo code is executed in the following steps below:

Input the number of c and m for function $J_m(\mu, V)$
Output c which function $J_m(\mu, V)$ equal minimum value

BEGIN
1. Input the cluster number, c, the weight exponent, m
   Initialize V, V $\in R^{sxc}$, j = 0
2. Repeat until $J_m(\mu, V)$ is no longer decreasing or $(||\mu^{(j+1)} - \mu^{(j)}|| \le \varepsilon)$
   Compute the object function $J_m(\mu, V)$ according to Equation (3)
   j := j + 1
   Update membership function μ_{ij} according to Equation (4)
   Update fuzzy centroid v_i according to Equation (7) and $\mu^{(j)}$
3. Show the result of fuzzy clustering

END

The threshold value ε is very small. In this case, the threshold value was set equal to 0.00001. The FCM always converges to the minimum of J_m but the results depend on randomly choosing value of μ_{ij}, and calculate of v_i at first loop. If the μ_{ij} is different, the local is different. If the centroid at first loop includes special element, the result may be mismatched with real center. Because in the set of object data which are similar character the results always exacter than in the set of object data which is totally different characters.

2.3 An Energy-Efficient Data Transfer Model of Wireless Sensor Networks Based on the Coalitional Game Theory

Data transfer model of WSNs is very important on the energy consumption. To reduce energy consumption some nodes in a WSN form a coalition by transferring data coordinately instead of transferring alone [3]. Game theory is the study of mathematical models of conflict and cooperation between individual decision makers. Coalitional game is usually used to increase the coalitions' payoff and has been widely used in communication networks [3]. Payoff function is presented the remained energy and the cost of transfer data to sink. In [3], the author applies Shapley value to measure the payoff, uses Nash equilibrium to determine the approximate data transfer strategies of the formed coalitions and uses a Markov process as the model of the forming the coalitions.

3 A High Energy Efficiency Approach Based on Fuzzy Clustering Topology for Long Lifetime in Wireless Sensor Networks

It is assumed that there are two kinds of nodes: N1 and N2. N1 nodes have twice the energy and twice the radius coverage of N2 nodes. In [6], Kaur divided area where nodes cover into grids such as 4×4, 3×3, and 2×2. After the area is divided into grids, the circular area is formed into clusters of nodes. The circular area is drawn around the boundary of coverage area. The maximum number of 4×4 grids in the circle can be calculated by Eq. (8):

$$G = \frac{\text{area of circle}}{\text{area of 4x4 grid}} - \frac{\text{circumference of circle}}{\text{diagonal of 4x4 grid}} = \frac{\pi R^2}{64x^2} - \frac{2\pi R}{8x\sqrt{2}} \tag{8}$$

Where R is the radius of the circle area, x is the size of sub grid (see Fig. 1.).

N1 nodes are located at the center of 4×4 grid, the distance between N2 nodes and N1 nodes are not far. Therefore, the energy consumption could be saved.

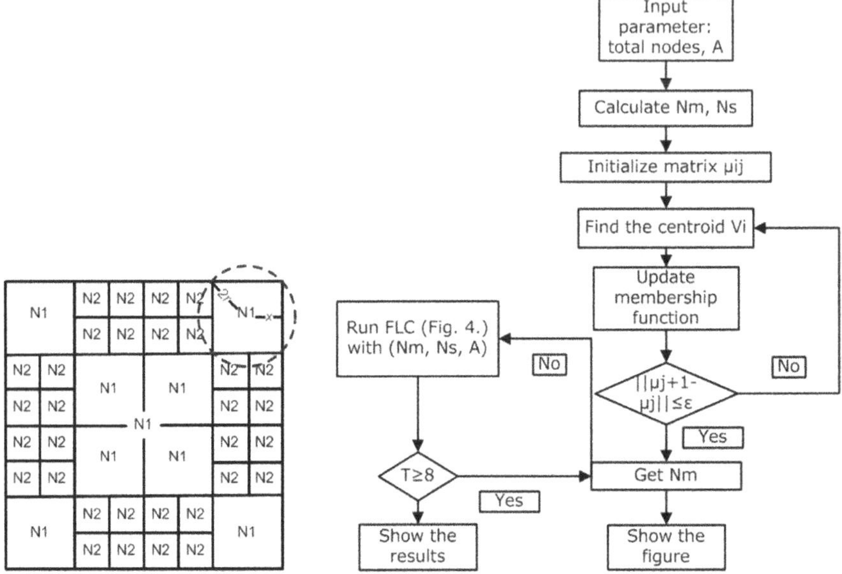

Fig. 1. Description of 4×4 grid

Fig. 2. The flowchart discovering optimal clusters

In case, the N1 nodes cannot be placed at the center of grids, because the distribution of nodes depends on the terrain of the area [6]. Thus, the real center may be variations in the results be different. In real case, the N1 nodes should be placed at the nearest the center of grids and N2 nodes are placed at the boundary. In WSNs can be individually distributed the grids not only one type of grid such as 3×3 grid but also 3 types of grid in the area.

After calculating the numbers of girds and the numbers of N1 nodes, N2 nodes, the FCM has been applied to partition the nodes into *c* clusters. The N1 nodes are located at the center of clusters and play a role of the master nodes and the N2 nodes play a role of the slave nodes. The flow chart of discovering optimal clusters is described by a flowchart in Fig. 2 above.

This simulation initializes 120 nodes in a 160×170 m^2 area, the number of master nodes is Nm is up to 7 and the number of slave nodes is Ns (see Fig. 3). If the number of nodes is too small and the area is too large, the energy consumption of nodes can be easily exhausted and then the lifetime of the node is decreased. However, the proposed approach suggests a solution to solve this problem.

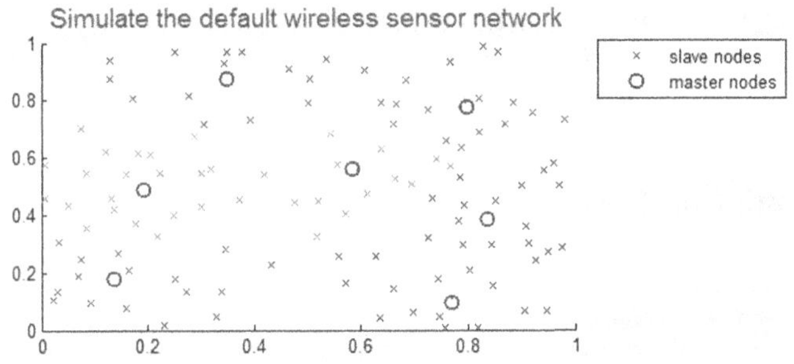
Simulate the default wireless sensor network
slave nodes
master nodes
0
0.2
0.4
0.6
0.8
1

Fig. 3. Illustrates the simple wireless sensor network of 120 nodes

By FLC, T as show in the flow chart is controlled by the administrator. Firstly, the input includes the number of master nodes (Nm), the number of slave nodes (Ns), and the area (A), the output is the network lifetime (T). The linguistic values of Nm, Ns and A are "small, medium, high". The linguistic values of T are "short, average and long". The FLC structure is shown as follows:

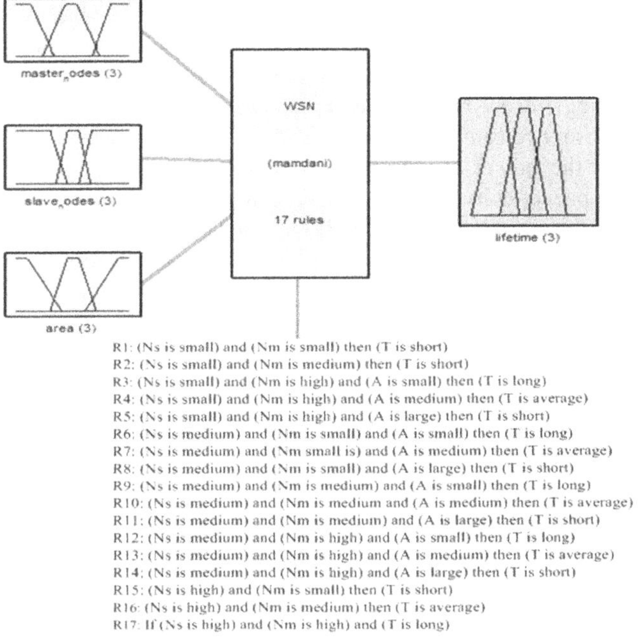
master_nodes (3)
slave_nodes (3)
area (3)
WSN
(mamdani)
17 rules
lifetime (3)
R1: (Ns is small) and (Nm is small) then (T is short)
R2: (Ns is small) and (Nm is medium) then (T is short)
R3: (Ns is small) and (Nm is high) and (A is small) then (T is long)
R4: (Ns is small) and (Nm is high) and (A is medium) then (T is average)
R5: (Ns is small) and (Nm is high) and (A is large) then (T is short)
R6: (Ns is medium) and (Nm is small) and (A is small) then (T is long)
R7: (Ns is medium) and (Nm small is) and (A is medium) then (T is average)
R8: (Ns is medium) and (Nm is small) and (A is large) then (T is short)
R9: (Ns is medium) and (Nm is medium) and (A is small) then (T is long)
R10: (Ns is medium) and (Nm is medium and (A is medium) then (T is average)
R11: (Ns is medium) and (Nm is medium) and (A is large) then (T is short)
R12: (Ns is medium) and (Nm is high) and (A is small) then (T is long)
R13: (Ns is medium) and (Nm is high) and (A is medium) then (T is average)
R14: (Ns is medium) and (Nm is high) and (A is large) then (T is short)
R15: (Ns is high) and (Nm is small) then (T is short)
R16: (Ns is high) and (Nm is medium) then (T is average)
R17: If (Ns is high) and (Nm is high) and (T is long)

Fig. 4. The FLC with 17 rules

The Fuzzy rule base is the most important part of FLC. There are totals 27 rules but this work presents optimally into 17 rules. Once inputting parameters and creating base rules are done, FLC is run by hand to show the results of T. The value of Nm, Ns or A can be changed by dragging and typing in working design. The relationship between output T and input parameters depend on linguistic values and Fuzzy rule base. The administrator can compare the results of T to choose the width of area or number of master nodes fitting with real situations.

4 Simulation Results

4.1 Objective of Simulation

The objective of this work is to prolong the network lifetime. In [10], Moslem presented 4 definitions of network lifetime from different sources. In this paper, the definition of network lifetime (T) for sensor network is based on the time when the first node dies. The simulation function is to verify the performance of T.

4.2 Simulation Scenarios

There are two scenarios in this simulation to evaluate the lifetime performance. Scenario 1: The plan of the administrator is to distribute randomly 100 sensor nodes in the 500×100m2 area. It is assumed that all nodes are active. The administrator would like to prolong lifetime in a constant area. By changing the number of master nodes and slave node, the administrator reduces the energy consumption and increase lifetime T. In this scenario, all transmissions are in single-hop mode and single event reporting. The single event reporting is in that the neighboring sensors report events to CH. Thus, sensors consume energy for sending, receiving and transmitting data. It is assumed that the power consumption of idle nodes is very small insignificant enough. In [10], packet delivery of each node can be modeled as an energy consumption model. The consumed energy of a 100-bit packet equals to 1μJ (e_t). The initial energy of each node equals 12J, a node produces a packet every 10 time units (t_0), cluster C achieves $T = t_0 * N$ when

$$E_t \leq E_o - e_t * n_b \quad + \sum_{j=2}^{N} E_j \tag{9}$$

Where E_t is the total energy consumption, E_j is the remain node energy at j^{th} packet, E_o is initial energy of a node, n_b is the total number of bytes in a packet, e_t is the consumption energy by delivering a 1-byte packet, N is the total number of packets of a node.

The total energy consumption is given as same as in [10]:

$$E_t = E_n + E_c = (\gamma_j P_t + \sum_{i \in subtree\ C_j} N_i)(m_2 \parallel s_1 - s' \parallel^4 + c_2) + e_r(P_r + \sum_{i \in subtree\ C_j} N_i) \tag{10}$$

Where E_n is denoted the energy depleted by the sensor for intercluster transmission, E_n is very small, this simulation assumes E_n equal to 0, E_c is the consumed energy in intercluster transmission by a node, N_i denotes the number of packets in level i+1,

γ_j denotes ratio, P_t denotes total number of packets that CH transmits to the Sink, m_2 and c_2 stand for the value of transmission parameter, P_r is the total number of packets received by CH, $s_1 = (x_1, y_1)$ is location of sensor nodes, s' is location of CH, $\|.\|$ denotes the Euclidean distance.

Similar in scenario 1, in scenario 2, the administrator change density by changing the area. When A is changed increase or decrease, the T is also changed. According to the simulation, administrator can test the real coverage area correspond to the T.

4.3 Simulation Results

In this simulation, the administrator verifies through computer simulations using Matlab. Notices that the simulation result is based on FLC and not involve the approximations are made in the analytical results to find a simple lifetime. Following the scenario 1, 100 nodes are divided into of 6 master nodes and 94 slave nodes. If Nm is increased from 6 to 10 nodes, the T is increased 55%. In [3], Tianying presented one the solution to prolong lifetime is used an energy-efficient data transfer model of WSNs based on Coalitional game theory. In this model, data transfer activity of sensor nodes is a game activity and the cooperating sensor nodes get more benefits than its work alone. Fig. 5 shows the difference between methods of Tianying and the proposed method in this paper:

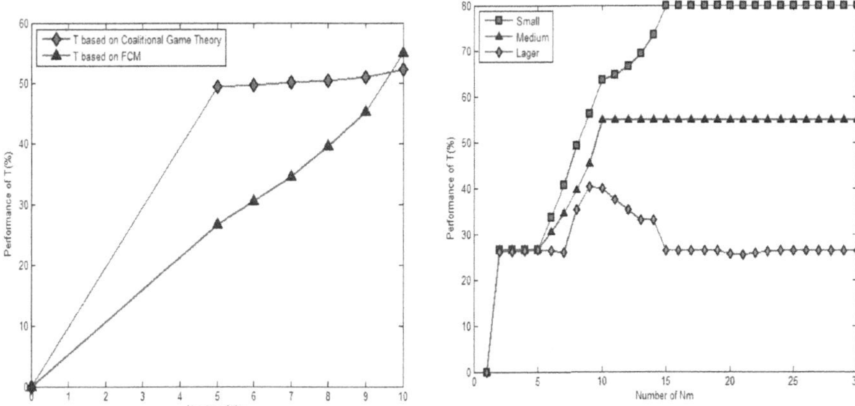

Fig. 5. Comparison of network lifetime based on Coalitional game theory and based on FCM

Fig. 6. Comparison of area and network lifetime

Comparison scenario 1 with in scenario 2, the administrator can change the density of nodes by changing area. When the areas are changing, the CH numbers are also changed. Affected by CH number, the energy consumption and lifetime are changed. The area decreases 40% surface area, the lifetime of nodes increases 25%. The administrator increase 60% surface area, the network lifetime also decreases 28.5%. Fig. 6 shows the effect of density in network lifetime (T).

5 Conclusion

In this paper, the proposed solution has not only covered the determining the number of a clustering in wireless sensor networks, but also is made the lifetime of sensor nodes 55% longer. The simulation was built with the Matlab environment because of its architecture rules is clearer rather than other software such as C++, Java and NS2. The program helps the administrator to make a decision of design the network effetely before employing real networks, in case the total nodes and the area are exactly known. Another important feature of this paper is the fuzzy c-mean algorithm deal with the problem in wireless sensor networks in clusters classification. Using this algorithm in the simulation model integrates effetely into transmission distance and the number of nodes to distribute a long-life cluster for more effective power consumption. In the future, this Fuzzy c-mean algorithm can be extended not only to determine the number of nodes but also to optimize the number of clusters in sensor networks.

References

1. Begonya, O., Luis, A., Christos, V.: Highly Reliable Energy-Saving MAC for Wireless Body Sensor Networks in Health care Systems. IEEE Journal on Selected Areas in Communications, 553–564 (2009)
2. Shu, H., Liang, Q., Gao, J.: Wireless Sensor Network Life time Analysis Using Interval Type-2 Fuzzy Logic Systems. IEEE Transactions on Fuzzy Systems, 416–427 (2008)
3. Tianying, W., Kun, Y., Weiyi, L., Jin, X.: An Energy-efficient Data Transfer Model of Wireless Sensor Networks Based on the Coalitional Game Theory. In: 2011 Eighth International Conference on Fuzzy Systems and Knowledge Discovery (FSKD), pp. 1354–1358 (2011)
4. Jang, J.-S.R., Sun, C.T., Mizutani, E.: Neuro-Fuzzy and Soft Computing A Computational Approach to Learning and Machine Intelligence. Prentice-Hall (1997)
5. Rafik, A.A., Witold, P.: Fundamentals of a Fuzzy-Logic-Based Generalized Theory of Stability. IEEE Transactions on Systems, Man, and Cybernetics – Part B: Cybernetics, 971–988 (2009)
6. Kaur, T., Baek, J.: A Strategic Deployment and Cluster-Header Selection for Wireless Sensor Networks. IEEE Transactions on Consumer Electronics, 1890–1897 (2009)
7. Horng, M.-F., Chen, Y.-T., Chu, S.-C., Pan, J.-S., Liao, B.-Y.: An Extensible Particles Swarm Optimization for Energy-Effective Cluster Management of Underwater Sensor Networks. In: Pan, J.-S., Chen, S.-M., Nguyen, N.T. (eds.) ICCCI 2010, Part I. LNCS (LNAI), vol. 6421, pp. 109–116. Springer, Heidelberg (2010)
8. Wang, Q.: Packet Traffic: A Good Data Source for Wireless Sensor Network Modeling and Anomaly Detection. IEEE Network, 15–21 (2011)
9. Verma, P., Yadava, R.D.S.: Fuzzy C-means Clustering Based Uncertainty Measure for Sample Weighting Boosts Pattern Classification Efficiency. In: Computational Intelligence and Signal Processing (CISP), pp. 31–35 (2012)
10. Moslem, N., Masoud, A.: Lifetime Analysis of Random Event-Driven Clustered Wireless Sensor Networks. IEEE Transactions on Mobile Computing, 1448–1458 (2011)
11. Chen, L., Gong, D.G., Wang, S.R.: Nearest Neighbor Classification by Partially Fuzzy Clustering. In: 26th International Conference on Advanced Information Networking and Applications Workshops, pp. 789–794 (2012)

Author Index